SHAKE UP SCIENCE 5

Pearson Education Limited
Edinburgh Gate
Harlow
Essex CM20 2JE
England
and Associated Companies throughout the world.

www.pearsonelt.com

First published 2016
ISBN: 978-1-2921-4482-5
Set in Futura LT Pro, Feltpen Com, Bauhaus Std, ITC Benguiat Gothic Std, ITC Zapf Dingbats Std, Avenir LT Pro

Acknowledgments
Picture Credits
The publisher would like to thank the following for their kind permission to reproduce their photographs:

(Key: b-bottom; c-center; l-left; r-right; t-top)

Student's Book: 123RF.com: 4b, 10b, 16tl, 18cl (bottom), 18bl, 20bl, 21tr, 22 (sun), 35br, 37c, 39t, 40tr, 40cl, 43b, 45t, 45b, 45bc, 48, 49t, 49tc, 49b, 51b, 52tc, 52tr, 63b, 64tl, 67tc, 67bc, 71t, 76tl, 82b, 100tl, 100tr, Gjermund Alsos 68, ampics 9c, Tobias Arhelger 18br, Kitch Bain 101b, Le Do 24br, Jacek Fulawka 17, Eric Isselee 106r, Ritu Jethani 107l, 111b, Kajornyot Krunkitsatien 4cl, 12tr, M.G. Mooij 85b, Gennadiy Poznyakov 102b, Thuansak Srilao 40tc, Kheng Guan Toh 78, Carlo Villa 90l, Monika Wisniewska 91t; **Alamy Images:** Jose Luis Pelaez Inc 90t; **Corbis:** Markus Botzek 83l, 87b, Stephen Frink 7t; **Fotolia.com:** Africa Studio 32t, alon 10t, 15b, Aaron Amat 65cr, andreusK 47b, Komarov Andrey 65b, 75t, anrymos 79b, antiksu 46b, BillionPhotos.com 28cr, Andrew Buckin 93br, Canaryluc 106cl, Dmitry Chulov 18tr, creativenature.nl 23t, Les Cunliffe 92t, Delphotostock 56, denio109 28cl, 30tl, destina 65t, Chris Dorney 65cl, Jaimie Duplass 83br, Elenarts 58t, ffphoto 30bl, fidelio 36cr, Susan Flashman 19r, forcdan 19l, freila 31t, hadzi3bart 52tl, highwaystarz 29t, 39b, Ad Hominem 53b, Igor 72bl, Vlad Ivantcov 69b, 70cr, jordi2r 71b, kalichka 21b, 22 (slug), Rita Kochmarjova 103t, 111t, kryvan 41t, ksena32 38tc, Oksana Kuzmina 13t, Henry Larsson 9t, Guillaume Le Bloas 88b, learchitecto 101t, marko 10c, Marla 72bc, Martinan 80b, Vasily Merkushev 46t, muuraa 38bc (l), nikolay100 72tc, Olandsfokus 25b, olllinka2 72c, Patryssia 31cr, piai 59, pixelrobot 72br (top), pressmaster 30tr, rdonar 24tr, Silvano Rebai 47t, Kimberly Reinick 38tl, 38tr, Rkphoto 13c, rodimovpavel 72t, romurundi 29c, Rzoog 57b, Server180 5c, Ljupco Smokovski 37b, sommai 28tc (r), Alex Stokes 72br, Dmytro Sukharevskyy 28c (r), sytnik 97c, mariusz szczygieł 92b, James Thew 58b, 89t, Timmary 30 (background), 38b, Alex Traksel 28b, valery121283 28tc (l), 28tr, 36bl, 36br, Alexandr Vasilyev 88c, Viacheslav Blizniuk 31b, vinr 73tr, Alain Wacquier 28tl, wckiw 33b, Winston Link 96cl, yanlev 29b, yellowj 93t, 99c; **Imagemore Co., Ltd:** 6b; **Imagestate Media:** John Foxx Collection 11b, 106l; Pearson Education Ltd: 70l, Gareth Boden 70br, Martyn F Chillmaid 67t, 95t, **Pearson Education, Inc.** 80c, Cheuk-king Lo 105c, Mohd Suhail. Pearson India Education Services Pvt. Ltd 55, Oxford Designers & Illustrators Ltd 80t, Pearson Education Asia Ltd 72cr, Rafal Trubisz 37l; **SF Glenview Photo Studio:** 14, 26, 50, 62, 74, 86, 98, 110; **Shutterstock.com:** Africa Studio 89b, AlenKadr 38br, Shelby Allison 35t, Andresr 7b, antpkr 84, Noam Armonn 69t, 75c, Andrey Armyagov 96t, Nagy-Bagoly Arpad 18tl, Can Balcioglu 70t, Andrey N Bannov 16tc, Basheera Designs 34, Roxana Bashyrova 64b, Belizar 25t, 27b, Berci 92c, Anthony Berenyi 97b, BMJ 20bl (c), 27c, BMJ. 20c, 21tc (r), 22 (plants), Steve Bower 20tr/c, 20bc (r), 22 (deer), Mark Bridger 16cl, 21tc (l), 22 (owl), Leonello Calvetti 44t, Tony Campbell 43t, Jacek Chabraszewski 102t, Coprid 93 (jug), David Crockett 72cl, 75bl, CyberKat 4tr, Daboost 96br, Gilles DeCruyenaere 22 (insect), Zhu Difeng 97t, 99b, Nikolay Dimitrov 18cr (bottom), Pichugin Dmitry 81b, 87c, Dennis Donohue 8t, Enciktat 18cr (top), Evgeny Kovalev spb 18cl, exopixel. 67tl, federicofoto 40cr, Bill Florence 4cr, Martin Fowler 11tl, Gelpi JM 77t, 87t, glenda 93bl, godrick 63c, Greenphile 67bl, Shane Gross 16b, Gtranquillity 38bc (r), Guentermanaus 43c, Mountain Hardcore 64tr, Shawn Hempel 5t, 15t, Jan van der Hoeven 82c, Terekhov Igor 67br, 105br, imging 95b, Eric Isselee 106cr, Chad A. Ivany 13b, JL Jahn 18cl (top), 27t, Doug James 18cr, Jhaz Photography 91b, 99t, Dmitry Kalinovsky 94, Jane Kelly 88tr, Brian Kinney 107r, kurhan 100c, kwest 45tc, 51c, Lagui 6tr, Lakeview Images 33c, Elena Larina 79t, Olivier Le Queinec 88cr, Martin Lehmann 12b, Larisa Lofitskaya 36bc (r), S.R. Maglione 23b, Piotr Marcinski 44b, Margouillat photo 73br, Mircea Maties 93 (coin), Maxx-Studio 77b, Antony McAulay 60, 61, Sue McDonald 103b, Steve McWilliam 16tr, 21tl, Dudarev Mikhail 104, Mopic 82t, Lukiyanova Natalia 93 (rings), Dmitry Naumov 73bc, NCG 40b, Yulia Nikulyasha Nikitina 88tc, Nito 93 (scissors), Olgysha 4tl, ollirg 49bc, Chepurnova Oxana 93 (boots), Pablo Hidalgo 40tl, 51t, Slavoljub Pantelic 100cl, Rick Parsons 88cl, Paul Matthew Photography 100tc, Paul Reeves Photography 16cr, Edyta Pawlowska 33t, Anita Patterson Peppers 81t, peresanz 54, 63t, Photosani 52b, Photovolcanica.com 40c, photowings 76b, Targn Pleiades 56 (background), Zeljko Radojko 76tr, Raftel 42, Redchanka. 11tr, Aida Ricciardiello 28c (l), 32b, Randy Rimland 4tc, Tony Rix 16c, Saiko3p 73bl, Sander van Sinttruye 53t, SaraJo 22 (rabbit), Ilin Sergey 9b, Serhiy Shullye 5b, Scenic Shutterbug 20t, 20br, 22 (lion), Slavapolo 6t, Smileus 85t, Ljupco Smokovski 88tl, 100b, Ljupco Smokovski. 105t, spaxiax 36bc (l), spaxiax. 31cl, Alex Staroseltsev. 105bl, StevenRussellSmithPhotos 22 (deer mouse), Alexey Stiop 57t, StudioSmart 96bl, tarasov 35bl, 93 (tyre), TessarTheTegu 22 (squirrel), Sirikorn Thamniyom 64tc, Mogens Trolle 8b, Suzanne Tucker 36tr, tusharkoley 41b, Ulga 76tc, Vectorlib.com 96cr, Vilainecrevette 11cr, Warren Price Photography 12tl, Ivonne Wierink 100cr, J.K. York 24bl, Yurchyks 67tr

Flash Cards: 123RF.com: alhovik, Ritu Jethani, Kajornyot Krunkitsatien, Gennadiy Poznyakov, Kheng Guan Toh; **Alamy Images**: Jose Luis Pelaez Inc; **Corbis**: Markus Botzek; **Fotolia.com**: Africa Studio, Aaron Amat, BillionPhotos.com, denio109, destina, Chris Dorney, Elenarts, ffphoto, forcdan, freila, Ad Hominem, Vlad Ivantcov, Andrea Izzotti, kalichka, Rita Kochmarjova, kryvan, Henry Larsson, marko, Vasily Merkushev, olllinka2, Patryssia, piai, Server180, sommai, Dmytro Sukharevskyy, Mariusz Szczygieł, James Thew, Timmary, valery121283, Alain Wacquier, yellowj, Mara Zemgaliete; **Imagestate Media**: John Foxx Collection; **Pearson Education Ltd**: Martyn F Chillmaid, Cheuk-king Lo, Pearson Education Asia Ltd, Rafal Trubisz; **Shutterstock.com**: Africa Studio, Shelby Allison, Andresr, antpkr, Noam Armonn, Nagy-Bagoly Arpad, Can Balcioglu, Basheera Designs, BMJ, Steve Bower, Mark Bridger, Jacek Chabraszewski, ChameleonsEye, Gilles DeCruyenaere, Zhu Difeng, Dennis Donohue, Evgeny Kovalev spb, exopixel., Konstantin Faraktinov, fotostory, Gelpi JM, Terekhov Igor, JL Jahn, Jhaz Photography, Dmitry Kalinovsky, Robyn Mackenzie, Mopic, Dmitry Naumov, ollirg, Edyta Pawlowska, Anita Patterson Peppers, peresanz, photobar, Photovolcanica.com, PlusONE, Aida Ricciardiello, saddako, SaraJo, Galushko Sergey, Scenic Shutterbug, Ljupco Smokovski, spaxiax, Alex Staroseltsev, StevenRussellSmithPhotos, TessarTheTegu, Warren Price Photography, Yurchyks

Worksheets: 123RF.com: 27al; **Fotolia.com:** nickolae 99btr, nicolasprimola 15a; **Pearson Education Ltd:** Gareth Boden 27ar; **Shutterstock.com:** Cico 51a, Critterbiz 51b, Maddrat 99bl, MillaF 87a, Morphart Creation 99bbr, Juriah Mosin 111a; **Sozaijiten:** 63b

All other images © Pearson Education

Every effort has been made to trace the copyright holders, and we apologize in advance for any unintentional omissions. We would be pleased to insert the appropriate acknowledgment in any subsequent edition of this publication.

Science Consultant
Mark Sander

Cover images: *Front:* **Fotolia.com:** mekcar l; **Getty Images:** Boarding1Now r; *Back:* **Fotolia.com:** alon l, hadzi3bart c; **Shutterstock.com:** Mopic r

Contents

Content and Language Integrated Learning (CLIL)

Increasingly, students around the world who don't speak English at home are learning content subjects such as science through the medium of English, meaning that English language learning is taking place at the same time as the learning of content.

Benefits include:

- exposure to and acquisition of English language in context, encouraging a more natural language learning process
- meaningful use of the English language, with students motivated to use English to find out more about real-world topics that interest them
- increased English fluency through using the language for a variety of purposes and in a number of different ways
- faster and higher-level development of skillswork, especially reading and writing
- preparation for future studies and the international workplace.

Varied support for English language learners is provided throughout the teaching notes, including additional background information, suggestions for suitable language-learning activities, as well as strategies and techniques for developing skillswork.

Using the materials

Teaching and learning situations can differ widely, and, with this in mind, the series has been devised to allow teachers the flexibility to customize according to their requirements. Following a modular approach, each lesson can work as a self-standing unit of content, and teachers can pick and choose to fulfill their own curriculums. Fast-track routes can be followed in situations where less time is allocated for the teaching of primary science through English. More information about fast-track routes can be found online.

In addition to the wide range of reinforcement and extension activities provided through the ActiveTeach, an optional **Workbook** is also available. The **Workbook** has been especially tailored for the requirements of English language learners and provides:

- activities relating to each lesson's key vocabulary and concepts
- targeted practice of already known grammar
- comprehensive development of science-related reading and writing skills
- a progression through receptive understanding to productive ability
- an emphasis on real-world application and students' own experience.

Student's Book

Start examining the Big Question.

Define learning goals for the unit.

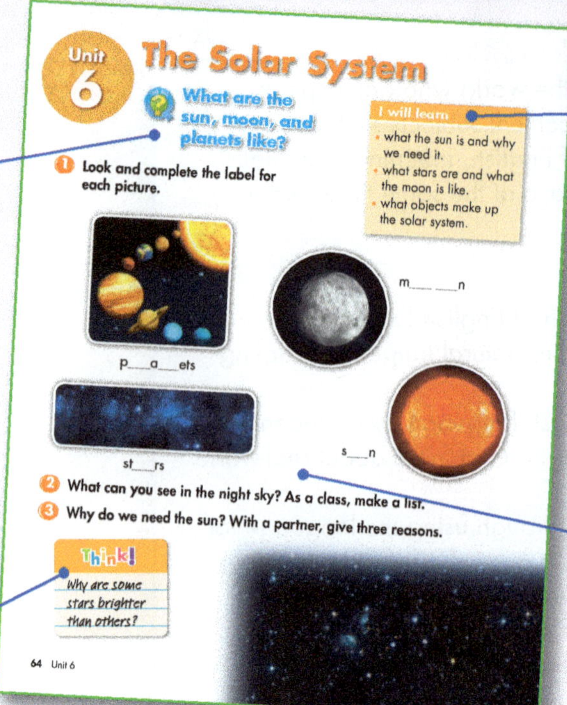

Activate previous knowledge and introduce the topic.

Engage critical thinking and begin to unfold the Big Question.

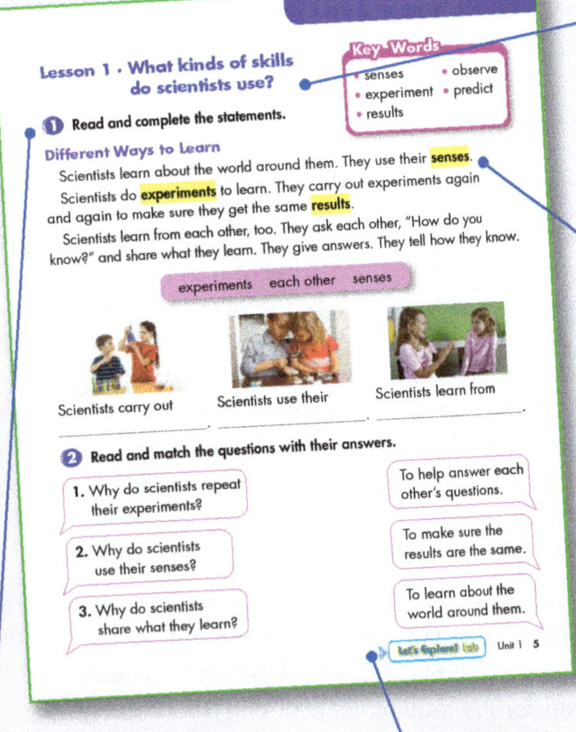

Bring science to life with clearly defined questions, real-world contents, and scientific facts.

Learn key words through texts and definitions in a glossary at the end of each Student's Book.

Experiment in class or online; recording observations in Student's Books gives a sense of ownership. Expand thinking with *Activity Cards* on the ActiveTeach.

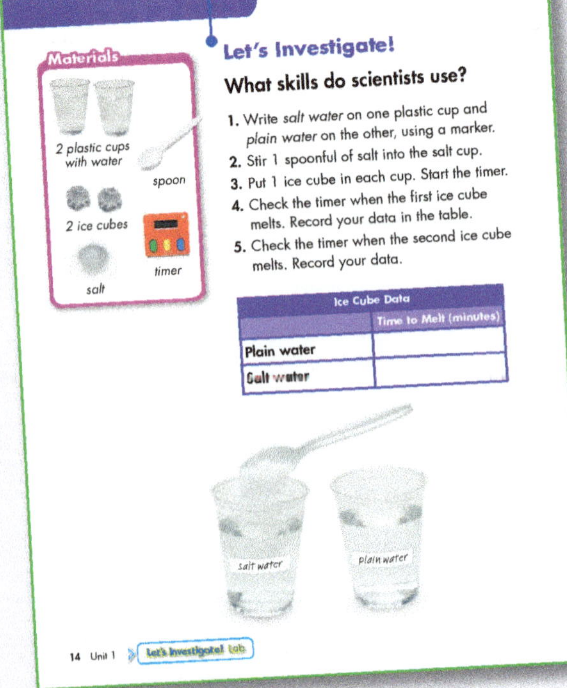

Think, read, and write like a scientist to make learning personal, relevant, and engaging. Explicit instruction brings science concepts to life.

Link to digital activities to explore topics before reading.

8 Read and underline what your heart and lungs do.

Lungs and Heart

Your lungs take in air when you breathe. Oxygen is in the air. You need oxygen to live.

Your heart pumps blood to all parts of your body. The blood picks up oxygen from the lungs. The blood carries the oxygen from your lungs to all parts of your body. Your heart works together with your lungs to keep you healthy.

Your heart and lungs work for your entire life. Your heart beats about 90 times per minute. When you're an adult, your heart beats about 80 times per minute. Adults breathe more slowly, too!

Flash Lab

Sound of a Heartbeat

Get a cardboard tube. Put one end over someone's heart. Place your ear on the other end. Listen. Tell what you hear.

9 Draw what this boy's lungs look like when he breathes in to play the trumpet.

Unit 4 47

Do quick activities in the classroom.

5 Read, look, and match each item to its recycling bin.

Recycle

People can recycle. Recycle means to change something so it can be used again. Paper, plastic, metal, and glass can be recycled.

Paper Glass Metal Plastic

6 With a partner, circle the parts of the toys that use recycled materials.

Go Green

Think of an item you normally throw in the trash or recycle. Design something new you can make from it and draw a plan in your notebook.

Unit 5 59

Provoke thought about how to protect Earth.

Do fun experiments with the family.

7 Read, look, and draw an arrow to show the direction the spoon and the train are moving in.

How Magnets Move Objects

A magnet can move some things without touching them. Look at the picture below. The spoon is moving, but the magnet is not touching the spoon. The force of the magnet pulls the spoon.

8 Draw an object that the magnet could move and an arrow to show the direction it would move in.

At-Home Lab

Get a variety of different sized metal objects and a magnet. Use the magnet to pull each object in turn and see how many of them you can move along the floor.

Lesson 3 Check Got it? 60-Second Video Unit 9 109

Review the main points of each lesson before taking the *Got it? Quiz*.

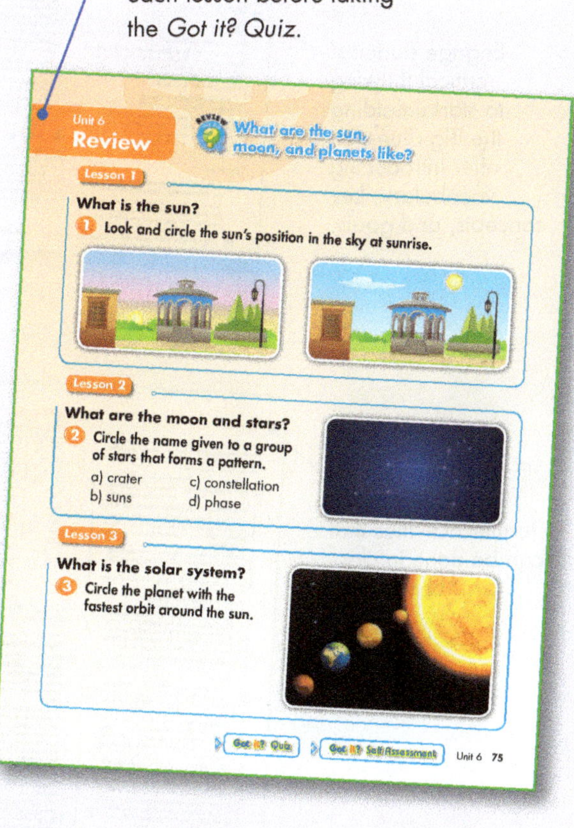
Unit 6 Review

What are the sun, moon, and planets like?

Lesson 1

What is the sun?

1 Look and circle the sun's position in the sky at sunrise.

Lesson 2

What are the moon and stars?

2 Circle the name given to a group of stars that forms a pattern.

a) crater c) constellation
b) suns d) phase

Lesson 3

What is the solar system?

3 Circle the planet with the fastest orbit around the sun.

Got it? Quiz Got it? Self-Assessment Unit 6 75

Assess progress at the end of each lesson and unit.

Review each lesson quickly and concisely.

Teacher's Book

Follow the 5-E methodology (pages xii-xiii) across each level's activities.

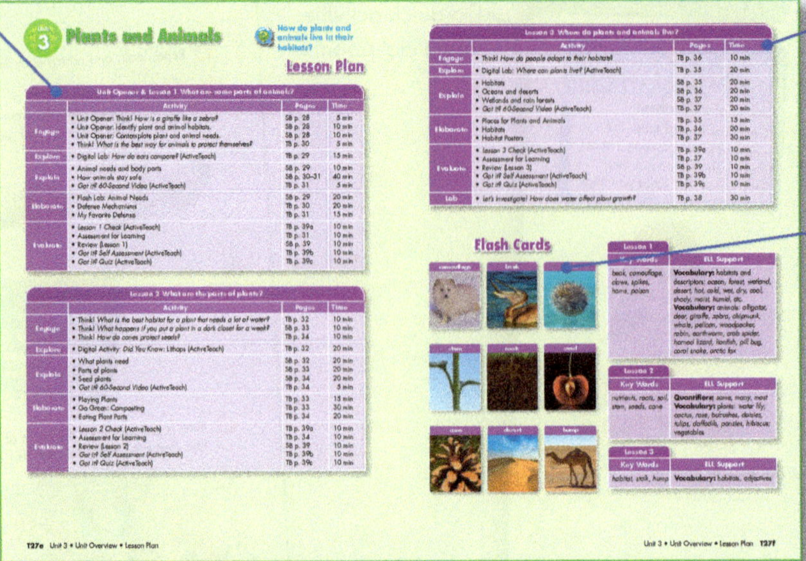
Plants and Animals
Lesson Plan
Flash Cards

Plan your lessons by selecting the activities that best suit your classroom needs, with an estimated time for each activity.

Select Flash Cards for use during the lessons.

Refer to objectives, vocabulary, and learning resources.

Plants and Animals
Introduce the Big Question

View the annotated Student's Book page for reference.

Engage students' critical thinking to start unfolding the Big Question after introducing vocabulary, key concepts, and goals.

Help students explore the topics in *Let's Explore!* Labs or *Explore My Planet!* Activities, hands-on or online.

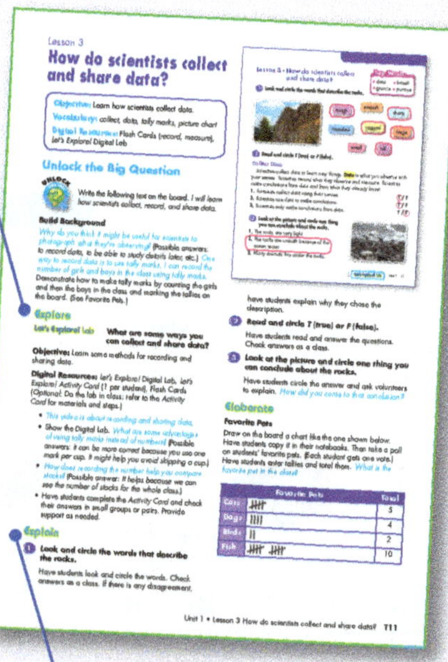
Lesson 3
How do scientists collect and share data?

Reinforce understanding of the scientific concepts through core content and activities.

Extend the scientific concepts through further activities that can be done in class or virtually.

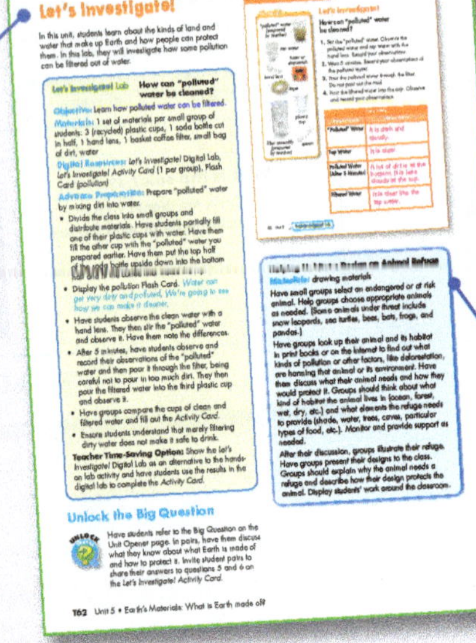
Let's Investigate!
Unlock the Big Question

Explore more ideas relating to the unit topic through additional creative activities.

Deepen students' knowledge and encourage students to elaborate on topics in creative ways.

Link to *Lesson Checks*, *Got it? 60-Second Videos*, as well as Assessment for Learning activities.

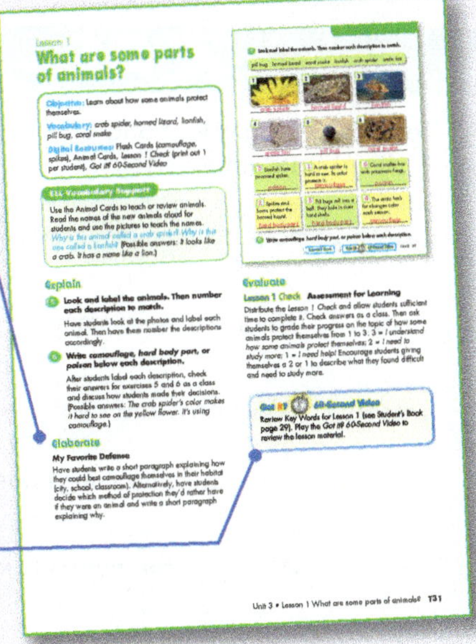

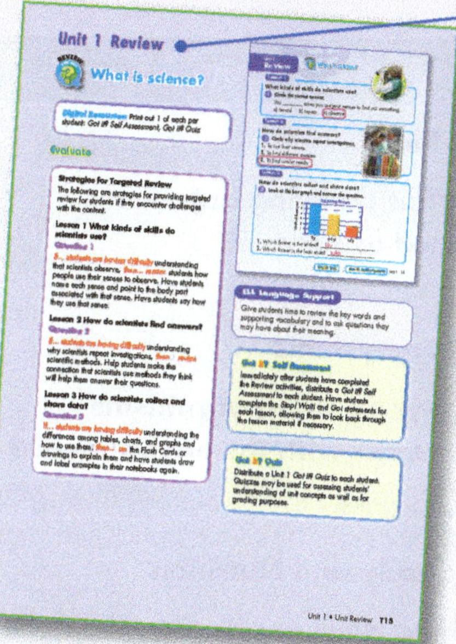

Address challenges students may have while reviewing the unit material and link to *Got it? Self Assessments* and *Got it? Quizzes*.

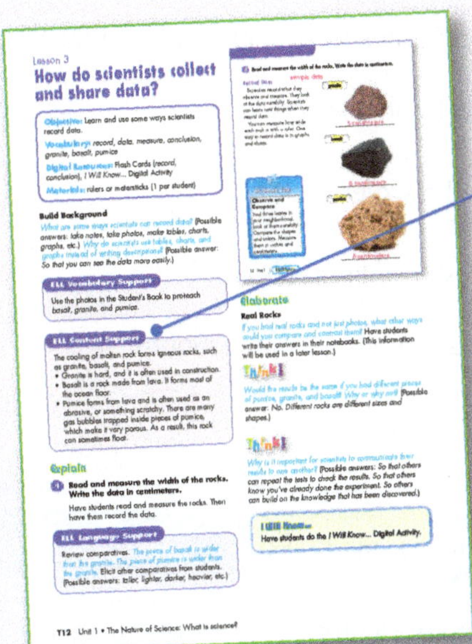

Use Study Guides to summarize the main points in each lesson and review the Big Question.

Support English Language Learners through background information, suitable activities, and skillswork strategies.

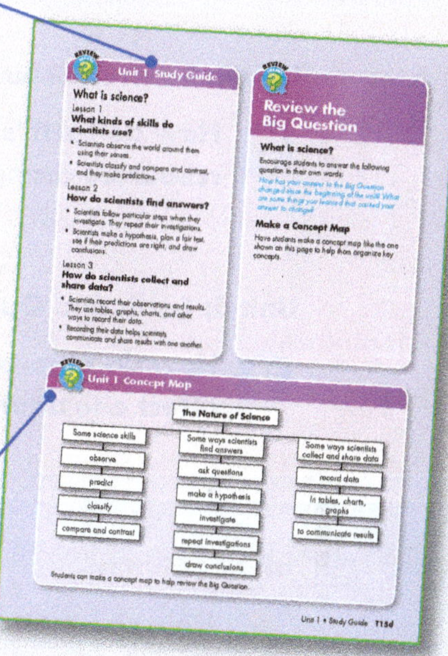

Review main unit concepts using concept maps downloaded from the ActiveTeach.

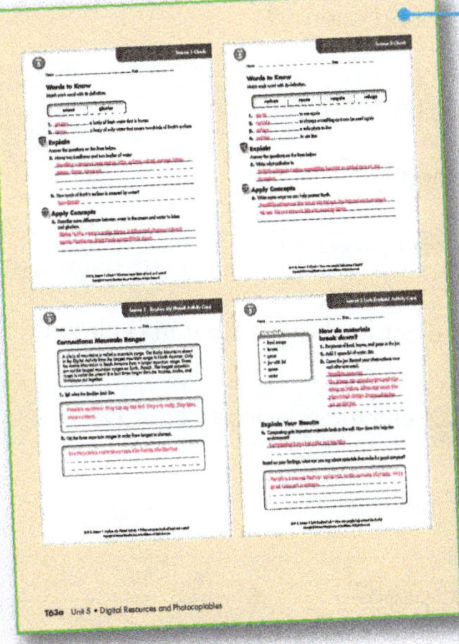

Check understanding and do exploratory activities using cards downloaded from the ActiveTeach.

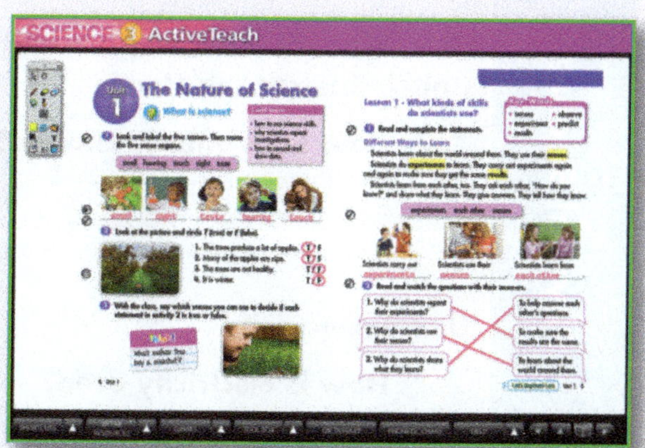

Access the digital activities, Flash Cards, and all printable resources, including *Activity Cards* and *Quizzes*, in the ActiveTeach.

Scope and Sequence

	Units	Lessons
Life Science	**Unit 1: Plants** **What do living organisms need to survive?**	Lesson 1: What plant and animal characteristics are inherited? Lesson 2: How do animals respond to the environment?
	Unit 2: Ecosystems **How do living organisms interact with the environment?**	Lesson 1: What are ecosystems? Lesson 2: What are food chains and food webs? Lesson 3: How do living things affect the environment?
	Unit 3: Body and Nutrition **How does my diet affect my health?**	Lesson 1: What are the nutrients in my food? Lesson 2: What are healthy and unhealthy diets?
Earth Science	**Unit 4: Earth's Resources** **How do Earth's resources change?**	Lesson 1: How can Earth's surface change rapidly? Lesson 2: Where is Earth's water? Lesson 3: What is the water cycle?
	Unit 5: Earth and Our Universe **How do objects in space affect one another?**	Lesson 1: How do star patterns change? Lesson 2: What are the phases of the moon? Lesson 3: What is the solar system?
Physical Science	**Unit 6: Matter** **How can matter be described and measured?**	Lesson 1: How is matter measured? Lesson 2: What are mixtures? Lesson 3: How does matter change?
	Unit 7: Energy and Heat **How does energy change?**	Lesson 1: What is sound energy? Lesson 2: What is light energy? Lesson 3: What is heat?
	Unit 8: Electricity **How is electricity used?**	Lesson 1: What is static electricity? Lesson 2: How do electric charges flow in a circuit? Lesson 3: How does electricity transfer energy?
	Unit 9: Motion **How can motion be described and measured?**	Lesson 1: What is motion? Lesson 2: What is speed?

I will learn...	Key Words
• what plant and animal characteristics are inherited.	• *characteristics, offspring, heredity, inherit, competition, camouflage, advantage*
• how animals respond to the environment.	• *behavior, stimulus, instinct, migration, protection, hibernation*
• what ecosystems are.	• *ecosystem, tundra, rain forest, desert, grassland, wetland, habitat, population*
• what food chains and food webs are.	• *energy, food chain, food web, resources*
• how living things affect the environment.	• *competition*
• what nutrients are in my food.	• *diet, calorie, carbohydrate, glucose, starch, fiber, cholesterol, amino acids, vegetarian*
• what are healthy and unhealthy diets.	• *grains, dairy, serving size, malnutrition, undernourished, anorexia nervosa, deficient, overnourished, obesity*
• how Earth's surface can change rapidly.	• *plates, volcano, fault, earthquake, focus, epicenter*
• where Earth's water is.	• *fresh water, glacier, ice cap, groundwater*
• what the water cycle is.	• *water vapor, evaporation, condensation, precipitation, water cycle*
• how star patterns change.	• *star, constellation, astronomer*
• about the phases of the moon.	• *eclipse, lunar eclipse, solar eclipse*
• about the solar system.	• *solar system, gravity, ellipse, planet, asteroid, comet, inner planets, outer planets*
• how matter is measured.	• *mass, volume*
• how to separate matter.	• *mixture, filtration, evaporation, condensation, solution, solute, solvent, solubility*
• how matter changes.	• *physical change, chemical change*
• what sound energy is.	• *sound, vibration, sound wave, frequency, wavelength, pitch, volume, amplitude*
• what light energy is.	• *refraction, reflection, absorption*
• what heat is.	• *heat, conduction, convection, radiation*
• what static energy is.	• *atom, electric charge, static electricity, electric force, lightning*
• how electric charges flow in a circuit.	• *electric current, circuit, battery, conductor, insulator, series circuit, parallel circuit*
• how electricity transfers energy.	• *resistor, filament, phantom energy*
• what motion is.	• *motion, relative motion, reference point, force, balanced forces, gravity, mass, weight*
• what speed is.	• *speed, velocity, direction, acceleration, average speed*

5E-Methodogy

Shake Up Science is based on the 5E-Methodology: Engage, Explore, Explain, Elaborate, Evaluate.

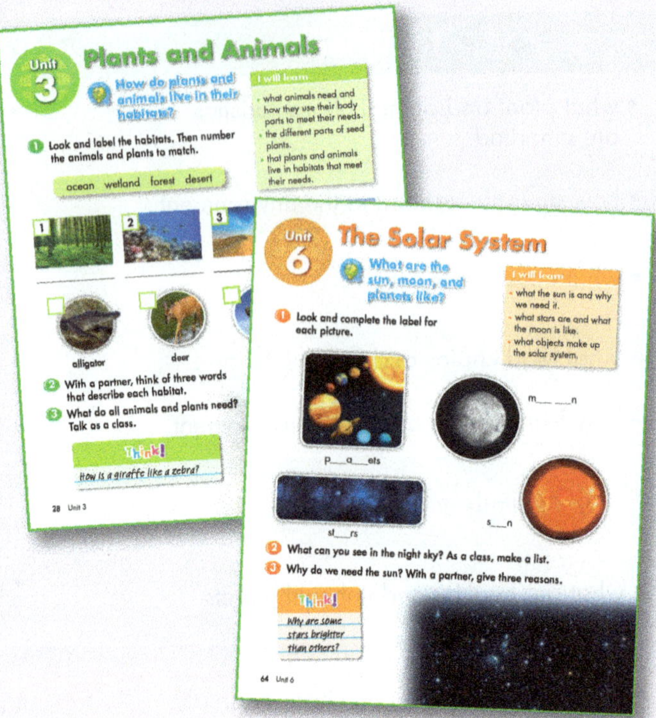
Unit 3

Plants and Animals

How do plants and animals live in their habitats?

I will learn

- what animals need and how they use their body parts to meet their needs.
- the different parts of seed plants.
- that plants and animals live in habitats that meet their needs.

1 Look and label the habitats. Then number the animals and plants to match.

ocean wetland forest desert

alligator deer

2 With a partner, think of three words that describe each habitat.

3 What do all animals and plants need? Talk as a class.

Think!

How is a giraffe like a zebra?

28 Unit 3

Unit 6

The Solar System

What are the sun, moon, and planets like?

I will learn

- what the sun is and why we need it.
- what stars are and what the moon is like.
- what objects make up the solar system.

1 Look and complete the label for each picture.

p__a__ets m____n st__rs s__n

2 What can you see in the night sky? As a class, make a list.

3 Why do we need the sun? With a partner, give three reasons.

Think!

Why are some stars brighter than others?

64 Unit 6

Engage

On the first page of every unit, students are introduced to the Big Question, the question that will guide their learning throughout the unit. On this page, students are encouraged to start engaging with the topic. *Think!* questions in the Student's Book and additional questions in the Teacher's Book help students to think critically and unfold the Big Question.

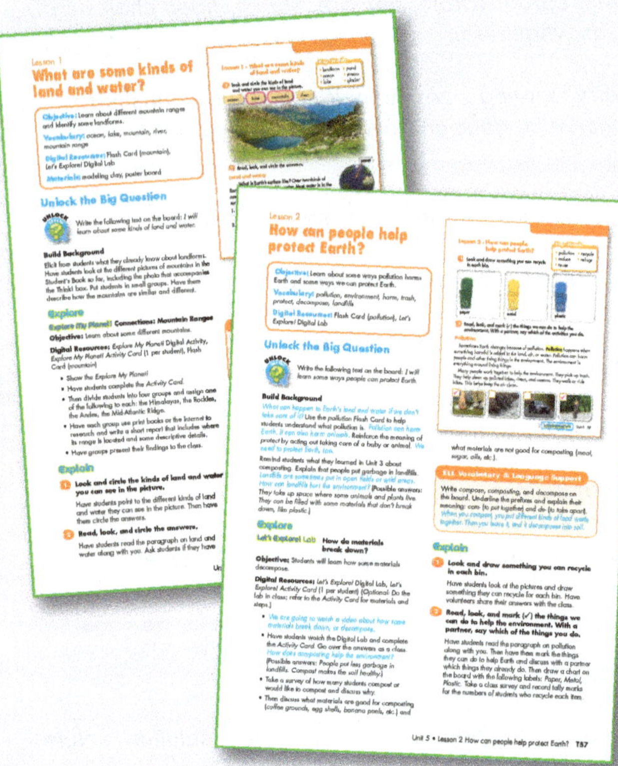
Lesson 1

What are some kinds of land and water?

Unlock the Big Question

Explore

Explain

Lesson 2

How can people help protect Earth?

Unlock the Big Question

Build Background

Explore

Explain

ELL Vocabulary & Language Support

Unit 5 • Lesson 2 How can people help protect Earth? T57

Explore

In every lesson, students have the opportunity to explore the main concepts before they start reading core content. This is done through the *Let's Explore!* Lab or *Explore My Planet!* Activity. *Let's Explore!* Labs offer opportunities to explore, hands-on, an idea relevant to the lesson. *Explore My Planet!* Activities expand on unit concepts through a short presentation of related content followed by a reflection activity. Teachers may opt to do the activities in class or show them via the ActiveTeach. Related *Activity Cards* that can be printed from the ActiveTeach help further reinforce students' activity-based learning.

Explain

After exploring a topic, students build their understanding. Students read a variety of texts that explain scientific concepts. Core content is accompanied by activities for students to do—individually, in pairs, in small groups, or as a class—to enhance understanding of the concepts. ELL Content Support boxes provide background information or activities teachers may wish to include in their lesson plans. *I Will Know...* Digital Activities can be used to reinforce and practice the main ideas. *Got it? 60-Second Videos* provide a comprehensive review of the scientific concepts covered in each lesson. At the end of every unit, *Let's Investigate!* Labs consolidate unit concepts through hands-on experiments. Teachers can opt to do the Labs hands-on in class or to show the digital materials instead. Students reflect on their learning by answering questions in their Student's Books. Related *Activity Cards* that can be printed from the ActiveTeach help further extend students' reflection.

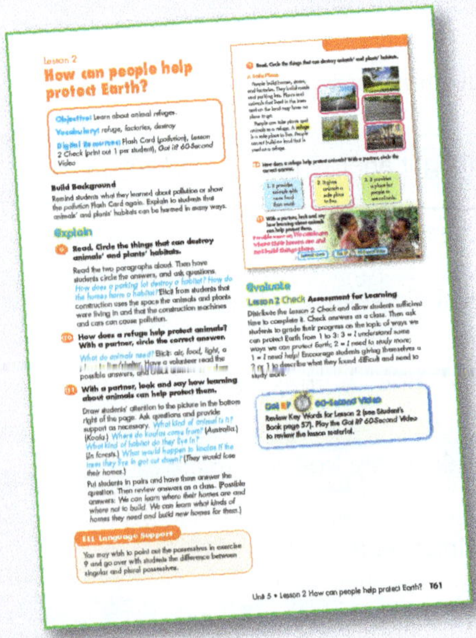
Lesson 2

How can people help protect Earth?

Build Background

Explain

Evaluate

ELL Language Support

Unit 5 • Lesson 2 How can people help protect Earth? T61

Elaborate

Throughout the units are a number of activities designed to deepen students' knowledge of the topic in fun and creative ways. In addition, class projects and other kinds of creative activities are offered at the end of every unit. Students are given various opportunities to write, perform skits, make murals, give presentations, do research, compose poems, and design inventions, among other things.

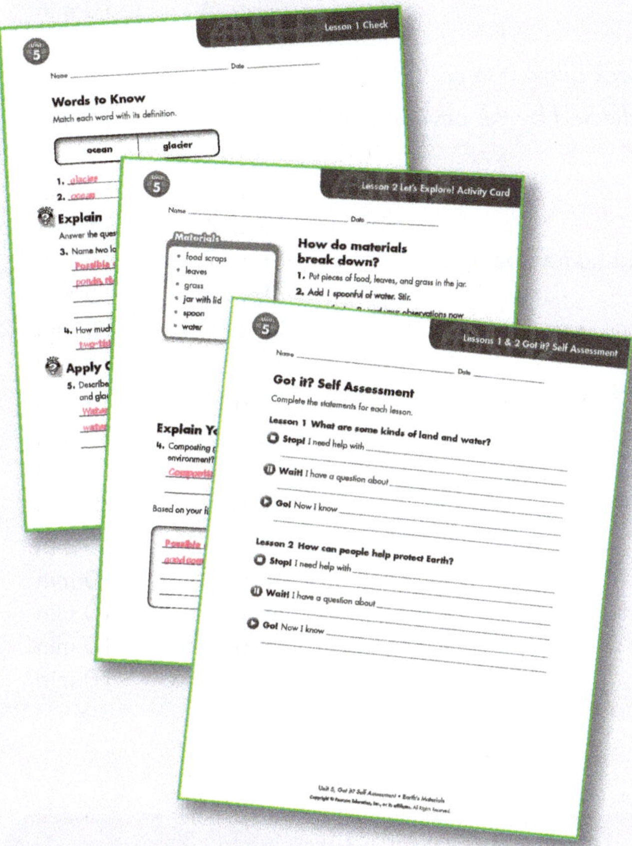

Evaluate

There are numerous instances for evaluating learning and progress. At the end of each lesson, students can watch the *Got it? 60-Second Video* to review key concepts, do a *Lesson Check*, and complete a unit review, with targeted review strategies to address challenges students may have with content. *Got it? Self Assessments* help students to assess their progress and to judge what they need to study further. There are also *Got it? Quizzes* to help evaluate understanding of each unit. Unit-specific Study Guides and Concept Maps provide clear summaries and additional tools for evaluation and review.

21st Century Skills

The 21st Century skills of critical thinking, collaboration, communication, and creativity are methodically developed across the digital and print components for each level of *Shake Up Science*. In an increasingly globally competitive workforce, it is more critical than ever to prepare students for the careers of tomorrow.

Communication Opportunities

Activities in *Shake Up Science* are highly participative and require students to collaborate and share their ideas. There are a number of opportunities in each unit for students to communicate in pairwork, groupwork, or whole class activities, to give presentations, and to express themselves through writing.

Accountability and Self-Direction

Lesson Checks at the end of each lesson and *Got it? Self Assessments* at the end of each unit encourage students to self-evaluate and to make their own judgments about what they need to review.

Critical Thinking and Problem Solving

Shake Up Science systematically cultivates students' skills of critical thinking and problem solving, with a science-related Big Question to lead each unit's learning, lots of exposure to scientific methodology, and *Think!* boxes relating to real-life topics.

Digital Literacy

Digital activities work hand-in-hand along with the print materials to help engage students and expand their understanding of scientific concepts as well as for review and feedback. Digital activities can be used in various ways in class. Throughout the series, internet research activities serve to complement the digital package, providing structured guidance for student-led exploration of the key scientific concepts.

Unit 1 Plants and Animals

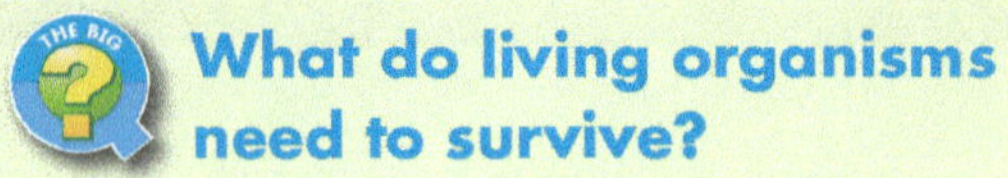

Lesson Plan

Unit Opener & Lesson 1 What plant and animal characteristics are inherited?			
	Activity	**Pages**	**Time**
Engage	• Unit Opener: Think! *What helps a coconut travel across the water?* • Unit Opener: Discuss how plants and animals protect themselves. • Unit Opener: Discuss why dogs usually bark. • Think! *How is a zebra's pattern like a fingerprint?* • Think! *What kinds of plants does the giraffe's neck allow it to eat...?*	SB p. 4 SB p. 4 SB p. 4 SB p. 6 SB p. 8	5 min 10 min 10 min 5 min 5 min
Explore	• Digital Lab: *How can some characteristics be affected by the environment?* (ActiveTeach)	TB p. 5	15 min
Explain	• Characteristics of plants and animals • Inherited characteristics of plants and animals • Inherited characteristics of peacock flounder and human beings • Competition and advantage • How peppered moths evolved to survive • *Got it? 60-Second Video* (ActiveTeach)	SB p. 5 SB p. 6 SB p. 7 SB p. 8 SB p. 9 TB p. 9	15 min 15 min 15 min 15 min 15 min 5 min
Elaborate	• Science Notebook: Describing Animals • More about the Peacock Flounder • Science Notebook: Longer Necks • Flash Lab: Dimpled Cheeks	TB p. 6 TB p. 7 TB p. 8 SB p. 9	15 min 20 min 10 min 20 min
Evaluate	• *Lesson 1 Check* (ActiveTeach) • Assessment for Learning • Review (Lesson 1) • *Got it? Self Assessment* (ActiveTeach) • *Got it? Quiz* (ActiveTeach)	TB p. 15a TB p. 9 SB p. 15 TB p. 15b TB p. 15b	10 min 10 min 10 min 10 min 10 min

Lesson 2 How do animals respond to the environment?			
	Activity	**Pages**	**Time**
Engage	• Think! *What advantages do insects that look like plants have?* • Think! *How does hibernation help some animals survive?*	SB p. 11 TB p. 12	5 min 5 min
Explore	• Digital Activity: *Misconception: Echolocation* (ActiveTeach)	TB p. 10	15 min
Explain	• Animal behaviors caused by stimuli • Animal instincts • Migration, protection, and hibernation • Behaviors that develop as a result of training • *Got it? 60-Second Video* (ActiveTeach)	SB p. 10 SB p. 11 SB p. 12 SB p. 13 TB p. 13	15 min 15 min 15 min 15 min 5 min
Elaborate	• Science Notebook: Animal Behaviors • Instinctive Animal Behavior Posters • At-Home Lab: Migrating Animals • Science Notebook: Hibernation • Science Notebook: My Learned Behavior	TB p. 10 TB p. 11 SB p. 12 TB p. 12 TB p. 13	15 min 20 min 15 min 15 min 15 min
Evaluate	• *Lesson 2 Check* (ActiveTeach) • Assessment for Learning • Review (Lesson 2) • *Got it? Self Assessment* (ActiveTeach) • *Got it? Quiz* (ActiveTeach)	TB p. 15a TB p. 13 SB p. 15 TB p. 15b TB p. 15b	10 min 10 min 10 min 10 min 10 min
Lab	• *Let's Investigate! How can some fish float?* (ActiveTeach)	SB p. 14	30 min

Flash Cards

Lesson 1

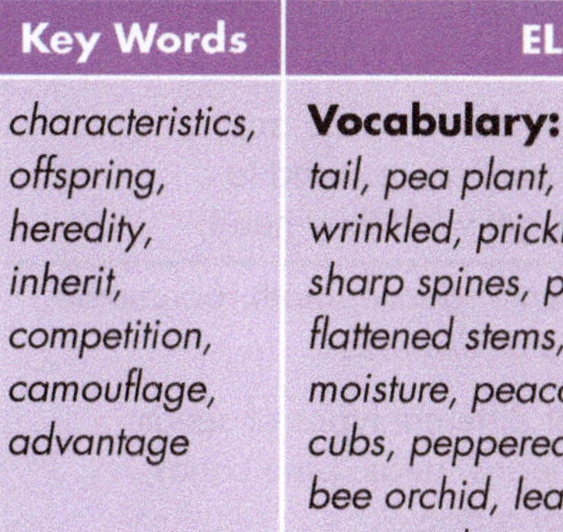
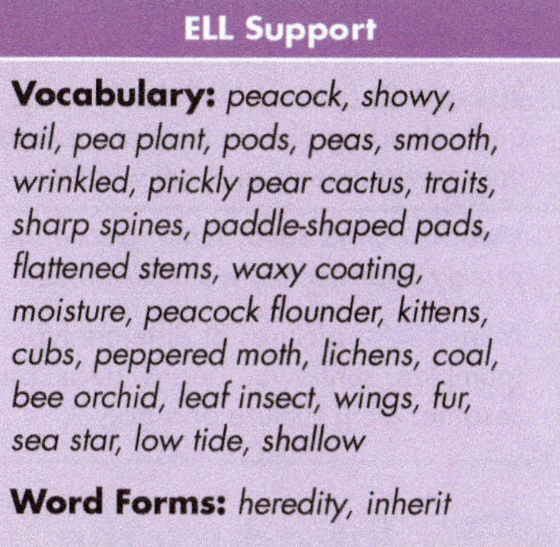

Key Words	ELL Support
characteristics, offspring, heredity, inherit, competition, camouflage, advantage	**Vocabulary:** *peacock, showy, tail, pea plant, pods, peas, smooth, wrinkled, prickly pear cactus, traits, sharp spines, paddle-shaped pads, flattened stems, waxy coating, moisture, peacock flounder, kittens, cubs, peppered moth, lichens, coal, bee orchid, leaf insect, wings, fur, sea star, low tide, shallow* **Word Forms:** *heredity, inherit*

Lesson 2

Key Words	ELL Support
behavior, stimulus, instinct, migration, protection, hibernation	**Vocabulary:** *shell, snow monkeys, response, geese, flocks, porcupine, quills, threatened* **Animal Vocabulary:** *sea star, goose, porcupine, marmot, monarch butterfly, geese, lion cub, white-crowned sparrow*

Unit 1 Plants and Animals

Unit Objectives

Lesson 1: Students will explain that plants and animals inherit characteristics that may help them survive and reproduce.

Lesson 2: Students will demonstrate an understanding of how animals respond to their environments and get what they need.

Vocabulary: *seed, turtle, goose, moth, porcupine, prickly pear cactus, water dispersal*

Materials: pictures of different types of seeds (pumpkin, sunflower, beans, peas, pine cones, sesame, etc.), picture of a dog

Introduce the Big Question

What do living organisms need to survive?

Build Background Display pictures of different types of seeds. *What is a seed? What are seeds for?* Have students brainstorm. Guide them to conclude that seeds are necessary for flowering plants to reproduce. On the board, draw a fern to remind students that not all plants produce seeds.

Engage

Think!

What helps a coconut travel across the water?

Point to the photo on the bottom right and have students identify what it is. *Did you know that coconuts are seeds? How do you think this coconut got into the water?* Ask volunteers to share their ideas.

1 Look and label.

Point to the pictures and allow students to say any words they already know. Ask students to work in pairs and write the words. Review the answers by pointing to the pictures for students to say the words.

2 Look at the animals and plants in the pictures above. How does each plant or animal protect itself? With a partner, make a list of your ideas.

In pairs, students discuss how each plant or animal protects itself. Review the answers with the whole class. (Possible answers: *Turtles hide in their shells; Geese can fly away; Moths can camouflage themselves against dark backgrounds; Porcupines have sharp quills; Cactuses have sharp spines.*)

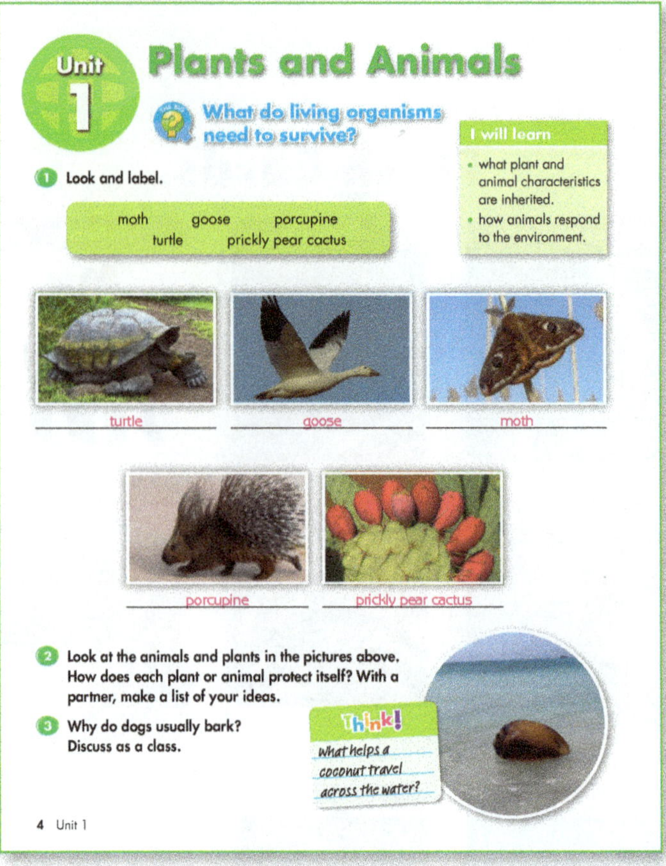

Unit 1 Plants and Animals

What do living organisms need to survive?

I will learn

- what plant and animal characteristics are inherited.
- how animals respond to the environment.

1 Look and label.

moth goose porcupine turtle prickly pear cactus

turtle

goose

moth

porcupine

prickly pear cactus

2 Look at the animals and plants in the pictures above. How does each plant or animal protect itself? With a partner, make a list of your ideas.

3 Why do dogs usually bark? Discuss as a class.

Think!

What helps a coconut travel across the water?

4 Unit 1

3 Why do dogs usually bark? Discuss as a class.

Display the picture of a dog. *Who has a pet dog? What do most dogs do?* Divide the class into small groups and have them list what most dogs do. Write students' ideas on the board. Then have the class brainstorm why dogs bark. (Possible answers: *Dogs may bark to say hello to their owners, to request attention, to show they are excited, hungry, thirsty, anxious,* etc.)

Think! Again!

Revisit the question *What helps a coconut travel across the water?* Divide the class into small groups and have them discuss. Ask volunteers to share their answers. Use board drawings to explain that coconuts are hollow in the center and have thick shells, called husks. *How do you think a coconut's hollow center and thick husk help it travel?* (Possible answer: *The hollow center helps it float, and the husk provides protection.*) *Where do you think this coconut is going?* Guide students to conclude that coconuts can float across the water until they wash up on a shore, where they can grow into new coconut trees.

Lesson 1

What plant and animal characteristics are inherited?

Objective: Learn what plant and animal characteristics are inherited.

Vocabulary: *peacock, showy, tail, pea plant, stems, leaves, flowers, pods, peas, characteristics, qualities, organism, wrinkled, smooth, parents, pass on to, offspring, heredity*

Digital Resources: Flash Cards (*characteristics, offspring*), *Let's Explore!* Digital Lab

Materials: selected Animal Cards, pictures of different animals with their babies, four bags per pair of students, 4 sets of cards per pair of students: stems (*tall, short*), flowers (*red, white*), pods (*green, yellow*), peas (*smooth, wrinkled*)

Unlock the Big Question

Write the following text on the board: *I will learn that plants and animals inherit characteristics that may help them survive and reproduce.*

Build Background On the board, draw a healthy plant in a pot close to a window and another plant with drooping leaves close to a wall. In small groups, have students discuss the differences between both plants and how the environment affects them. Discuss answers as a class.

Explore

Let's Explore! Lab **How can some characteristics be affected by the environment?**

Objective: Understand how the environment can affect plant and animal characteristics.

Digital Resources: *Let's Explore!* Digital Lab, *Let's Explore! Activity Card* (1 per student), Environmental Effect Cards Part 1 and Part 2 (1 set per group)

- Use the Flash Card to pre-teach *characteristics*.
- Show the Digital Lab.
- Demonstrate the activity, with students' help, by picking one A Card and matching it with a B Card first and then with a C Card.
- Have groups match the cards and display them together on their tables.
- Have students complete the *Activity Card* and check their answers in small groups or pairs. Provide support as needed.

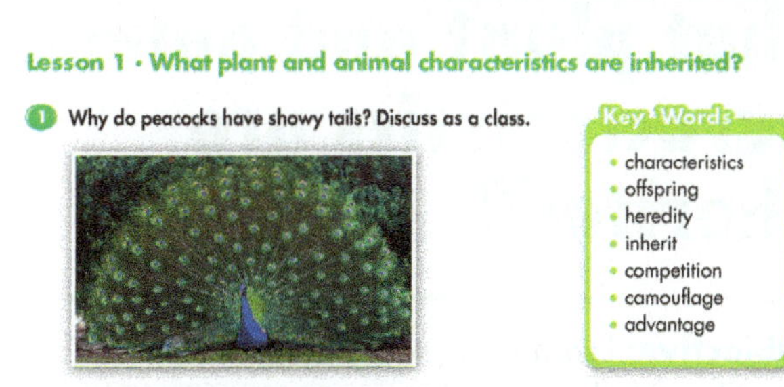

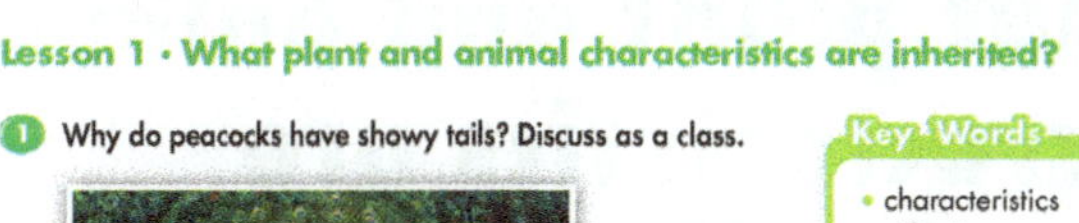

Lesson 1 • What plant and animal characteristics are inherited?

1 **Why do peacocks have showy tails? Discuss as a class.**

Key Words
- characteristics
- offspring
- heredity
- inherit
- competition
- camouflage
- advantage

2 **Read and underline the different characteristics of Mendel's pea plants.**

Characteristics of Living Things

In the middle of the nineteenth century, a monk named Gregor Mendel was hard at work in his garden. He noticed that his pea plants were not all exactly alike. All of the pea plants had stems, leaves, flowers, pods, and peas. But they also had some differences in their characteristics. **Characteristics** are the qualities an organism has. Some of the plants were tall, while others were short. Some had purple flowers, while others had white ones. The pods were green or yellow. The peas themselves were smooth or wrinkled.

pea plant

The pea plants were like their parents because of characteristics passed on to them. But Mendel found that the offspring did not always look exactly like their parents. Sometimes they had different characteristics. Some **offspring** even had different characteristics than other plants with the same parents. Mendel asked himself why. Many years later, his work became the basis for the scientific study of **heredity**, or the passing of characteristics from parents to offspring.

3 **What characteristics do most pea plants have? With a partner, make a list.**

Let's Explore! Lab Unit 1 5

Explain

1 Why do peacocks have showy tails? Discuss as a class.

Allow volunteers to describe the peacock in the picture. *Only male peacocks have showy tails. Their tails are brightly colored and attractive. Why do you think that is?* Read the question aloud. Pair students to discuss the answer. Invite pairs to share their ideas with the class. (Possible answer: *Showy tails help peacocks attract mates.*)

2 Read and underline the different characteristics of Mendel's pea plants.

Elicit the names of the parts of a pea plant and write them on the board: *stems, leaves, pods, peas.* Write the following questions on the board: *What did Gregor Mendel grow in his garden? What did he find out?* Ask students to read the first paragraph to find the answers. Write the word *characteristics* on the board and elicit its definition. Use the *offspring* Flash Card to explain that all living things receive characteristics or qualities from their parents. Have students read and underline the different characteristics Mendel's pea plants showed. Finally, elicit from students the importance of Mendel's work.

3 What characteristics do most pea plants have? With a partner, make a list.

Remind students that, although pea plants have different characteristics, most pea plants share some characteristics. Have pairs list these characteristics.

Lesson 1

What plant and animal characteristics are inherited?

Objective: Learn how plants and animals can inherit some characteristics.

Vocabulary: *inherited, prickly pear cactus, survive, environment, inherit, traits, parents, offspring, sharp spines, paddle-shaped pads, flattened stems, waxy coating, hold in, moisture*

Digital Resources: Flash Card (*offspring*), *I Will Know...* Digital Activity

Materials: picture of a fingertip, pictures of a horse, zebra, and a peacock and a peahen

Build Background Display the *offspring* Flash Card and have students discuss the similarities between the lion and the cub. As a class, discuss what characteristics make them look alike.

Explain

4 Read and write the characteristic that helps the prickly pear cactus survive in a dry environment.

Use the picture of the prickly pear cactus to pre-teach *sharp spines, paddle-shaped pads, flattened stems, waxy coating,* and *moisture*. Have students describe the environment where a prickly pear cactus lives. Students read and write a characteristic that helps this plant survive in a dry environment.

5 Read and compare zebras and horses. Write two ways they are the same and two ways they are different.

Display the pictures of a horse and a zebra. Have students read and write two ways they are the same and two ways they are different.

Elaborate

Science Notebook: Describing Animals

Have students write a description of an animal in their Science Notebooks. Provide language support as needed. Divide the class into pairs. Have students take turns reading their descriptions for their partners to guess what animal they described.

Males and Females

Display the pictures of a peacock and a peahen. Have students describe the differences between them. *Peacocks and peahens look quite different because males and females inherit different characteristics from their parents.* Ask students

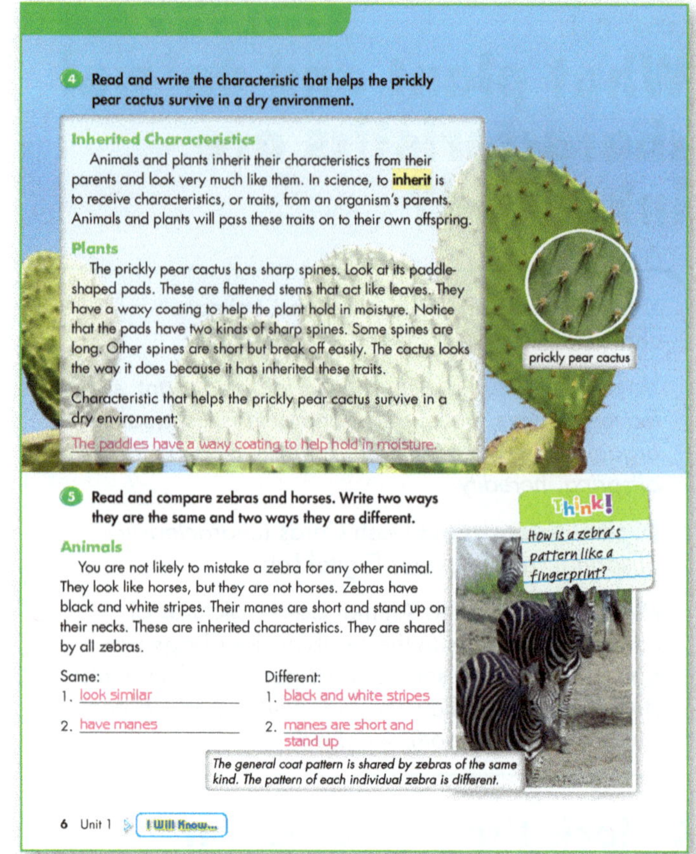

4 Read and write the characteristic that helps the prickly pear cactus survive in a dry environment.

Inherited Characteristics

Animals and plants inherit their characteristics from their parents and look very much like them. In science, to **inherit** is to receive characteristics, or traits, from an organism's parents. Animals and plants will pass these traits on to their own offspring.

Plants

The prickly pear cactus has sharp spines. Look at its paddle-shaped pads. These are flattened stems that act like leaves. They have a waxy coating to help the plant hold in moisture. Notice that the pads have two kinds of sharp spines. Some spines are long. Other spines are short but break off easily. The cactus looks the way it does because it has inherited these traits.

Characteristic that helps the prickly pear cactus survive in a dry environment:

The paddles have a waxy coating to help hold in moisture.

prickly pear cactus

5 Read and compare zebras and horses. Write two ways they are the same and two ways they are different.

Animals

You are not likely to mistake a zebra for any other animal. They look like horses, but they are not horses. Zebras have black and white stripes. Their manes are short and stand up on their necks. These are inherited characteristics. They are shared by all zebras.

Same:
1. look similar
2. have manes

Different:
1. black and white stripes
2. manes are short and stand up

Think!

How is a zebra's pattern like a fingerprint?

The general coat pattern is shared by zebras of the same kind. The pattern of each individual zebra is different.

6 Unit 1 I Will Know...

to research on the Internet another animal species whose males and females look different. Have students make a poster explaining the main differences.

ELL Vocabulary Support

Write the words *heredity* and *inherit* on the board. Write the following sentence frames on the board and have students complete them.

In science, to __________ is to receive characteristics from an organism's parents. Mendel's work became the basis for the scientific study of ________.

Think!

How is a zebra's pattern like a fingerprint?

Display the picture of a fingertip and have students discuss what they know about fingerprints. Read the question aloud and discuss the answers with the students. (Answer: *Each zebra's pattern is unique.*)

I Will Know...

Have students do the *I Will Know...* Digital Activity.

Lesson 1

What plant and animal characteristics are inherited?

Objective: Learn how the peacock flounder and human beings inherit some characteristics.

Vocabulary: *peacock flounder, flat, pattern, match, background, traits, human beings, height*

Digital Resources: Flash Card (*inherit*)

Materials: students' family photos

Build Background On the board, draw a fish, part by part, and have students guess what it is. Write the word *fish* on the board. Pair students and have them brainstorm, for two minutes, the characteristics that most fish have. Elicit fish characteristics and write them on the board. (Possible answers: *They can swim; They have scales; They have fins;* etc.)

Explain

6 Read and underline three inherited characteristics of the peacock flounder.

Point to the picture for students to describe. Have students read and underline three inherited characteristics of the peacock flounder. Check answers as a class. Ask *How does changing color help the peacock flounder survive? It can blend into its surroundings so that it is less visible to predators.*

7 Read and write three characteristics you may have inherited from your parents. Then share your answers with a partner.

Ask students to read the text on their own and to write three characteristics they may have inherited from their parents. Have students share their answers with a partner. Display the *inherit* Flash Card and have students list the characteristics the children in the picture inherited from their parents.

Think!

Do humans inherit all their characteristics from their parents? Why or why not?

Have small groups brainstorm. Then discuss as a class. (Possible answer: *No. Some characteristics are unique to each individual.*) *Sometimes people with tall parents do not grow to be as tall as their parents. Why might this happen?* (Possible answer: *They might not inherit that characteristic.*)

6 Read and underline three inherited characteristics of the peacock flounder.

Did you look twice at the fish in the photo? Something does not look quite right. The peacock flounder has both eyes on one side of its body! This flat fish is unusual in another way, too. The peacock flounder can change its color and pattern to match its background. This allows it to surprise the animals it eats as they swim by. It also hides itself from animals that would eat it. This fish looks and acts the way it does because it has inherited these traits.

7 Read and write three characteristics you may have inherited from your parents. Then share your answers with a partner.

Human Beings

People also inherit many characteristics from their parents. A person's parents may be very tall, so that person may grow to be very tall also. Height is not the only inherited characteristic. Some characteristics, such as hair and eye color, are also inherited. However, this is not always the case. Sometimes a child may grow up to be taller or shorter than his or her parents or have a different hair color.

Characteristics I may have inherited from my parents:

1. Possible answers: height,
2. hair color,
3. eye color

Unit 1 7

Elaborate

More about the Peacock Flounder

What makes the peacock flounder different from other fish? Elicit the fish's characteristics and write them on the board. *Why does the peacock flounder look and act the way it does? Because it has inherited these traits.* Tell students that these fish have other characteristics that also make them different. Write the following questions on the board for students to research: *Where do peacock flounder live? What do they eat? Why are they called that? What makes baby flounder different from their parents? How do adult flounder swim?* Once students have researched, divide the class into trios and have them share their information. Then ask each group to illustrate the information and label a poster that shows the peacock flounder's characteristics. Have each group present its poster to the class.

Science Notebook: Characteristics I Inherited from My Parents

Have each student bring a photograph of their family and write in their Science Notebooks the characteristics they inherited from each of their parents. Pair students and have them share their photos and discuss their characteristics.

Lesson 1

What plant and animal characteristics are inherited?

Objective: Learn how animals with different characteristics compete.

Vocabulary: *parents, offspring, advantages, give birth, kittens, cubs, competition, resources, eyesight, sense of smell, male, pass characteristics on to offspring*

Materials: pictures of puppies from the same litter, pictures of a lion and a giraffe, selected Animal Cards

Build Background Display the pictures of a lion and a giraffe. Divide the class into two groups, A and B. In three minutes, group A will write as many lion characteristics as they can and group B as many giraffe characteristics as they can. Check answers as a class. The winning team will be the one that listed more characteristics.

Think!

What kinds of plants does the giraffe's neck allow it to eat more easily than other animals?

Point to the picture of the two giraffes. Encourage students to say where giraffes live and what they eat.

Explain

8 Read and circle the animal in each situation that has the advantage. Then compare your answers with a partner.

Read the text out loud. Say the word *competition*. Have a volunteer read the sentence that defines the word. Display a picture of puppies from the same litter. *What might these puppies compete for? Food!* Have students discuss which puppies have better chances of survival and why. (Possible answer: *The ones that are bigger or stronger because they can get more food.*) Ask students to read each situation and circle the animal that has the advantage. Have pairs compare their answers. Encourage volunteers to explain the reasons for their choices.

9 Read. How did giraffes' necks get so long? Discuss as a class and write the answer.

Look at the giraffes in the picture. How do you think giraffes' necks got so long? Write students' predictions on the board. Once students read the text, have them share their ideas with the class.

8 Read and circle the animal in each situation that has the advantage. Then compare your answers with a partner.

Parents, Offspring, and Advantages

You know that baby animals look somewhat like their parents. Cats give birth to kittens, and lions give birth to lion cubs. Sometimes, offspring from the same parents can look different from each other. They may have different characteristics than other organisms of the same type. It may be easier or more difficult for the offspring with different characteristics to compete. **Competition** occurs when two or more living things need the same resources in order to survive.

1. Two lion cubs are running after a rabbit. Which lion cub catches the rabbit?
   a. The lion cub that is hungrier.
   b. The lion cub that is faster.
   c. The lion cub that is bigger.

2. Two dogs are looking for a hidden piece of meat. Which dog finds the meat?
   a. The dog with better eyesight.
   b. The dog with the bigger mouth.
   c. The dog with the better sense of smell.

9 Read. How did giraffes' necks get so long? Discuss as a class and write the answer.

One example that shows competition is in giraffes. Male giraffes use their long necks to fight with other males. The winner of the fight is more attractive to female giraffes. This male reproduces. The longer and stronger a male giraffe's neck is, the better chance he has to pass these characteristics on to offspring. Over time, giraffes inherit longer and stronger necks.

Giraffes' necks got so long because the giraffes with longer necks survived and passed this characteristic to their offspring.

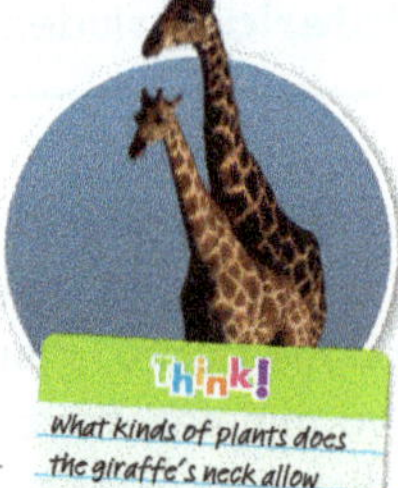

Think! What kinds of plants does the giraffe's neck allow it to eat more easily than other animals?

8 Unit 1

Think! Again!

Revisit the question from the beginning of the class: *What kinds of plants does the giraffe's neck allow it to eat more easily than other animals?* (Possible answer: *The leaves of tall trees.*)

Elaborate

Science Notebook: Longer Necks

Have students consider why a female giraffe would prefer male giraffes with longer necks rather than those with shorter necks. Have students write their ideas in their Science Notebooks. Discuss students' responses as a class.

Giraffe Facts Competition

Have students research on the Internet three interesting facts about giraffes. Then divide the class into small groups and have them share their information. The winning team will be the one that collects more facts. (Possible answers: *They are the tallest mammals in the world. They only sleep between ten minutes and two hours a day. They sleep standing up. Their tongues can be up to 45 cm long. They have four stomachs. A giraffe's heart can be 60 cm long and weigh more than ten kg. The spot pattern of each individual giraffe is different. A male giraffe can weigh about 1,400 kilograms. They are not aggressive animals. Fights between males last only a few minutes, and they hardly ever hurt each other.*)

Lesson 1

What plant and animal characteristics are inherited?

Objective: Learn how peppered moths evolved to have a dark color.

Vocabulary: *peppered moth, camouflage* (n), *background lichens, coal, die off, survive, advantage, compete*

Digital Resources: Flash Card (*camouflage*), *Lesson 1 Check* (print out 1 per student), *Got it? 60-Second Video*

Build Background Use board drawings to pre-teach *peppered moth, camouflage,* and *lichen.*

Explain

10 Read and circle *T* (true) or *F* (false). With a partner, correct the false statements.

Invite students to read the paragraph and circle the answers. Then pair students to correct the false statements. Check answers as a class.

ELL Content Support

The peppered moth is one of the best-known examples of evolution by natural selection and is often referred to as Darwin's moth. During the Industrial Revolution, the coal that was burned produced soot that darkened the trees in the industrial areas of England. Naturalists noted that the light form of the moth was more common in the countryside, while the dark moth prevailed in the sooty regions. The conclusion was that the darker moths had some sort of survival advantage in the newly darkened landscape.

11 Read again and put the steps of the evolution of the peppered moth in order (1–4).

Elicit from students how peppered moths survived before coal use increased in England. *What color did peppered moths use to be before coal use increased in England? Light gray!* Have students put the steps of the evolution of the peppered moth in order.

Elaborate

Animals That Use Camouflage

Divide the class into small groups. Have them research on the Internet animals that use camouflage. Ask students to choose one animal and make posters that illustrate how the animal uses camouflage to survive.

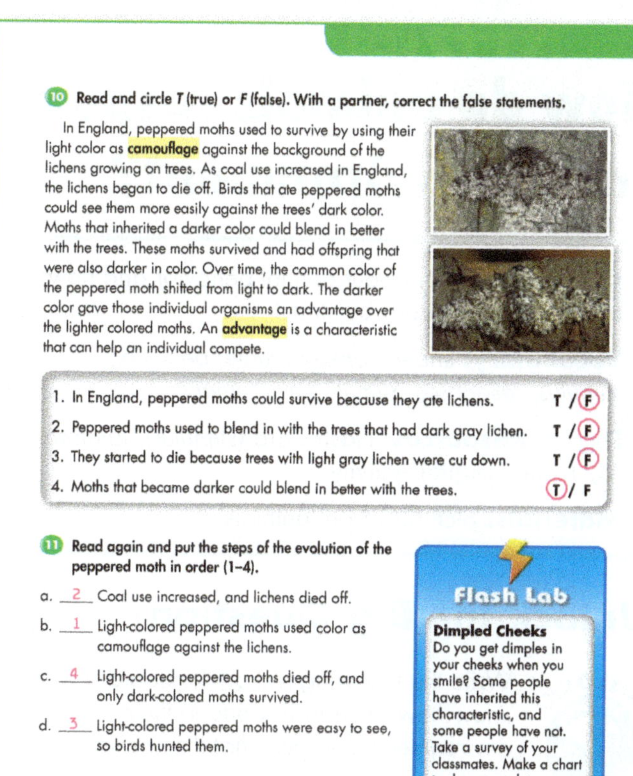

10 **Read and circle *T* (true) or *F* (false). With a partner, correct the false statements.**

In England, peppered moths used to survive by using their light color as **camouflage** against the background of the lichens growing on trees. As coal use increased in England, the lichens began to die off. Birds that ate peppered moths could see them more easily against the trees' dark color. Moths that inherited a darker color could blend in better with the trees. These moths survived and had offspring that were also darker in color. Over time, the common color of the peppered moth shifted from light to dark. The darker color gave those individual organisms an advantage over the lighter colored moths. An **advantage** is a characteristic that can help an individual compete.

1. In England, peppered moths could survive because they ate lichens. T / (F)
2. Peppered moths used to blend in with the trees that had dark gray lichen. T / (F)
3. They started to die because trees with light gray lichen were cut down. T / (F)
4. Moths that became darker could blend in better with the trees. (T) / F

11 **Read again and put the steps of the evolution of the peppered moth in order (1–4).**

a. __2__ Coal use increased, and lichens died off.
b. __1__ Light-colored peppered moths used color as camouflage against the lichens.
c. __4__ Light-colored peppered moths died off, and only dark-colored moths survived.
d. __3__ Light-colored peppered moths were easy to see, so birds hunted them.

Flash Lab

Dimpled Cheeks
Do you get dimples in your cheeks when you smile? Some people have inherited this characteristic, and some people have not. Take a survey of your classmates. Make a chart to show your data.

Lesson 1 Check | Got it? 60-Second Video | Unit 1 9

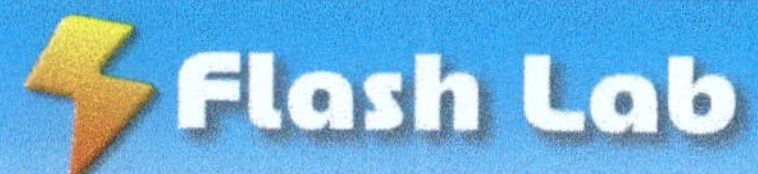

Flash Lab

Dimpled Cheeks

Explain to students what a dimple is. *Do you get dimples in your cheeks when you smile? Some people have inherited this characteristic, and some people have not.* Have students take a survey of their classmates and make a pie chart that shows the number of students with and without dimples.

Evaluate

Lesson 1 Check Assessment for Learning

Distribute the *Lesson 1 Check* and guide students as they complete it. Check answers as a class. Then ask students to grade their progress on the topic of inherited characteristics from 1 to 3: 3 = *I understand inherited characteristics in plants and animals;* 2 = *I need to study more;* 1 = *I need help!* Encourage students giving themselves a 1 or 2 to describe what they found difficult and what they need to study more.

Got it?  **60-Second Video**

Review the Key Words for Lesson 1 (see Student's Book page 5). Play the *Got it? 60-Second Video* to review the lesson material.

Lesson 2

How do animals respond to the environment?

Objective: Learn about animal behavior and how behaviors are caused by stimuli.

Vocabulary: *snow monkeys, turtle, behavior, stimulus, responses, stimuli, environment*

Digital Resources: Flash Card (*stimulus*), *Explore My Planet!* Digital Activity

Materials: pictures of pet animals

Unlock the Big Question

Write the following text on the board: *I will know how animals respond to their environments and get what they need.*

Build Background Display pictures of pet animals. Ask volunteers who have pets to describe how their pets behave when they are hungry. Write students' ideas on the board.

Explore

Explore My Planet! Misconception: Echolocation

Objective: Students will learn how bats use echolocation to locate prey.

Digital Resources: *Explore My Planet!* Digital Activity, *Explore My Planet! Activity Card* (1 per student)

- Show the *Explore My Planet!* Ask students to look at the picture and describe it.
- Read the *Explore My Planet!* with students. Remind students that animals use their senses to gather information about their environment. Have students list the five senses. Then ask them to name two senses bats use to get information about their surroundings. (Possible answers: *sight* and *hearing*)
- Ask students to work independently or in pairs to complete the *Activity Card*.
- Provide support as needed. Check answers as a class.

Explain

1 How are the monkeys in the picture responding to their environment? Discuss as a class.

Call students' attention to the picture of the monkeys. Have students describe the monkeys' surroundings.

Lesson 2 • How do animals respond to the environment?

1 How are the monkeys in the picture responding to their environment? Discuss as a class.

2 Read. Why would a turtle hide in its shell? With a partner, list two examples.

Key Words

- behavior
- stimulus
- instinct
- migration
- protection
- hibernation

Animal Behaviors

Have you ever tried to touch a turtle? If so, you may have seen a typical behavior of turtles. When a turtle feels threatened, it may pull its head inside its shell. This behavior protects the turtle from other animals.

Behaviors are the ways that animals act. Every behavior is caused by a stimulus. A **stimulus** is something that causes a reaction in a living thing. Some behaviors are responses to stimuli in the environment. When a turtle pulls its head inside its shell, it is reacting to something it has heard, seen, or smelled in its environment. Other behaviors are responses to stimuli inside an animal. For example, hunger is a stimulus that causes animals to look for food and eat.

Snow monkeys in the winter.

Turtle hiding in its shell.

A turtle would hide its head in its shell because…

1. Possible answers: it heard something.
2. it smelled something.

3 How do you respond to the following stimuli? Write your answers and compare them with a partner.

Possible answers:

1. When the weather is cold, I start to shiver.
2. When the weather is hot, I start to sweat and drink more water.
3. When I am scared, I feel my heart beat faster.

10 Unit 1 Explore My Planet!

Then ask students to explain what the surroundings suggest about the climate in which these monkeys live. Students may say that the monkeys are huddling for body warmth in the cold temperatures.

2 Read. Why would a turtle hide in its shell? With a partner, list two examples.

Point to the picture of the turtle for students to describe. Have pairs find two examples of a stimulus that might cause a turtle to pull its head into its shell.

3 How do you respond to the following stimuli? Write your answers and compare them with a partner.

A stimulus is something that causes a reaction in a living thing. What do you do when somebody tickles you? Write on the board *When somebody tickles me, I laugh. What is the stimulus that makes you laugh? Tickling!* Have students complete the sentences and compare answers with a partner.

Elaborate

Science Notebook: Animal Behaviors

Discuss animal behaviors students have observed. Have students identify what caused each behavior. For example, students may have observed a dog bark (*behavior*) when the doorbell rings (*stimulus*). Have students write in their Science Notebooks the sentence frame *A dog may bark when* ________. and complete it with as many options as they can. Check answers as a class.

Lesson 2

How do animals respond to the environment?

Objective: Learn about animal instincts.

Vocabulary: *bee orchid, leaf insect, inherit, physical characteristics, wings, fur, sea star, low tide, shallow, suck, pant*

Digital Resources: *I Will Know...* Digital Activity

Materials: selected Animal Cards or pictures of different animals

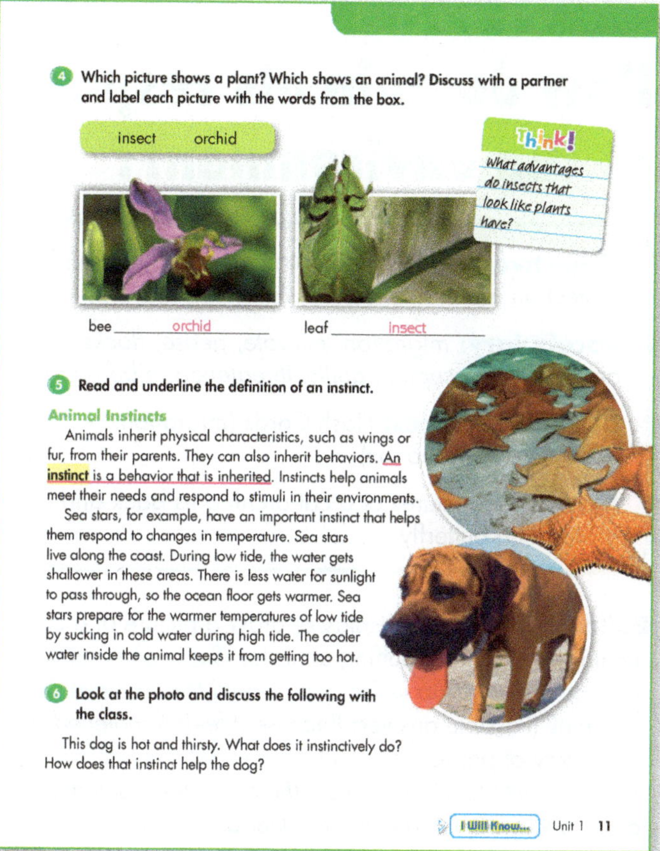

4 Which picture shows a plant? Which shows an animal? Discuss with a partner and label each picture with the words from the box.

insect orchid

Think!
What advantages do insects that look like plants have?

bee orchid

leaf insect

5 Read and underline the definition of an instinct.

Animal Instincts

Animals inherit physical characteristics, such as wings or fur, from their parents. They can also inherit behaviors. An instinct is a behavior that is inherited. Instincts help animals meet their needs and respond to stimuli in their environments.

Sea stars, for example, have an important instinct that helps them respond to changes in temperature. Sea stars live along the coast. During low tide, the water gets shallower in these areas. There is less water for sunlight to pass through, so the ocean floor gets warmer. Sea stars prepare for the warmer temperatures of low tide by sucking in cold water during high tide. The cooler water inside the animal keeps it from getting too hot.

6 Look at the photo and discuss the following with the class.

This dog is hot and thirsty. What does it instinctively do? How does that instinct help the dog?

I Will Know... Unit 1 11

Build Background Display pictures of different animals on the board. Taking turns, volunteers write below each picture the physical characteristics each animal inherited from its parents.

Explain

4 Which picture shows a plant? Which shows an animal? Discuss with a partner and label each picture with the words from the box.

Call students' attention to the two pictures at the top of the page. Invite pairs to discuss which picture shows an animal. Explain that some plants and animals disguise themselves by blending in with their surroundings in order to hide from predators or prey.

Think!

What advantages do insects that look like plants have?

Review with students what they already know about camouflage. Discuss answers to the question with the class.

5 Read and underline the definition of an instinct.

Ask students to read and underline the definition of an instinct. Guide them to conclude that, not only do animals inherit physical characteristics, but they also inherit behaviors. Use board drawings to explain how sea stars use instinctive behavior to survive.

ELL Content Support

Are only basic behaviors instinctive?

Students may think that only basic behaviors, such as a baby grabbing an object that touches its hand, are instinctive. Many instinctive behaviors are more involved. Spiders instinctively spin webs using different types of silk. The threads in the middle of the web are sticky, so they are more likely to trap prey. The threads on the outside are not adhesive, which allows the spider to move along them easily to reach the prey.

6 Look at the photo and discuss the following with the class.

Have pairs describe and predict what is happening to the dog in the picture. *Dogs cannot sweat through their skin like we do. When their body temperature rises, they pant to make air circulate through their bodies to cool down. Panting is an instinctive behavior. What else can dogs do instinctively when they are hot and thirsty?* Discuss answers with the class.

Elaborate

Instinctive Animal Behavior Posters

Distribute an Animal Card to each student. Have students draw and color their animals on sheets of construction paper. Then ask each student to research three instinctive behaviors of their animal on the Internet. Have students draw and write how those behaviors help it survive in its environment. Display the posters on the classroom walls and ask students to present them to the class.

I Will Know...

Have students do the *I Will Know...* Digital Activity.

Lesson 2

How do animals respond to the environment?

Objective: Learn that hibernation, migration, and protection are examples of instinctive behavior.

Vocabulary: *migration, migrate, geese, flocks, protection, porcupine, quills, threatened, hibernation*

Digital Resources: Flash Cards (*migration, protection, hibernation*)

Materials: pictures of a cat arching its back and of a Monarch butterfly

Build Background Display or draw a picture of a cat arching its back and puffing up its fur. *Suppose a cat arches its back and puffs up its fur. Why would the cat have that reaction?* (Possible answer: *Because it feels threatened and it is a way of protecting itself.*) *How might this behavior help the cat?* (Possible answer: *The cat makes itself appear larger and more threatening to other animals.*)

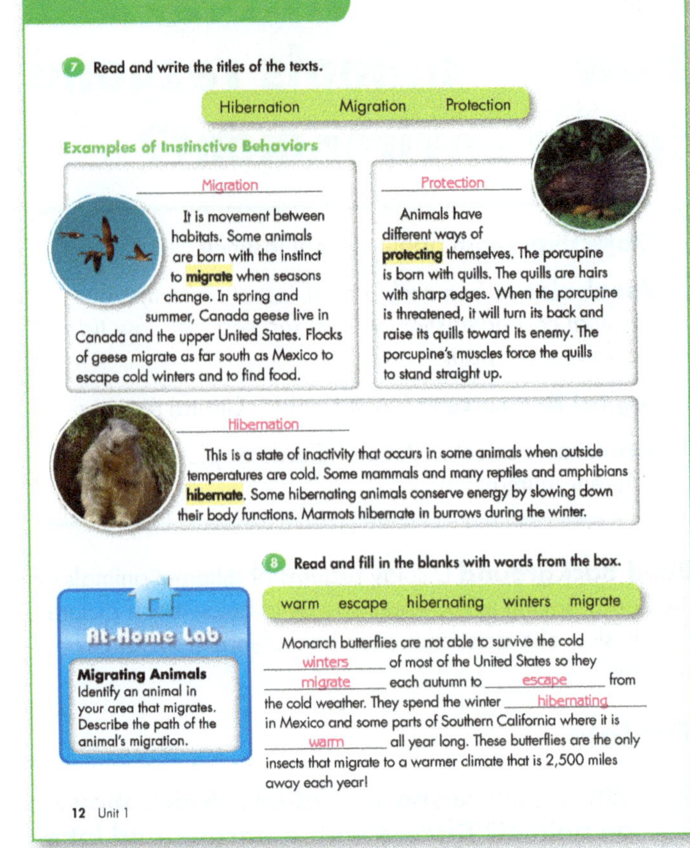

7 Read and write the titles of the texts.

Hibernation Migration Protection

Examples of Instinctive Behaviors

Migration

It is movement between habitats. Some animals are born with the instinct to **migrate** when seasons change. In spring and summer, Canada geese live in Canada and the upper United States. Flocks of geese migrate as far south as Mexico to escape cold winters and to find food.

Protection

Animals have different ways of **protecting** themselves. The porcupine is born with quills. The quills are hairs with sharp edges. When the porcupine is threatened, it will turn its back and raise its quills toward its enemy. The porcupine's muscles force the quills to stand straight up.

Hibernation

This is a state of inactivity that occurs in some animals when outside temperatures are cold. Some mammals and many reptiles and amphibians **hibernate**. Some hibernating animals conserve energy by slowing down their body functions. Marmots hibernate in burrows during the winter.

8 Read and fill in the blanks with words from the box.

warm escape hibernating winters migrate

Monarch butterflies are not able to survive the cold winters of most of the United States so they migrate each autumn to escape from the cold weather. They spend the winter hibernating in Mexico and some parts of Southern California where it is warm all year long. These butterflies are the only insects that migrate to a warmer climate that is 2,500 miles away each year!

At-Home Lab

Migrating Animals
Identify an animal in your area that migrates. Describe the path of the animal's migration.

12 Unit 1

Explain

7 Read and write the titles of the texts.

Cats' arching their backs is another example of instinctive behavior. We are going to read about three more examples of instinctive behavior.
Cover the names of the *migration*, *protection*, and *hibernation* Flash Cards and display the cards. Have students predict what kind of instinctive behavior they think these animals may have. Have students read and label each text.

ELL Content Support

Exploit the opportunity to review animal vocabulary seen in this unit. Use board drawings for students to guess what you draw: *goose/geese, porcupine, marmot, monarch butterfly, peacock/peahen, mane, lion cub, wings, fur, sea star,* etc.

8 Read and fill in the blanks with words from the box.

Display the Monarch butterfly picture. *What kind of instinctive behavior do you think Monarch butterflies use to survive?* Have students read and complete the text with the words from the box.

Elaborate

Monarch Butterfly Migration Routes

Have students research on the Internet the flight paths of Monarch butterflies. Ask them to use a map to draw the routes.

Migrating Animals

Tell students that they can use field guides at their library or on the Internet to help them identify an animal that migrates in their area. Encourage students to use a map, globe, or atlas to locate and trace the path of the animal's migration.

Science Notebook: Hibernation

Have students make a Frayer model for hibernation in their Science Notebooks. Tell students to write the word *Hibernation* in the center circle on the page and its definition at the top left. Have students list some of the characteristics of hibernation at the top right. In the lower left, have students find and fill in examples, such as deer mice, prairie dogs, chipmunks, grizzly bears, and hedgehogs. The lower right is for non-examples, such as pandas.

ELL Content Support

True Hibernation
Most animals do not truly hibernate. In true hibernation, body temperature drops to almost 0 °C, metabolism slows down, and heart rate drops. A true hibernator may appear to be dead. It must wake up every few days to eat, however. Ground squirrels and bats are examples of true hibernators.

Lesson 2

How do animals respond to the environment?

Objective: Learn how some behaviors may develop as a result of training.

Vocabulary: *training, skunk, hunt, prey* (n), *learned behavior, pride, pounce, sparrow*

Digital Resources: Flash Card (*offspring*), *Lesson 2 Check* (print out 1 per student), *Got it? 60-Second Video*

Build Background Ask the lesson question again and allow students to say what they have learned so far about how animals respond to the environment.

Explain

9 Read and underline the sentence that tells how the lion cub learns to hunt its prey.

Display the *offspring* Flash Card. Have students discuss what the lion cub eats while it is a baby and what it will have to learn as it gets older. Students read and underline how the lion cub learns to hunt its prey.

10 With a partner, name two behaviors that human babies might learn from their parents.

Point to the picture and encourage students to describe what is happening. (*The mother is showing the baby girl how to brush her teeth.*) Pairs discuss two behaviors that human babies might learn from their parents. Check answers as a class. Guide students to conclude that babies learn many things by observing and imitating their parents.

11 Read. If this adult sparrow cannot complete its song, what can you conclude? Discuss as a class.

Have students look at the sparrow in the picture and discuss how sparrows learn to sing. Read the question aloud before students read the text. Have students discuss why an adult sparrow might not be able to complete its song. (Answer: *Because it was separated from its parents when it was young.*)

9 Read and underline the sentence that tells how the lion cub learns to hunt its prey.

Learned Behavior

Not all behaviors are instinctual. Some behaviors develop as a result of training or changes in experience. Young animals learn many things as they interact with the environment. A dog that attacks a skunk may get sprayed with a bad-smelling liquid. The dog may learn to keep away from skunks.

Human babies learn many things by observing their parents. Young animals do, too. Lion cubs learn to hunt by watching older lions. A pride, or group of lions, often hunts together. Zebras are common prey for lions. A herd of zebras keeps safe from attack by staying together. When a zebra is separated from the herd, the lions will chase it toward a group of lions that is hiding. The lions will then pounce on their prey. A lion cub learns to pounce on its prey by pouncing on its mother's twitching tail. Learning the pouncing behavior helps the lion cub survive and get the food it needs.

10 With a partner, name two behaviors that human babies might learn from their parents.

11 Read. If this adult sparrow cannot complete its song, what can you conclude? Discuss as a class.

Learning and Instinct Combined

Some behaviors are partly instinctive and partly learned. The white-crowned sparrow inherits the ability to recognize the song its species sings. But knowing how to sing the song is not inherited. Sparrows must learn the song from their parents. Scientists have found that young sparrows that are separated from their parents never learn to sing the complete song.

Humans inherit the ability to learn much more than animals can learn. For example, humans inherit the ability to learn language. But we are not born knowing English, Spanish, or Chinese. We must learn the words used in our language.

Unit 1 13

Elaborate

Science Notebook: My Learned Behavior

Have students work in small groups and brainstorm different ways people learn. For example, young children might learn how to tie their shoes by first having a family member show them (observation) and then by trying it themselves (hands-on learning) until they can tie their shoes themselves. Then have students list basic behaviors they have learned, such as eating with a spoon or riding a bike, who taught them, and how they learned the behaviors.

Evaluate

Lesson 2 Check Assessment for Learning

Distribute the *Lesson 2 Check* and guide students as they complete it. Check answers as a class. Then ask students to grade their progress on the topic of how animals respond to the environment from 1 to 3: 3 = *I understand about how animals respond to the environment;* 2 = *I need to study more;* 1 = *I need help!* Encourage students giving themselves a 1 or 2 to describe what they found difficult and what they need to study more.

Got it?  **60-Second Video**

Review Key Words for Lesson 2 (see Student's Book page 10). Play the *Got it? 60-Second Video* to review the lesson material.

Let's Investigate!

In this unit, students learn what plant and animal characteristics are inherited and how animals respond to the environment. In this lab, students will observe how some fish use a swim bladder to float.

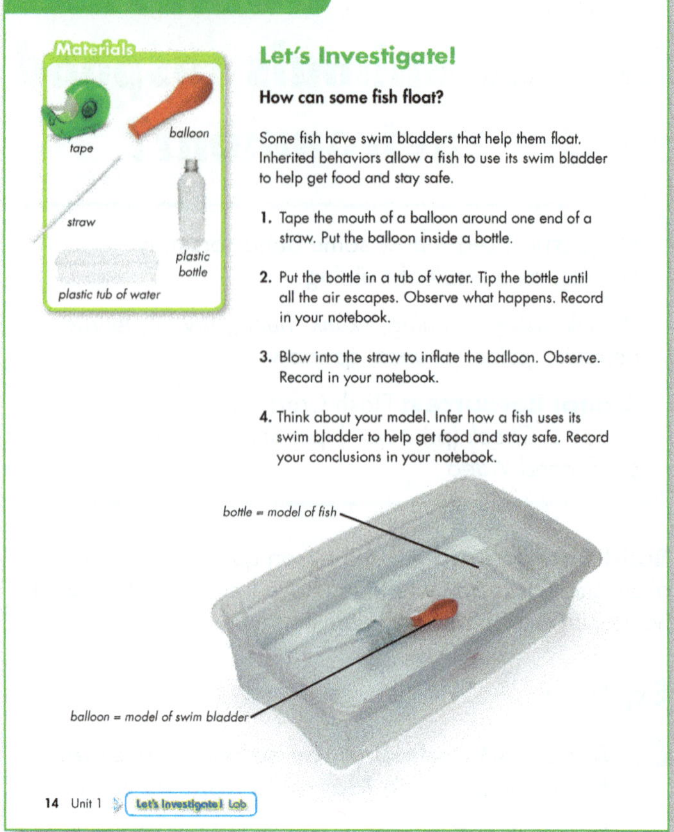

Materials

tape

balloon

straw

plastic bottle

plastic tub of water

Let's Investigate!

How can some fish float?

Some fish have swim bladders that help them float. Inherited behaviors allow a fish to use its swim bladder to help get food and stay safe.

1. Tape the mouth of a balloon around one end of a straw. Put the balloon inside a bottle.
2. Put the bottle in a tub of water. Tip the bottle until all the air escapes. Observe what happens. Record in your notebook.
3. Blow into the straw to inflate the balloon. Observe. Record in your notebook.
4. Think about your model. Infer how a fish uses its swim bladder to help get food and stay safe. Record your conclusions in your notebook.

bottle = model of fish

balloon = model of swim bladder

14 Unit 1 Let's Investigate! Lab

Let's Investigate! Lab **How can some fish float?**

Objective: Students will make a model of a swim bladder to demonstrate how fish float and sink.

Materials: 1 set of materials per small group of students: clear tape, balloon (15 cm), flexible plastic straw, clear plastic bottle (500 mL), rectangular plastic tub, water (to fill tub half full)

Digital Resources: *Let's Investigate!* Digital Lab, *Let's Investigate! Activity Card* (1 per group)

- Divide students into small groups and distribute materials.
- Ask students to tape the mouth of a balloon around one end of a straw and put the balloon inside the bottle.
- Have students put the bottle in a tub of water and tip the bottle until all the air escapes.
- Ask students to record their observations in their notebooks.
- At the end of the activity, have students share their observations with the class. Guide them to conclude that the swim bladder is an inherited characteristic that allows many types of fish to survive in water.

Teacher Time-Saving Option: Show the *Let's Investigate!* Digital Lab as an alternative to the hands-on lab activity.

Class Project: A Day in the Life

Materials: construction paper (1 sheet per group), markers

Distribute materials. Divide the class into small groups. Ask students to choose one of the animals that they read about in the unit and write and illustrate a cartoon strip about a day in the life of this animal. Students should describe the behaviors, both instinctual and learned, that help the animal to escape predators and/or function as a predator. Display the cartoon strips on the classroom walls for the class to read.

Unlock the Big Question

Have students refer to the Big Question on the Unit Opener page. In pairs, have them recall what they have learned about what plants and animals need to survive. Invite student pairs to share their answers to question 5 on the *Let's Investigate! Activity Card.*

Unit 1 Review

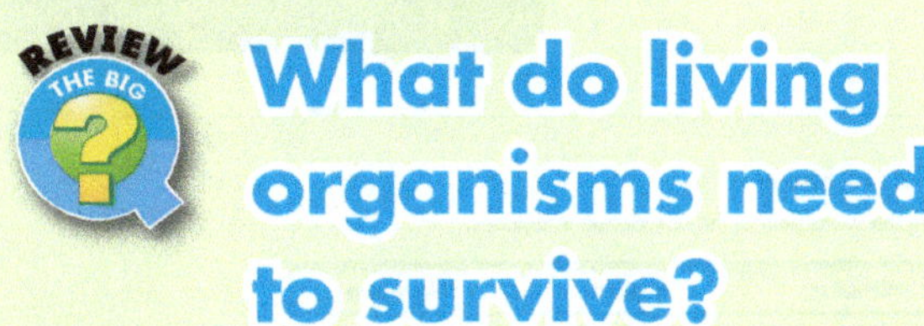

What do living organisms need to survive?

Digital Resources: Print out 1 of each per student: *Got it? Self Assessment*, *Got it? Quiz*

Evaluate

Strategies for Targeted Review

The following are strategies for providing targeted review for students if they encounter challenges with the content.

Lesson 1 What plant and animal characteristics are inherited?

Question 1

If... students are having difficulty completing the sentences, then... direct students to Lesson 1. Encourage them to look back at the texts where the words appear and read them in context.

Question 2

If... students are having difficulty remembering characteristics baby giraffes inherit from their parents, then... direct students to page 8 and have them look at the information about giraffes at the bottom of the page.

Lesson 2 How do animals respond to the environment?

Question 3

If... students are having difficulty deciding how to match the information, then... direct students to look back over Lesson 2 and find the words in the text to help understand them.

Question 4

If... students are having difficulty identifying the instinctive and learned behaviors, then... direct students to look back over Lesson 2 and find the definitions of these types of behaviors. Then elicit examples.

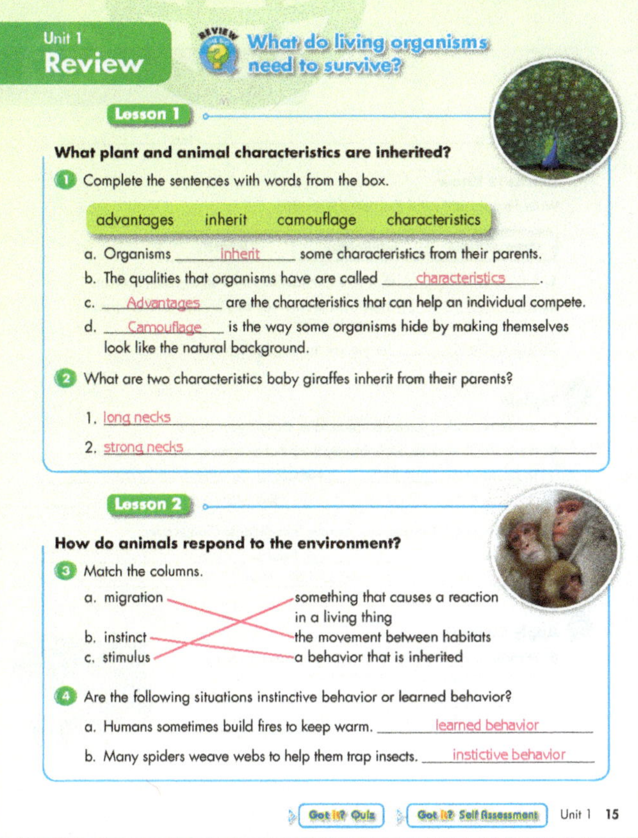

Unit 1 Review

What do living organisms need to survive?

Lesson 1

What plant and animal characteristics are inherited?

1 Complete the sentences with words from the box.

advantages inherit camouflage characteristics

a. Organisms ___inherit___ some characteristics from their parents.
b. The qualities that organisms have are called ___characteristics___.
c. ___Advantages___ are the characteristics that can help an individual compete.
d. ___Camouflage___ is the way some organisms hide by making themselves look like the natural background.

2 What are two characteristics baby giraffes inherit from their parents?

1. long necks
2. strong necks

Lesson 2

How do animals respond to the environment?

3 Match the columns.

a. migration — the movement between habitats
b. instinct — a behavior that is inherited
c. stimulus — something that causes a reaction in a living thing

4 Are the following situations instinctive behavior or learned behavior?

a. Humans sometimes build fires to keep warm. ___learned behavior___
b. Many spiders weave webs to help them trap insects. ___instictive behavior___

Got it? Quiz Got it? Self Assessment Unit 1 15

ELL Language Support

Before students start working on the Review activities, have them read each question aloud along with you.

Got it? Self Assessment

Immediately after students have completed the Review activities, distribute a *Got it? Self Assessment* to each student. Have students complete the *Stop! Wait! and Go!* statements for each lesson, allowing them to look back through the lesson material if necessary.

Got it? Quiz

Distribute a Unit 1 *Got it? Quiz* to each student. Quizzes may be used for assessing students' understanding of unit concepts as well as for grading purposes.

Lesson 1 Check

Name ______________________ Date ____________

Words to Know

Write the word next to the description it matches.

characteristics	advantage	inherit

1. inherit ______ to receive characteristics from an organism's parents
2. advantage ______ a characteristic that can help an individual compete in its environment
3. characteristics ______ the qualities that an organism has

Explain

Write whether each statement is true or false. Explain your choice.

4. A pea plant's environment is the only thing that determines its characteristics.
   This statement is false because a pea plant inherits some of its characteristics, such as the color of its flowers, from its parents.
5. One advantage of the peacock flounder is its ability to change color.
   This statement is true because changing color helps the peacock flounder match its background and catch food.

Apply Concepts

6. Study the picture. What are three characteristics the baby koala probably inherited from its parent?
   Possible answer: It probably inherited its black nose, gray and white fur, and four legs.

Lesson 2 Check

Name ______________________ Date ____________

Words to Know

Write the word next to the description it matches.

stimulus	instinct	migration

1. instinct ______ a behavior that is inherited
2. migration ______ the movement between habitats
3. stimulus ______ something that causes a reaction in a living thing

Explain

Write whether each statement is true or false. Explain your choice.

4. An animal's behavior is either learned or instinctive.
   This statement is false because some behaviors are partly instinctive and partly learned.
5. A dog doing tricks is an example of a learned behavior.
   This statement is true because a dog must be trained to do tricks. It is not born knowing how to do them.

Apply Concepts

6. Many spiders weave webs to help them trap insects. Why is it important for this behavior to be an instinct?
   Possible answer: It is important because spiders are not raised by parents that teach them how to weave webs to catch food. Spiders need webs as soon as they are born to catch food.

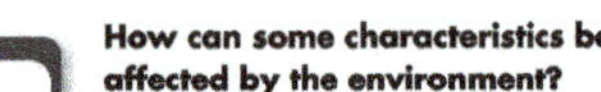

Lesson 1 Let's Explore! Activity Card

Name ______________________ Date ____________

Materials

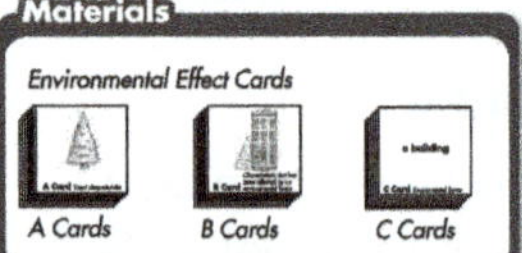

Environmental Effect Cards

A Cards *B Cards* *C Cards*

How can some characteristics be affected by the environment?

Many characteristics are inherited. Some are affected by the environment. A Cards show living things as they often appear. B Cards show how the living things may appear depending on the environment. C Cards tell what factors affected the living things.

1. Observe the living thing on an A Card. Match it with a B Card.
2. Find the matching C Card.
3. Repeat for each A Card. Compare your matches with others. Explain any differences.

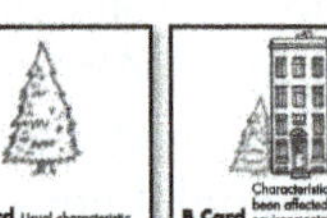

Explain Your Results

4. Pick an A Card. Explain how the characteristic was affected by the environment.
   Possible answer: The usual shape of an evergreen tree will be flattened on one side if a building is too close to the tree as it grows.

What would happen if a living thing could not change its characteristics when its environment changed?

Possible answer: The living thing would have a hard time growing, finding food, or surviving.

Lesson 2 Explore My Planet! Activity Card

Name ______________________ Date ____________

Misconception: Echolocation

Have you ever heard that bats cannot see?

A common misconception about bats is that they are blind. Some bats use something called echolocation to locate prey, such as insects. Echolocation uses sound energy. Bats make a sound and then hear the echo as it bounces off an object, such as a delicious mosquito. Then the bat knows how far away and in what direction the insect is flying. Because bats can find prey in the dark, many people have assumed they were blind. But all types of bats have eyes that can see. They can see only in black, white, and shades of gray.

1. How can some bats get information from their environments?
   Possible answer: Some bats can use echolocation or sight.

Explain one reason a bat's echolocation is useful.

Possible answer: It is useful because the bat can hunt at night.

Unit 1

Let's Investigate! Activity Card

Name ______________________ Date ____________

Analyze and Conclude

5. What would happen to the fish if its swim bladder filled with water?

Possible answer: The fish would sink.

Unit 1

Lessons 1 & 2 Got it? Self Assessment

Name ______________________ Date ____________

Got it? Self Assessment

Complete the statements for each lesson.

Lesson 1 What plant and animal characteristics are inherited?

Stop! I need help with ______________________

Wait! I have a question about ______________________

Go! Now I know ______________________

Lesson 2 How do animals respond to the environment?

Stop! I need help with ______________________

Wait! I have a question about ______________________

Go! Now I know ______________________

Unit 1

Got it? Quiz

Name ______________________ Date ____________

Got it? Quiz

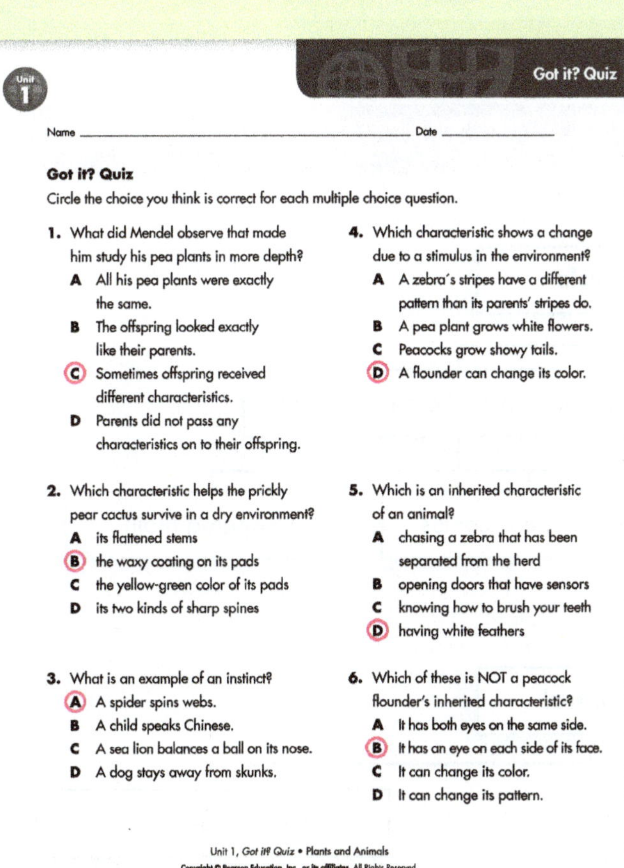

Circle the choice you think is correct for each multiple choice question.

1. What did Mendel observe that made him study his pea plants in more depth?
- **A** All his pea plants were exactly the same.
- **B** The offspring looked exactly like their parents.
- **(C)** Sometimes offspring received different characteristics.
- **D** Parents did not pass any characteristics on to their offspring.

2. Which characteristic helps the prickly pear cactus survive in a dry environment?
- **A** its flattened stems
- **(B)** the waxy coating on its pads
- **C** the yellow-green color of its pads
- **D** its two kinds of sharp spines

3. What is an example of an instinct?
- **(A)** A spider spins webs.
- **B** A child speaks Chinese.
- **C** A sea lion balances a ball on its nose.
- **D** A dog stays away from skunks.

4. Which characteristic shows a change due to a stimulus in the environment?
- **A** A zebra's stripes have a different pattern than its parents' stripes do.
- **B** A pea plant grows white flowers.
- **C** Peacocks grow showy tails.
- **(D)** A flounder can change its color.

5. Which is an inherited characteristic of an animal?
- **A** chasing a zebra that has been separated from the herd
- **B** opening doors that have sensors
- **C** knowing how to brush your teeth
- **(D)** having white feathers

6. Which of these is NOT a peacock flounder's inherited characteristic?
- **A** It has both eyes on the same side.
- **(B)** It has an eye on each side of its face.
- **C** It can change its color.
- **D** It can change its pattern.

Unit 1

Got it? Quiz

Name ______________________ Date ____________

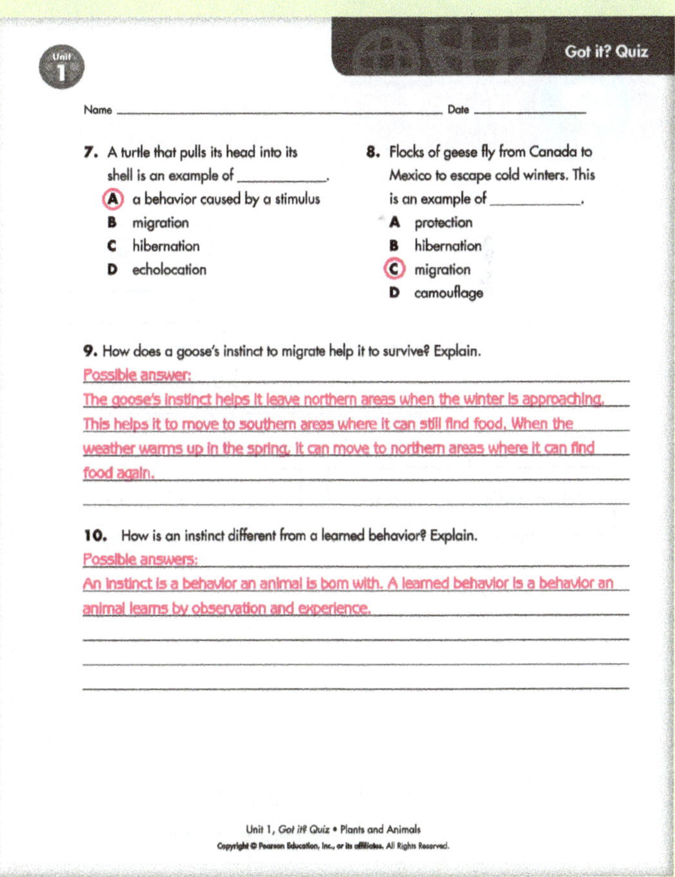

7. A turtle that pulls its head into its shell is an example of ____________.
- **(A)** a behavior caused by a stimulus
- **B** migration
- **C** hibernation
- **D** echolocation

8. Flocks of geese fly from Canada to Mexico to escape cold winters. This is an example of ____________.
- **A** protection
- **B** hibernation
- **(C)** migration
- **D** camouflage

9. How does a goose's instinct to migrate help it to survive? Explain.

Possible answer:
The goose's instinct helps it leave northern areas when the winter is approaching. This helps it to move to southern areas where it can still find food. When the weather warms up in the spring, it can move to northern areas where it can find food again.

10. How is an instinct different from a learned behavior? Explain.

Possible answers:
An instinct is a behavior an animal is born with. A learned behavior is a behavior an animal learns by observation and experience.

Unit 1 Study Guide

What do living organisms need to survive?

Lesson 1

What plant and animal characteristics are inherited?

- Organisms inherit some characteristics from their parents.
- Some characteristics may give an individual an advantage over other individuals.

Lesson 2

How do animals respond to the environment?

- Animal behaviors are responses to stimuli in the environment or stimuli within the animal. These responses can help animals survive.
- Animals inherit instinctive behaviors. Other behaviors are learned.

Review the Big Question

What do living organisms need to survive?

Have students use what they have learned from the unit to answer the question in their own words.

How has your answer to the Big Question changed since the beginning of the unit? What are some things you learned that caused your answer to change?

Make a Concept Map

Have students make a concept map like the one shown on this page to help them organize key concepts.

Unit 1 Concept Map

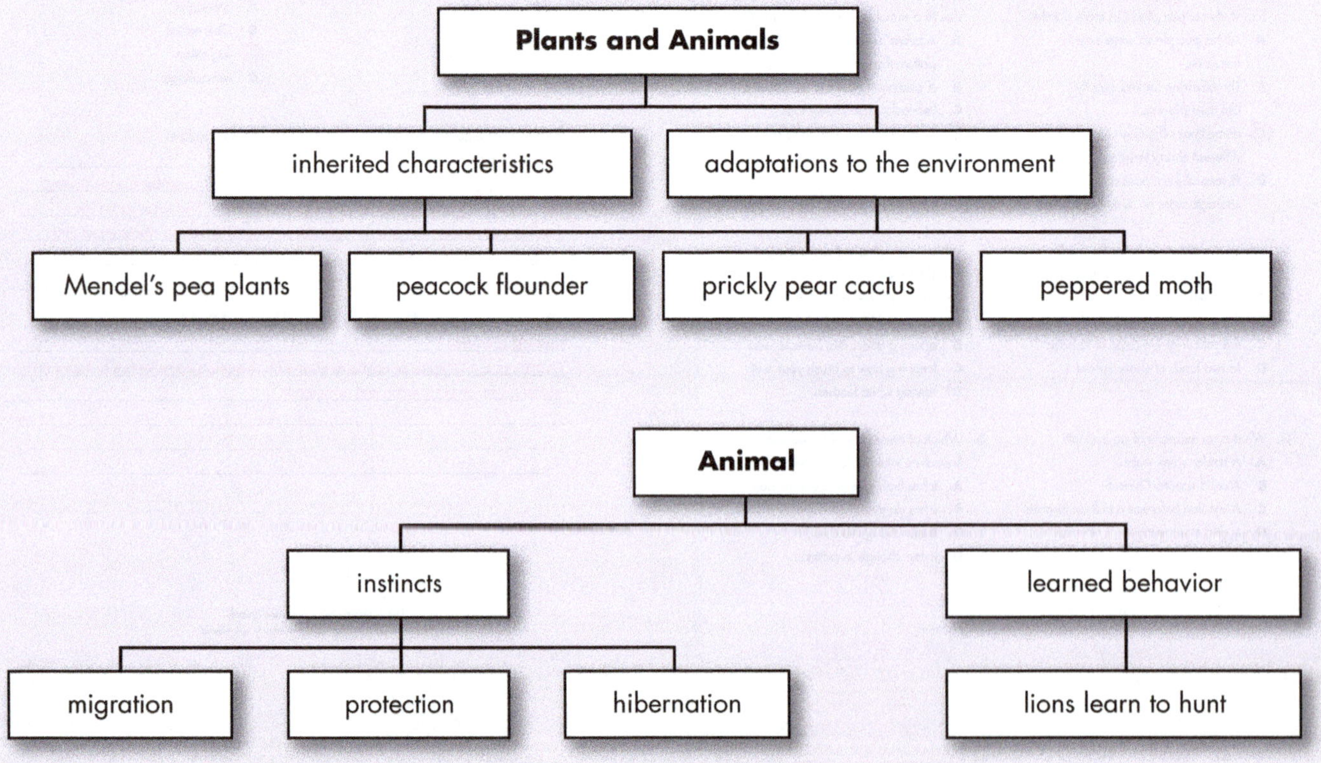

- **Plants and Animals**
  - inherited characteristics
    - Mendel's pea plants
    - peacock flounder
  - adaptations to the environment
    - prickly pear cactus
    - peppered moth

- **Animal**
  - instincts
    - migration
    - protection
    - hibernation
  - learned behavior
    - lions learn to hunt

Students can make a concept map to help review the Big Question.

Teacher's Notes

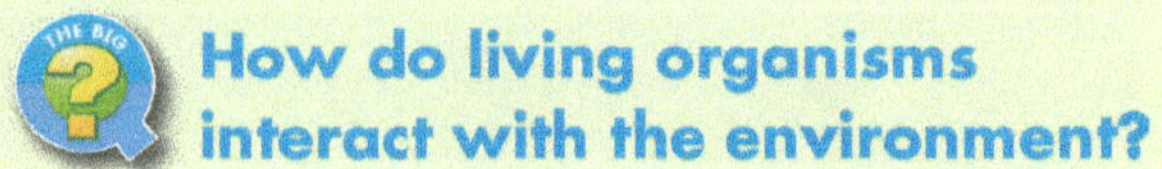

Lesson Plan

Unit Opener & Lesson 1 What are ecosystems?			
	Activity	**Pages**	**Time**
Engage	• Unit Opener: Think! *What does a manatee need to survive?*	SB p. 16	5 min
	• Unit Opener: List where some animals live.	SB p. 16	10 min
	• Unit Opener: Name what some animals eat.	SB p. 16	10 min
	• Think! *What are the primary factors that make ecosystems different?*	TB p. 18	10 min
	• Think! *What sort of structures does the platypus have that help it survive?*	SB p. 19	10 min
Explore	• Digital Activity: *Let's Blog: Alligator Farm* (ActiveTeach)	TB p. 17	15 min
Explain	• Parts of an ecosystem and kinds of ecosystems in North America	SB p. 17	15 min
	• Descriptions of five different ecosystems	SB p. 18	15 min
	• Living things within their ecosystems and the platypus' structures for survival	SB p. 19	15 min
	• *Got it? 60-Second Video* (ActiveTeach)	TB p. 19	5 min
Elaborate	• Alligators vs. Crocodiles	TB p. 17	20 min
	• Ecosystems in My Country	TB p. 18	20 min
	• Science Notebook: More about Ecosystems	TB p. 18	20 min
	• Science Notebook: Platypus Fun Facts	TB p. 19	15 min
Evaluate	• *Lesson 1 Check* (ActiveTeach)	TB p. 27a	10 min
	• Assessment for Learning	TB p. 19	10 min
	• Review (Lesson 1)	SB p. 27	10 min
	• *Got it? Self Assessment* (ActiveTeach)	TB p. 27b	10 min
	• *Got it? Quiz* (ActiveTeach)	TB p. 27c	10 min

Lesson 2 What are food chains and food webs?			
	Activity	**Pages**	**Time**
Engage	• Think! *Why are decomposers, such as the banana slug, important in a food chain?*	SB p. 21	5 min
	• Think! *In what ways are producers, consumers, and decomposers alike and different?*	TB p. 21	5 min
	• Think! *Would a mouse still be able to get food if all the insects in the ecosystem were gone?*	TB p. 22	5 min
Explore	• Digital Lab: *How do food webs show connections?* (ActiveTeach)	TB p. 20	20 min
Explain	• Flow of energy in energy pyramids and food chains	SB p. 20	15 min
	• Food chain in a forest ecosystem	SB p. 21	15 min
	• Food web in a forest ecosystem	SB p. 22	15 min
	• How populations in ecosystems can change naturally	SB p. 23	20 min
	• *Got it? 60-Second Video* (ActiveTeach)	TB p. 23	5 min
Elaborate	• Decomposers	TB p. 21	20 min
	• Food Web Posters	TB p. 22	20 min
	• Changes in Food Webs	TB p. 22	15 min
	• At-Home Lab: Decomposer's Delight	SB p. 23	15 min
Evaluate	• *Lesson 2 Check* (ActiveTeach)	TB p. 27a	10 min
	• Assessment for Learning	TB p. 23	10 min
	• Review (Lesson 2)	SB p. 27	10 min
	• *Got it? Self Assessment* (ActiveTeach)	TB p. 27b	10 min
	• *Got it? Quiz* (ActiveTeach)	TB p. 27c	10 min

Lesson 3 How do living things affect the environment?			
	Activity	**Pages**	**Time**
Engage	• Think! *Why might an invasive species grow better in a new environment than it does in its native environment?*	TB p. 24	5 min
Explore	• Digital Lab: *What happens when one part of an ecosystem is removed?* (ActiveTeach)	TB p. 24	20 min
Explain	• How organisms compete to survive and how plants cause change • How animals and humans cause change in the environment • *Got it? 60-Second Video* (ActiveTeach)	SB p. 24 SB p. 25 TB p. 25	15 min 15 min 5 min
Elaborate	• Dams Built by Humans • Science Notebook: Natural Disasters and Ecosystems	TB p. 25 TB p. 25	20 min 20 min
Evaluate	• *Lesson 3 Check* (ActiveTeach) • Assessment for Learning • Review (Lesson 3) • *Got it? Self Assessment* (ActiveTeach) • *Got it? Quiz* (ActiveTeach)	TB p. 27a TB p. 25 SB p. 27 TB p. 27b TB p. 27c	10 min 10 min 10 min 10 min 10 min
Lab	• *Let's Investigate! How do earthworms meet their needs in a model of an ecosystem?* (ActiveTeach)	SB p. 26	30 min

Flash Cards

Lesson 1

Key Words	ELL Support
ecosystem, tundra, rain forest, desert, grassland, wetland, habitat, population	**Vocabulary:** *climate, soil, sandy soil, fertile soil, arctic fox, mountain lion, banana slug, fir tree, coyote, grasshopper, prairie chicken, water lily, cypress tree, raccoon, alligator, shrub, living, nonliving, species, wildebeest, plains, platypus, webbed feet, fur* **Irregular Plurals:** *caribou–caribou, cactus–cactuses or cacti, bison–bison, mouse–mice, fish–fish, moose–moose, deer–deer, species–species*

Lesson 2

Key Words	ELL Support
energy, food chain, food web, resources	**Vocabulary:** *energy pyramid, herbivore, carnivore, decomposer, producer, consumer, photosynthesis, carbon dioxide, predator, prey, fungi, bacteria, plentiful, living space*

Lesson 3

Key Words	ELL Support
competition	**Vocabulary:** *balance (n), resources, sprout, log, loosestrife, kudzu, benefit (n/v), harm (n/v), beaver, stream, dam, birdhouse, backyard, chemicals, beetles, crops*

Ecosystems

Unit Objectives

Lesson 1: Students will describe the parts of ecosystems and some examples of ecosystems.

Lesson 2: Students will explain how energy flows in a food chain and a food web. Students will also learn how some organisms compete for resources.

Lesson 3: Students will describe how a sudden change to one group of organisms affects an environment.

Vocabulary: *deer, lizard, mouse, owl, mountain lion, grasshopper, forest, rain forest, desert, grassland, wetland, tundra, manatee*

Materials: selected Animal Cards, pictures of different ecosystems, such as a forest, rain forest, desert, grassland, etc.

Introduce the Big Question

How do living organisms interact with the environment?

Build Background Display pictures of different ecosystems on the board and elicit their names. Have students, one at a time, come to the front, take an Animal Card, and write the animal's name below the picture of the ecosystem it lives in. Explain that the same animal can live in different ecosystems.

Engage

What does a manatee need to survive?

Point to the photo on the bottom right and have students identify what it is. *What animal is it? It's a manatee! What does a manatee need to survive?* Ask volunteers to share their and their reasoning with the class.

1 Look and label.

Use the photos to elicit vocabulary and teach new words. Have students label the photos. Review the answers by pointing to the pictures for students to say the words.

2 Where does each of the animals above live? With a partner, make a list of your ideas.

Explain to students that each animal, depending on the species, may live in different habitats. In pairs, have students discuss where each animal lives. Draw a six-column chart with the names of the animals as headings and write the students' answers in the corresponding columns. (Possible answers: *Deer may live in forests, grasslands, and rain forests. Lizards may live in deserts, in forests, and in rain forests. Mice may live in forests, deserts, grasslands, and cities. Owls may live in forests, tundra, and deserts. Mountain lions, also known as cougars, pumas, or catamounts, mainly live in forests, but can also be found in wetlands. Grasshoppers may live in forests, rain forests, grasslands, wetlands, and deserts.*)

Unit 2 Ecosystems

How do living organisms interact with the environment?

I will learn
- what ecosystems are.
- what food chains and food webs are.
- how living things affect the environment.

1 Look and label.

mouse owl lizard grasshopper mountain lion deer

deer | lizard | mouse | owl | mountain lion | grasshopper

2 Where does each of the animals above live? With a partner, make a list of your ideas.

3 What does each of the animals eat? Discuss as a class.

Think! What does a manatee need to survive?

16 Unit 2

3 What does each of the animals eat? Discuss as a class.

Read the question out loud. *What does each of the animals eat?* Write students' ideas on the board. Have students discuss how the six animals can be connected. (Possible answer: *The six animals can live in the forest.*) Guide students to conclude other ways the animals can be related to one another. For example, mice may eat grasshoppers, and owls and mountain lions may eat mice.

Think! Again!

Revisit the question. Explain to students that manatees live in rivers, bays, canals, and coastal areas. Divide the class into small groups and have them discuss what they think manatees need to survive. Ask volunteers to share their answers. (Possible answers: *A manatee needs warm water to live in. It also needs underwater plants to eat.*)

Lesson 1

What are ecosystems?

Objective: Learn what an ecosystem is.

Vocabulary: *ecosystem, living things, nonliving things, environment, interact, tundra, rain forest, desert, grassland, wetland, climate, soil, sandy, fertile*

Digital Resources: Flash Card (*wetland*), *Explore My Planet!* Digital Activity

Materials: pictures of an alligator and a crocodile, a globe

Unlock the Big Question

Write the following text on the board: *I will learn the parts of ecosystems and some examples of ecosystems. I will also know how specific structures of organisms help them live in their habitats.*

Build Background Display the *wetland* Flash Card. Brainstorm the living and nonliving things that can be found in this kind of ecosystem and write them on the board. Have students draw a two-column chart in their notebooks with the headings *Living Things* and *Nonliving Things*. Have pairs write the words that are on the board in the corresponding columns. Check answers as a class.

ELL Content Support

Use the pictures of an alligator and a crocodile to elicit some of the main differences between them. Alligators have shorter and wider snouts than crocodiles, which have longer and narrower snouts.

Explore

Explore My Planet! Let's Blog: Alligator Farm

Objective: Students will learn about albino alligators kept in the St. Augustine alligator farm.

Digital Resources: *Explore My Planet!* Digital Activity, *Explore My Planet! Activity Card* (1 per student)

- Use a globe to show students that alligators live in the southeast of the United States and in China. Tell them that the St. Augustine Alligator Farm Zoological Park in Florida has more than 23 species of crocodiles and alligators that visitors can see.
- Show the *Explore My Planet!*
- Have students complete the *Activity Card* and check their answers in small groups or pairs. Provide support as needed.

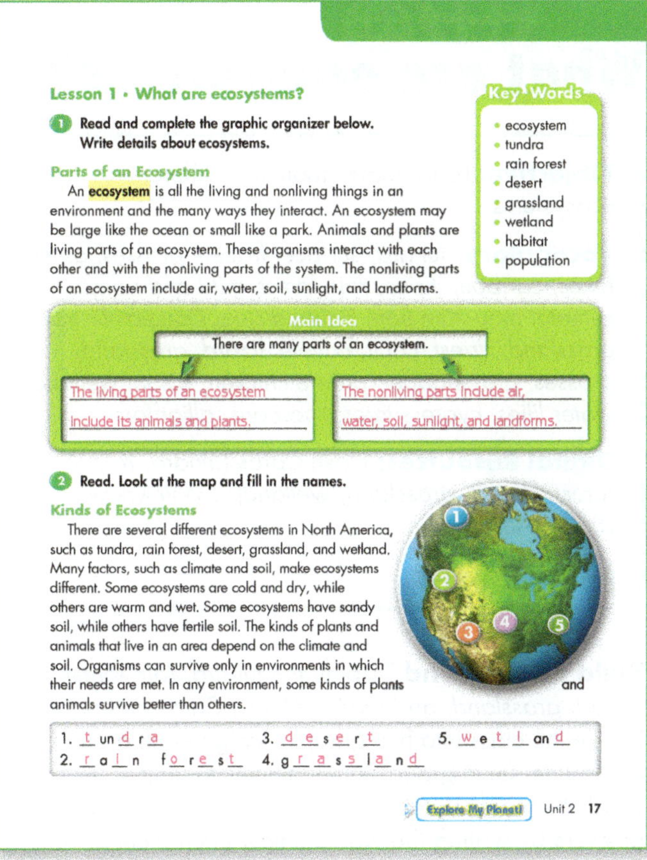

Lesson 1 • What are ecosystems?

Key Words
- ecosystem
- tundra
- rain forest
- desert
- grassland
- wetland
- habitat
- population

1 Read and complete the graphic organizer below. Write details about ecosystems.

Parts of an Ecosystem

An **ecosystem** is all the living and nonliving things in an environment and the many ways they interact. An ecosystem may be large like the ocean or small like a park. Animals and plants are living parts of an ecosystem. These organisms interact with each other and with the nonliving parts of the system. The nonliving parts of an ecosystem include air, water, soil, sunlight, and landforms.

Main Idea: There are many parts of an ecosystem.

The living parts of an ecosystem include its animals and plants.

The nonliving parts include air, water, soil, sunlight, and landforms.

2 Read. Look at the map and fill in the names.

Kinds of Ecosystems

There are several different ecosystems in North America, such as tundra, rain forest, desert, grassland, and wetland. Many factors, such as climate and soil, make ecosystems different. Some ecosystems are cold and dry, while others are warm and wet. Some ecosystems have sandy soil, while others have fertile soil. The kinds of plants and animals that live in an area depend on the climate and soil. Organisms can survive only in environments in which their needs are met. In any environment, some kinds of plants and animals survive better than others.

1. t un d r a
2. r a i n f o r e s t
3. d e s e r t
4. g r a s s l a n d
5. w e t l an d

Explore My Planet! Unit 2 17

Explain

1 Read and complete the graphic organizer below. Write details about ecosystems.

Read *Parts of an Ecosystem* with the class. *What is the text about? Ecosystems! What is the text's main idea?* Explain to students that graphic organizers can help them focus on the most important concepts within a text. Ask students to read the text again and complete the graphic organizer. Have pairs compare their answers.

2 Read. Look at the map and fill in the names.

Display the globe where all students can see it. Invite volunteers to point to different ecosystems they know about, like the Sahara Desert or the rain forest in Brazil. Then have them locate North America and say what kind of ecosystems they think there are in that part of the world. Have students read and complete the names.

Elaborate

Alligators vs. Crocodiles

Have students research on the Internet similarities and differences between alligators and crocodiles. Put students in pairs and have them make a Venn diagram that shows what these two species have and do not have in common. Allow volunteers to present their diagrams to the class.

Lesson 1

What are ecosystems?

Objective: Learn more about five different ecosystems.

Vocabulary: *tundra, surface, frozen, grasses, arctic foxes, caribou, thrive, rain forest, mild and rainy climate, mountain lion, banana slug, Douglas fir tree, grassland, driest, cacti, coyote, lizard, grassland, grasses, grasshopper, prairie chicken, bison, wetland, water lilies, cypress trees, raccoons, alligators*

Digital Resources: Flash Cards (*tundra, rain forest, desert, grassland, wetland*), *I Will Know...* Digital Activity

Materials: globe

Build Background Display the *tundra, rain forest, desert, grassland,* and *wetland* Flash Cards on the board. Divide the class into five groups. Assign each group one ecosystem. Give groups three minutes to write a list of words they associate with their ecosystems. Check their answers by writing the words below each picture. The winning team will be the one with the longest word list.

Explain

3 Read and number the pictures of the ecosystems (1–5) to match their descriptions.

Have students read and number the pictures. Ask students to scan the texts and underline the words that describe each ecosystem. Elicit the words and write them under the Flash Cards on the board.

4 Read again and write the names of the animals shown in the pictures.

Have students read again and write the names of the animals shown in the pictures. Elicit names of other animals mentioned in the texts. Ask questions to make sure students know what animals they are.

ELL Vocabulary Support

Write the following words on the board and elicit their plural forms: *caribou, cactus, bison, mouse, fish, moose,* and *deer.* Explain to students that these nouns have irregular plurals that are not made by adding -*s* or -*es* to the end of the word. (Answers: *caribou–caribou, cactus–cacti, bison–bison, mouse–mice, fish–fish, moose–moose, deer–deer.*) To give students practice, have them use the plural forms in sentences that describe which ecosystem the animals live in, for example, *Caribou and moose live in the tundra.*

3 Read and number the pictures of the ecosystems (1–5) to match their descriptions.

4

1 A **tundra** is a cold region. The ground beneath the surface is frozen all year. Some grasses can grow, but trees cannot. Arctic foxes, caribou, and other animals thrive in these areas.

caribou

3

2 Not all **rain forests** are tropical. The mild and rainy climate of the Pacific Northwest supports temperate rain forests. They are home to organisms such as mountain lions, banana slugs, and Douglas fir trees.

mountain lion

5

3 The driest ecosystem is a **desert**. Some plants and animals have adapted to the limited water supply. Cacti, coyotes, and lizards are desert organisms.

lizard

1

4 **Grasslands**, as their name suggests, are covered with grasses. They receive a medium amount of rain. Grasshoppers, prairie chickens, and bison are animals that live in grasslands in North America.

grasshopper

2

5 In **wetlands**, the ground is covered with water for at least part of the year. Water lilies and cypress trees grow in some wetlands. Different kinds of animals, such as insects, raccoons, and alligators, live in wetlands, too.

raccoon

4 Read again and write the names of the animals shown in the pictures.

18 Unit 2 I Will Know...

Elaborate

Ecosystems in My Country

Have students look at the globe and identify where they live and determine which ecosystems can be found in their country. In small groups, have students make a poster that clearly marks the geographic area of each ecosystem.

Science Notebook: More about Ecosystems

Have students choose and research one type of ecosystem in more depth. Ask them to write a one-page paper explaining why it is important to protect the living and nonliving things in their chosen ecosystem. Encourage students to think about how the ecosystem would be affected if one of the living or nonliving things in that ecosystem were removed.

Ask *What are the primary factors that make ecosystems different?* (Answer: *Climate and soil are the primary factors.*)

I Will Know...

Have students do the *I Will Know...* Digital Activity.

Lesson 1

What are ecosystems?

Objective: Learn about living things within their ecosystems.

Vocabulary: *habitat, address, shrub, survive, savannah, grassland, population, species, wildebeests, plains, platypus, survival, wings, webbed feet, fur*

Digital Resources: Flash Cards (*tundra, rain forest, desert, grassland, wetland*), *Lesson 1 Check* (print out 1 per student), *Got it? 60-Second Video*

Materials: globe, picture of a savannah

Build Background Display a globe where all students can see it. Have students locate Africa and say what kinds of ecosystems there are on that continent. Display a picture of a savannah and write the word *savannah* on the board. Have students say what savannahs are like. Write their ideas on the board. (Possible answers: *generally hot with dry and rainy seasons, with grasses, some trees, and large herds of animals*) Then have students say what kind of animals live there.

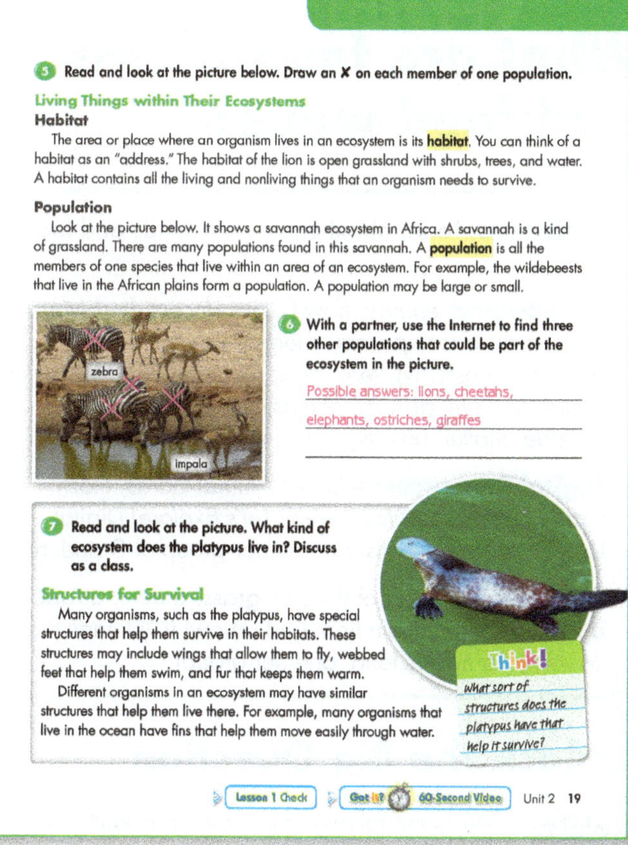

5 Read and look at the picture below. Draw an ✗ on each member of one population.

Living Things within Their Ecosystems

Habitat

The area or place where an organism lives in an ecosystem is its **habitat**. You can think of a habitat as an "address." The habitat of the lion is open grassland with shrubs, trees, and water. A habitat contains all the living and nonliving things that an organism needs to survive.

Population

Look at the picture below. It shows a savannah ecosystem in Africa. A savannah is a kind of grassland. There are many populations found in this savannah. A **population** is all the members of one species that live within an area of an ecosystem. For example, the wildebeests that live in the African plains form a population. A population may be large or small.

6 With a partner, use the Internet to find three other populations that could be part of the ecosystem in the picture.

Possible answers: lions, cheetahs, elephants, ostriches, giraffes

7 Read and look at the picture. What kind of ecosystem does the platypus live in? Discuss as a class.

Structures for Survival

Many organisms, such as the platypus, have special structures that help them survive in their habitats. These structures may include wings that allow them to fly, webbed feet that help them swim, and fur that keeps them warm.

Different organisms in an ecosystem may have similar structures that help them live there. For example, many organisms that live in the ocean have fins that help them move easily through water.

Explain

5 Read and look at the picture below. Draw an ✗ on each member of one population.

Ask students to read the first paragraph and say what *habitat* means. Ask them to read the second paragraph and underline the definition of *population*. Finally, have students read and draw an ✗ on each member of one population.

6 With a partner, use the Internet to find three other populations that could be part of the ecosystem in the picture.

In pairs, students research on the Internet three other populations that live on the African savannah. Check answers as a class.

7 Read and look at the picture. What kind of ecosystem does the platypus live in? Discuss as a class.

Point to the picture on the bottom right and have students name the animal. Read the text aloud and have students discuss what kind of ecosystem the platypus lives in.

Think!

Have students look at the platypus. *What sort of structures does the platypus have that help it survive?* Discuss as a class.

Elaborate

Science Notebook: Platypus Fun Facts

Have students research and write in their Science Notebooks five different facts about platypuses. In small groups, have students share their information. Check answers as a class.

Evaluate

Lesson 1 Check Assessment for Learning

Distribute the *Lesson 1 Check* and guide students as they complete it. Check answers as a class. Then ask students to grade their progress on the topic of ecosystems from 1 to 3: 3 = *I understand how living and nonliving things interact in different ecosystems*; 2 = *I need to study more*; 1 = *I need help!* Encourage students giving themselves a 1 or 2 to describe what they found difficult and what they need to study more.

Review the Key Words for Lesson 1 (see Student's Book page 17). Play the *Got it? 60-Second Video* to review the lesson material.

Lesson 2

What are food chains and food webs?

Objective: Learn about energy pyramids and food chains.

Vocabulary: *energy, flow* (v), *producers, herbivores, carnivores, decomposers, energy pyramid, consumer, upward, transfer* (v), *fern, deer, mountain lion, food chain, sunlight, absorb, photosynthesis, carbon dioxide, matter, pass* (v), *flow* (n)

Digital Resources: selected Animal Cards, Flash Cards (*food chain, food web*), *Let's Explore!* Digital Lab, Food Web Cards (one set per group of 8 students)

Materials: pictures of the sun, grass, a deer, a rabbit, a squirrel, a mountain lion, and an owl (*Optional*: ball of yarn, 35 meters long, one per group)

Unlock the Big Question

Write the following text on the board: *I will learn how energy flows in a food chain and a food web. I will know how some organisms compete for the same resources.*

Build Background On the board, display the pictures of the sun, grass, a rabbit, and a mountain lion at random. Ask students to help you make a food chain to show how energy flows. *Which is the first link of a food chain? Sunlight. What do each of these animals eat?* Place the sun picture on the left side of the board and, with the students' help, continue placing the rest of the pictures to the right of what the organisms eat or get energy from. Then display the pictures of the deer, squirrel, and owl. *Now tell me what these other animals eat.* Place the pictures to the right of what the animals eat.

Explore

Let's Explore! Lab How do food webs show connections?

Objective: Students will make a model of a food web and show how it makes connections.

Digital Resources: *Let's Explore!* Digital Lab, *Let's Explore! Activity Card* (1 per student) (*Optional*: Do the lab in class; refer to the *Activity Card* for materials and steps.)

- Use the *food chain* and *food web* Flash Cards and the food web diagram on the board to review or pre-teach the meanings of *food chain* and *food web*. *A food chain shows the transfer of energy from one organism to another by eating and being eaten. A food web is a system of overlapping food chains.*
- Review or pre-teach key vocabulary using of the Food Web Cards: *kelp, kelp crab, sea urchin, octopus, sea otter, sea star, sea gulls,* and *orca.*

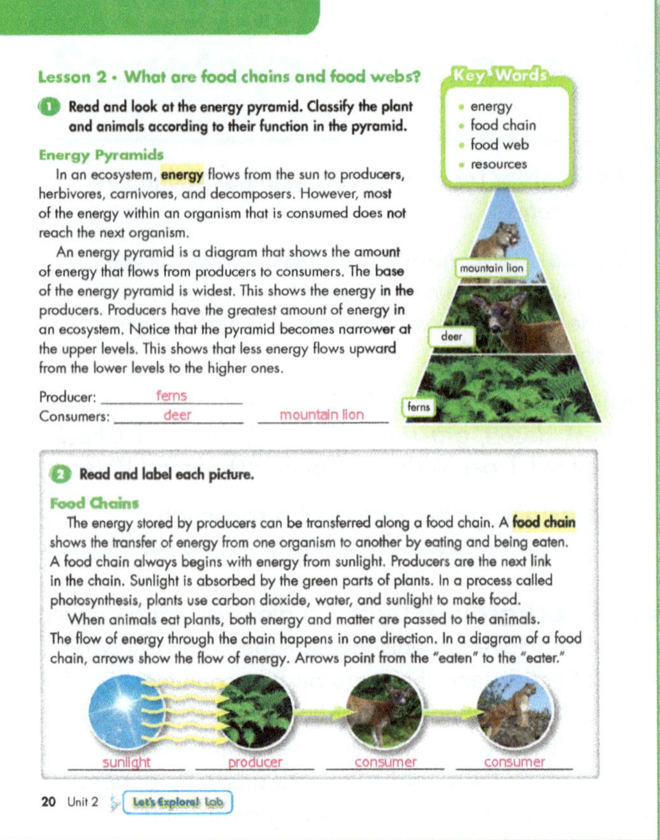

Lesson 2 • What are food chains and food webs?

1 Read and look at the energy pyramid. Classify the plant and animals according to their function in the pyramid.

Energy Pyramids

In an ecosystem, **energy** flows from the sun to producers, herbivores, carnivores, and decomposers. However, most of the energy within an organism that is consumed does not reach the next organism.

An energy pyramid is a diagram that shows the amount of energy that flows from producers to consumers. The base of the energy pyramid is widest. This shows the energy in the producers. Producers have the greatest amount of energy in an ecosystem. Notice that the pyramid becomes narrower at the upper levels. This shows that less energy flows upward from the lower levels to the higher ones.

Producer: ferns
Consumers: deer mountain lion

Key Words
- energy
- food chain
- food web
- resources

2 Read and label each picture.

Food Chains

The energy stored by producers can be transferred along a food chain. A **food chain** shows the transfer of energy from one organism to another by eating and being eaten. A food chain always begins with energy from sunlight. Producers are the next link in the chain. Sunlight is absorbed by the green parts of plants. In a process called photosynthesis, plants use carbon dioxide, water, and sunlight to make food.

When animals eat plants, both energy and matter are passed to the animals. The flow of energy through the chain happens in one direction. In a diagram of a food chain, arrows show the flow of energy. Arrows point from the "eaten" to the "eater."

sunlight producer consumer consumer

20 Unit 2 Let's Explore! Lab

- Show the Digital Lab. Drag and drop what each organism eats with the students' help.
- Ask students to work independently or in pairs to complete the *Activity Card.*
- Provide support as needed. Check answers as a class.

Optional: Divide the class into groups of eight. Give each group a copy of the Food Web Cards and a ball of yarn 35 meters long. Have students follow the instructions on the *Activity Card* and answer the questions.

Explain

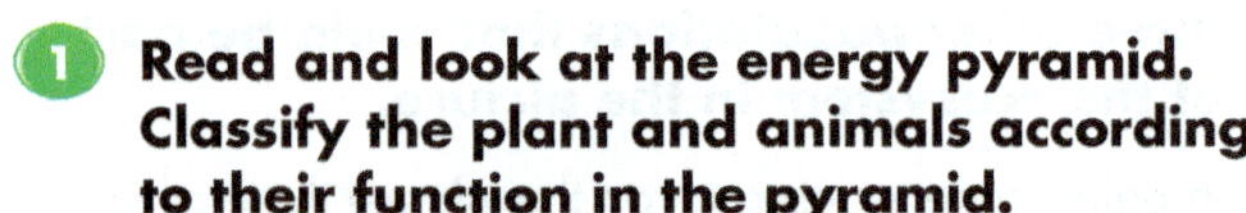

1 Read and look at the energy pyramid. Classify the plant and animals according to their function in the pyramid.

Draw a pyramid on the board and add lines to divide the pyramid into thirds. Label the bottom third *producer* and the middle and top thirds *consumer*. Have students look at the food web on the board and name a producer and two consumers that would be part of the same energy pyramid. Students read and classify the producer and consumers. Remind students that producers make their own food and consumers eat other living things.

2 Read and label each picture.

Look at the diagram. What does it show? A food chain! Have students read and say what a food chain is. Students read and label the pictures. *What is the beginning of all food chains? Energy from sunlight. What is the next link? Producers! And what is the third link? Consumers!*

Lesson 2

What are food chains and food webs?

Objective: Learn how energy flows in food chains.

Vocabulary: *energy flow, food chain, sunlight, ferns, mice, owl, forest, deer, predators, mountain lion, prey, decomposer, fungi, bacteria, waste, soil, reuse, matter*

Digital Resources: Flash Card (*food chain*), *I Will Know...* Digital Activity

Materials: flip-chart paper, markers, pictures of earthworms, cockroaches, houseflies, slugs, mushrooms, snails, woodlice, and beetles

Build Background Make a chart on the board or on flip-chart paper with the headings *Producers* and *Consumers* to help students categorize the information from the previous class.

Explain

3 Look and number each picture to show the flow of energy in the food chain.

Elicit some examples of food chains. Then have students look and number each picture to show the flow of energy in the food chain.

4 How does energy flow in this food chain? Discuss with a partner and write your answer.

Invite pairs of students to discuss and write how energy flows in the food chain from the previous activity. Check answers as a class.

5 Read and fill in the blanks with words from the box.

Read the text with students and ask them what it is about. (*How energy flows in a forest food chain.*) Then have students reread the text individually and fill in the blanks with words from the box. Check answers as a class.

Think!

Ask *Why are decomposers, such as the banana slug, important in a food chain?* Elicit answers from the class. (Possible answer: *Because they recycle energy from waste and dead organisms.*)

Elaborate

Decomposers

Display pictures of decomposers (earthworms, cockroaches, houseflies, slugs, mushrooms, snails, woodlice, beetles, etc.) and elicit their names. Write the following list on the board: *earthworms, cockroaches, houseflies, slugs, mushrooms,*

3 Look and number each picture to show the flow of energy in the food chain.

3 4 2 1

4 How does energy flow in this food chain? Discuss with a partner and write your answer.

This food chain begins with sunlight.
Then the ferns use the sunlight to make food. Then the mice eat the ferns. Finally, the owl eats the mice.

5 Read and fill in the blanks with words from the box.

mountain lions | dead | deer | plants | sun

In a forest ecosystem, a food chain might consist of the sun, plants, a deer, and a mountain lion. Mountain lions are predators. Predators get the energy they need by hunting and killing prey. Deer are prey for the mountain lion. The mountain lion uses the food energy that was stored in plants and passed to the deer.

Decomposers, such as fungi and bacteria, are at each level of a food chain. They get what they need from dead organisms and other waste. Decomposers return matter to the soil, air, and water. Producers can then reuse this matter.

The banana slug is a decomposer. It gets its energy from leaves, animal waste, and dead plant materials. Then it recycles what it eats into the soil.

Think!
Why are decomposers, such as the banana slug, important in a food chain?

I Will Know... Unit 2 21

snails, woodlice, and *beetles.* Divide the class into pairs. Assign one of the decomposers to each pair and have them research on the Internet their habitat, what they eat, and what food chain they belong to. Have students make a poster that illustrates and describes their decomposer. Invite volunteers to present their posters to the class.

ELL Content Support

Removal of One Component

A change in a living or nonliving part of an ecosystem upsets the balance in that ecosystem. For example, materials broken down by decomposers become nutrients in the soil that are needed by plants. If a group of decomposers dies out, the nutrients they would have put into the soil of the ecosystem will not be available for plants to use. The plants might weaken or die, and then there might be fewer plants for herbivores to eat.

Think!

Ask *In what ways are producers, consumers, and decomposers alike and different?* Elicit answers from the class. (Possible answer: *All organisms use energy to survive. However, different organisms get energy from different sources.*)

I Will Know...

Have students do the *I Will Know...* Digital Activity.

Lesson 2

What are food chains and food webs?

Objective: Learn how food chains combine to form a food web.

Vocabulary: *food source, overlap, food chains, combine, flow of energy, branch out, producer, consumer, predator, forest, prey, owl, mountain lion, rabbit, mice, squirrels, survive, disease, storms, pollution, hunting*

Digital Resources: selected Animal Cards, Flash Cards (*food chain, food web, tundra, rain forest, desert, grassland, wetland*)

Materials: pictures of the African savannah, a Mexican rain forest, an Australian coral reef, etc., poster board

Build Background Display the *food chain* and *food web* Flash Cards. Have students describe the pictures and explain the two concepts. Remind students that, in the *Let's Explore!* Digital Lab, they made a model food web. *Remember that a food chain shows how energy is transferred from one organism to another by eating and being eaten and that a food web is a system of overlapping food chains.*

Explain

6 Read and circle *T* (true) or *F* (false). With a partner, correct the false statements.

Invite students to read the paragraph and circle the answers. Then pair students to correct the false statements. Check answers as a class.

7 Look at the food web in a forest ecosystem. Circle a producer and draw an *X* on three consumers. Then draw arrows that show the flow from the sun as energy is transferred along the food chain through the producers to the consumers.

Point to the food web and have students say which ecosystem the food web belongs to. Ask volunteers to describe one food chain in the food web. (Possible answer: *Energy from the sun is transferred to plants, then to the rabbit, and then to the owl.*) Read the instructions aloud and have pairs do the activities. Check answers as a class.

Ask the following questions to check students' understanding: *For what same resource do the mountain lion and the owl compete?* (*The deer mouse.*) *Which animal is both a predator and prey?* (*The mouse; it is the only animal that both eats prey–insects–and is eaten by predators–mountain lions.*)

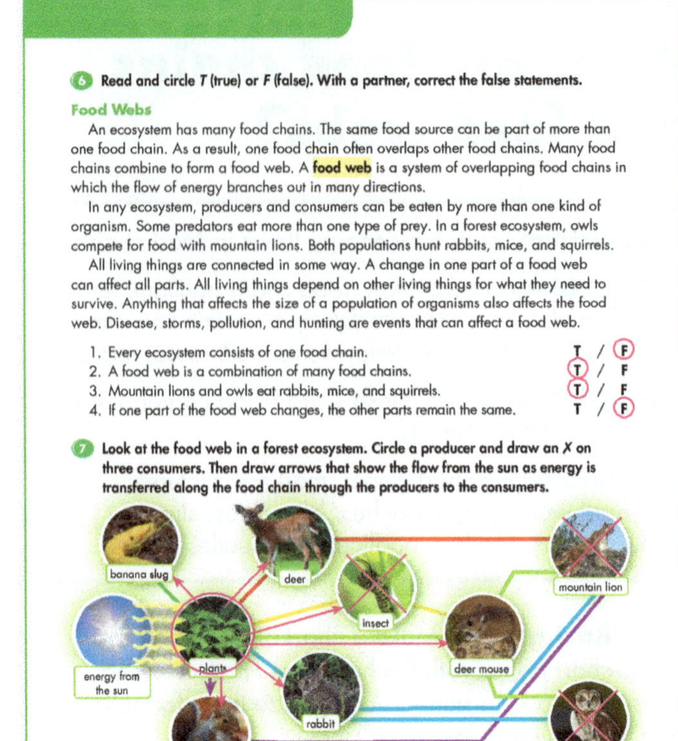

6 Read and circle *T* (true) or *F* (false). With a partner, correct the false statements.

Food Webs

An ecosystem has many food chains. The same food source can be part of more than one food chain. As a result, one food chain often overlaps other food chains. Many food chains combine to form a food web. A **food web** is a system of overlapping food chains in which the flow of energy branches out in many directions.

In any ecosystem, producers and consumers can be eaten by more than one kind of organism. Some predators eat more than one type of prey. In a forest ecosystem, owls compete for food with mountain lions. Both populations hunt rabbits, mice, and squirrels.

All living things are connected in some way. A change in one part of a food web can affect all parts. All living things depend on other living things for what they need to survive. Anything that affects the size of a population of organisms also affects the food web. Disease, storms, pollution, and hunting are events that can affect a food web.

1. Every ecosystem consists of one food chain. T / F
2. A food web is a combination of many food chains. T / F
3. Mountain lions and owls eat rabbits, mice, and squirrels. T / F
4. If one part of the food web changes, the other parts remain the same. T / F

7 Look at the food web in a forest ecosystem. Circle a producer and draw an *X* on three consumers. Then draw arrows that show the flow from the sun as energy is transferred along the food chain through the producers to the consumers.

22 Unit 2

Elaborate

Food Web Posters

Display the Flash Cards and pictures of different ecosystems. Divide the class into small groups. Have students research a food web from an ecosystem in a different part of the world, such as the African savannah, a Mexican rain forest, or an Australian coral reef. Ask students to record the information on poster board with a label and pictures for each animal, as well as arrows to show the flow of energy and matter. Invite groups to present their food webs to the class.

Discuss the following question as a class: *Would a mouse still be able to get food if all the insects in the ecosystem were gone?* (*Yes. The mouse could still get food from plants.*)

Changes in Food Webs

Discuss as a class how a forest fire could affect the food web pictured on this page. Write students' ideas on the board. (Possible answer: *If a forest fire killed many plants in this food web, there would be fewer animals. For example, the number of deer mice would decrease because they would not have enough food to eat.*) Have the same groups of students from the previous activity discuss how changing one part of the food web they chose could change other parts of the web.

Lesson 2

What are food chains and food webs?

Objective: Learn how populations in ecosystems can change naturally.

Vocabulary: *population, resources, deer mice, plentiful, living space, decrease* (v), *die, move out, available, grow*

Digital Resources: *Lesson 2 Check* (print out 1 per student), *Got it? 60-Second Video*

Materials: picture of a forest fire, map of the state or country where you live

Build Background Elicit a forest food chain and draw a diagram of it on the board. Display a picture of a forest fire. *How might a natural disaster, such as a forest fire or drought, affect a population?* (Possible answer: *The disaster might reduce the amount of available food, water, and living space, causing the population to decrease.*)

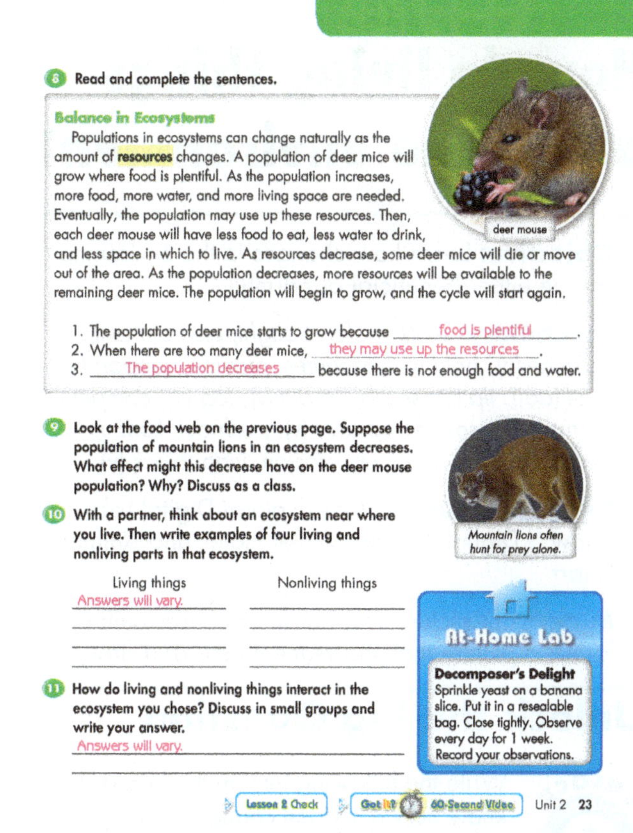

8 Read and complete the sentences.

Balance in Ecosystems

Populations in ecosystems can change naturally as the amount of **resources** changes. A population of deer mice will grow where food is plentiful. As the population increases, more food, more water, and more living space are needed. Eventually, the population may use up these resources. Then, each deer mouse will have less food to eat, less water to drink, and less space in which to live. As resources decrease, some deer mice will die or move out of the area. As the population decreases, more resources will be available to the remaining deer mice. The population will begin to grow, and the cycle will start again.

deer mouse

1. The population of deer mice starts to grow because food is plentiful.
2. When there are too many deer mice, they may use up the resources.
3. The population decreases because there is not enough food and water.

9 Look at the food web on the previous page. Suppose the population of mountain lions in an ecosystem decreases. What effect might this decrease have on the deer mouse population? Why? Discuss as a class.

Mountain lions often hunt for prey alone.

10 With a partner, think about an ecosystem near where you live. Then write examples of four living and nonliving parts in that ecosystem.

Living things	Nonliving things
Answers will vary.	

At-Home Lab

Decomposer's Delight
Sprinkle yeast on a banana slice. Put it in a resealable bag. Close tightly. Observe every day for 1 week. Record your observations.

11 How do living and nonliving things interact in the ecosystem you chose? Discuss in small groups and write your answer.

Answers will vary.

Lesson 2 Check | Got it? 60-Second Video | Unit 2 23

Explain

8 Read and complete the sentences.

Read the text with students. Ask *What happens when the amount of resources changes in an ecosystem? Populations change naturally. What population is described in the text? Deer mice!* Have pairs read the text again and complete the sentences.

9 Look at the food web on the previous page. Suppose the population of mountain lions in an ecosystem decreases. What effect might this decrease have on the deer mouse population? Why? Discuss as a class.

Have students discuss what might happen to the population of deer mice if the population of mountain lions decreases.

10 With a partner, think about an ecosystem near where you live. Then write examples of four living and nonliving parts in that ecosystem.

Display a map of your country. Have students name an ecosystem near where they live, and write it on the board. In pairs, students write examples of four living and nonliving things in that ecosystem.

11 How do living and nonliving things interact in the ecosystem you chose? Discuss in small groups and write your answer.

Divide the class into small groups. Have students discuss how living and nonliving things interact in the ecosystem they chose. Have volunteers present their ideas to the class.

At-Home Lab

Decomposer's Delight

Show students how to sprinkle yeast on a banana slice, put it in a resealable bag, and close it tightly. Students will observe it every day for one week and record their observations. Students should notice increasing yeast growth on the banana and that the banana will decompose somewhat. They may also notice that the bag has filled with a gas (CO_2) that has been produced.

Evaluate

Lesson 2 Check **Assessment for Learning**

Distribute the *Lesson 2 Check* and guide students as they complete it. Check answers as a class. Then ask students to grade their progress on the topic of food chains and food webs from 1 to 3: 3 = *I understand what food chains and food webs are*; 2 = *I need to study more*; 1 = *I need help!* Encourage students giving themselves a 1 or 2 to describe what they found difficult and what they need to study more.

Got it?  **60-Second Video**

Review Key Words for Lesson 2 (see Student's Book page 20). Play the *Got it? 60-Second Video* to review the lesson material.

Lesson 3

How do living things affect the environment?

Objective: Learn how changes in the environment can be either beneficial or harmful.

Vocabulary: *water hole, drought, environment, balance* (n), *hold* (v), *resources, provide, support* (v), *level* (adj), *tip* (v), *seeds, sprout* (v), *log* (n), *competition, harm* (v), *loosestrife, spread* (v), *push out, kudzu, beneficial, ground cover, erosion, vine, block* (v)

Digital Resources: *Let's Explore!* Digital Lab, *I Will Know...* Digital Activity, Water Holes and Animals Chart

Materials: picture of a house or houses built in a forest, a picture that shows the effects of a drought

Unlock the Big Question

Write the following text on the board: *I will know how a sudden change to one group of organisms affects an environment.*

Build Background Display the picture of the house built in a forest and have students discuss how the building of houses can affect the forest ecosystem.

Explore

Let's Explore! Lab What happens when one part of an ecosystem is removed?

Objective: Students will make and use a model to determine what happens when one component in an ecosystem is removed.

Digital Resources: *Let's Explore!* Digital Lab, *Let's Explore! Activity Card* (1 per student) (*Optional*: Do the lab in class; refer to the *Activity Card* for materials and steps.)

- Ask students what resources all living organisms need to survive. (Answers: *food, water,* and *living space*)
- Review or pre-teach the meanings of *water hole* and *drought.*
- Show the Digital Lab.
- Ask students to work independently or in pairs to complete the *Activity Card.*
- Provide support as needed. Check answers as a class.

I Will Know...

Have students do the *I Will Know...* Digital Activity.

Lesson 3 • How do living things affect the environment?

1 Read and underline the best title for the paragraph.

Key Words
• competition

The environment is like a balance. One side holds what lives in the environment. The other side holds resources that the environment provides. If the environment provides enough resources to support life, the balance is level.

Change often tips the balance. For example, tree seeds may sprout on a log and start to grow. The young trees need light and space to grow. The young trees are in competition. **Competition** occurs when two or more living things need the same resources in order to survive. Some trees get enough light. As their branches grow, they shade nearby plants. The environment changes. Other young trees may not get enough light to survive.

a. Balance in the Environment
b. Competition of Trees in the Forest
c. Changes in the Environment

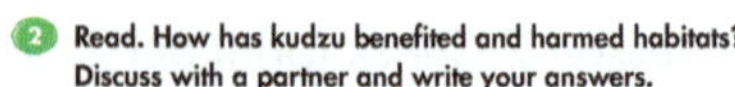

2 Read. How has kudzu benefited and harmed habitats? Discuss with a partner and write your answers.

Plants Cause Change

Changes can help some living things and harm others. For example, a plant called purple loosestrife was brought to the United States. No animals eat this plant. It is spreading to new places. That is great for the loosestrife! However, there is less space for other plants to grow. Some kinds of plants are completely pushed out of the environment.

kudzu

A plant called kudzu was originally brought to the United States from Japan. At first, kudzu was thought to be beneficial to the environment. Kudzu was used in gardens for its beauty and in many open spaces as a ground cover to prevent erosion. The plant, however, grows very well in the southeastern United States. Because the vines grow so well, they destroy forests by blocking sunlight that trees need.

Benefit: It prevents erosion in many open spaces.
Harm: It destroys forests by blocking sunlight.

24 Unit 2 Let's Explore! Lab I Will Know...

Explain

1 Read and underline the best title for the paragraph.

Have students look at the picture at the top right and describe it. Read the text aloud. Have pairs summarize in one sentence what the text is about. Write students' ideas on the board and have the class decide which sentence best summarizes the text. Then ask pairs to choose the best title for the paragraph.

2 Read. How has kudzu benefited and harmed habitats? Discuss with a partner and write your answers.

Have students scan the text and underline the names of the two plants described in the text. Explain that these two plants have caused positive and negative changes to some ecosystems. Read the question. Have pairs read, discuss, and write the answers.

Think!

Explain to students that purple loosestrife and kudzu are called invasive (or non-native) species. Ask *Why might an invasive species grow better in a new environment than it does in its native environment?* Invite students to discuss the question in small groups and share their ideas with the class. (Possible answers: *They have no competitors. They have no predators.*)

Lesson 3

How do living things affect the environment?

Objective: Learn how animals and humans cause change to the environment.

Vocabulary: *habitat, beavers, stream, shallow, pond, cut down trees, teeth, wood, dam, still water, shrubs, flooded, take away, flowing water*

Digital Resources: Flash Card (*food web*), *Lesson 3 Check* (print out 1 per student), *Got it? 60-Second Video*

Build Background Write the word *competition* on the board and have students look up its meaning in the glossary on page 113. Elicit examples of how living things discussed in this unit may compete. Write students' ideas on the board. (Possible answers: *Mountain lions and owls may compete for the same prey. Deer mice may compete for food, water, and living space. Young trees may compete for light and space to grow.*)

Explain

3 Read and underline two positive changes and circle two negative changes beavers cause in their environment.

Ask students to look at the picture of the beaver and read the caption. Have them discuss why they think beavers cut down trees and if the changes beavers cause to the environment are positive or negative. Then ask them to underline two positive changes and circle two negative changes.

4 Read and circle whether the changes are positive, negative, or both.

Point to the picture and encourage students to describe what happened. Explain that humans can also cause changes to the environment. Have pairs read and decide whether the changes mentioned are positive, negative, or both.

5 Discuss as a class the reasons for your choices.

Discuss with the students the reasons for their choices in the previous activity.

Elaborate

Dams Built by Humans

Tell students that, just like beavers, humans build dams that cause changes to the environment. Have students find out more about how dams change the environment wherever they are built. Students should present their findings to the class.

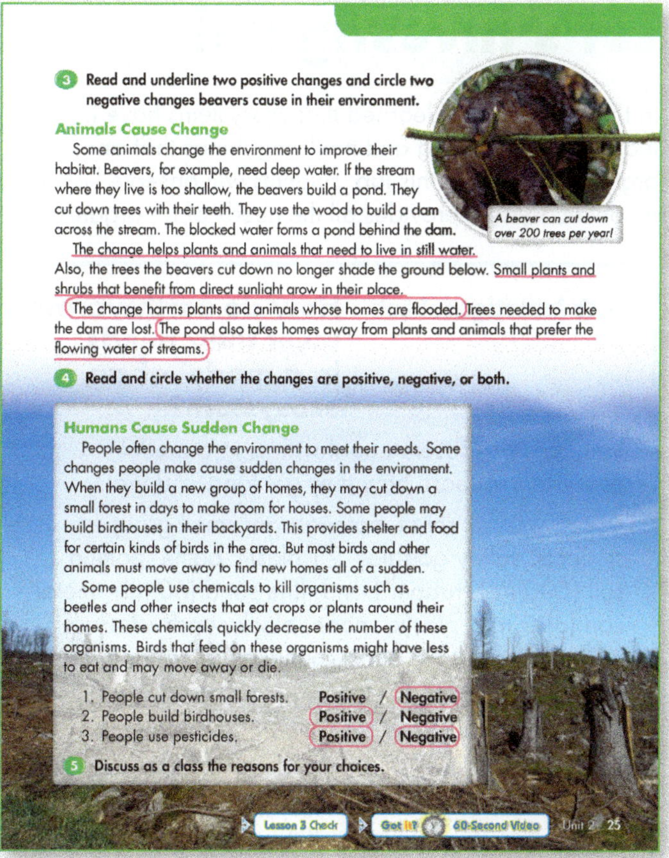

3 Read and underline two positive changes and circle two negative changes beavers cause in their environment.

Animals Cause Change

Some animals change the environment to improve their habitat. Beavers, for example, need deep water. If the stream where they live is too shallow, the beavers build a pond. They cut down trees with their teeth. They use the wood to build a dam across the stream. The blocked water forms a pond behind the dam. The change helps plants and animals that need to live in still water. Also, the trees the beavers cut down no longer shade the ground below. Small plants and shrubs that benefit from direct sunlight grow in their place.

The change harms plants and animals whose homes are flooded. Trees needed to make the dam are lost. The pond also takes homes away from plants and animals that prefer the flowing water of streams.

A beaver can cut down over 200 trees per year!

4 Read and circle whether the changes are positive, negative, or both.

Humans Cause Sudden Change

People often change the environment to meet their needs. Some changes people make cause sudden changes in the environment. When they build a new group of homes, they may cut down a small forest in days to make room for houses. Some people may build birdhouses in their backyards. This provides shelter and food for certain kinds of birds in the area. But most birds and other animals must move away to find new homes all of a sudden.

Some people use chemicals to kill organisms such as beetles and other insects that eat crops or plants around their homes. These chemicals quickly decrease the number of these organisms. Birds that feed on these organisms might have less to eat and may move away or die.

1. People cut down small forests. Positive / Negative
2. People build birdhouses. Positive / Negative
3. People use pesticides. Positive / Negative

5 Discuss as a class the reasons for your choices.

Lesson 3 Check Got it? 60-Second Video Unit 2 25

Science Notebook: Natural Disasters and Ecosystems

Have students write a one-page composition explaining how a natural disaster, such as a hurricane, drought, or flood, might change an ecosystem and how those changes might affect food webs in that ecosystem.

Evaluate

Lesson 3 Check Assessment for Learning

Distribute the *Lesson 3 Check* and guide students as they complete it. Check answers as a class. Then ask students to grade their progress on the topic of how living things affect the environment from 1 to 3: 3 = *I understand how living things affect the environment*; 2 = *I need to study more*; 1 = *I need help!* Encourage students giving themselves a 1 or 2 to describe what they found difficult and what they need to study more.

Got it? 60-Second Video

Review Key Words for Lesson 3 (see Student's Book page 24). Play the *Got it? 60-Second Video* to review the lesson material.

Let's Investigate!

In this unit, students learned that ecosystems have habitats that contain the living and nonliving things that an organism needs. In this lab, students will observe how a model ecosystem provides all the needs of an organism.

Let's Investigate! Lab **How do earthworms meet their needs in a model of an ecosystem?**

Objective: Observe earthworms in a model ecosystem to learn how they get food and water.

Materials: 1 set of materials per small group of students: safety goggles, 2 L plastic bottle, potting soil (500 mL), rubber band, fine sand (60 mL), metal can (300 mL), plastic spoon, black construction paper (2 sheets), 6 earthworms, masking tape, clear plastic cup, aluminum foil (30 cm square)

Digital Resources: *Let's Investigate!* Digital Lab, *Let's Investigate! Activity Card* (1 per group)

Advance Preparation: Prepare an earthworm bottle for each group. Cut off the top of a 2 L plastic bottle. Place a metal can upside down inside the bottle. Add 500 mL of moist potting soil around the metal can. The soil should be level with the top of the can. Alternatively, have each group prepare their earthworm bottle. Groups can also fill their cup with 60 mL of sand. Cut a 30 cm square of aluminum foil for each group.

- Divide students into small groups and distribute materials.
- Ask students to use a spoon to add a thin layer of sand and then six worms.
- Have students tape black paper around the bottle, cover the top with foil fastened with a rubber band, and wait for 24 hours.
- Students remove the paper and foil to observe the sand, dirt, and earthworms.
- Ask students to record their observations in their notebooks. Students will observe earthworms tunneling in the soil and possibly leaving waste in the sand.
- At the end of the activity, have students share their observations with the class.

Teacher Time-Saving Option: Show the *Let's Investigate!* Digital Lab as an alternative to the hands-on lab activity.

Let's Investigate!

How do earthworms meet their needs in a model of an ecosystem?

1. Obtain an earthworm bottle from your teacher. Use a spoon to add a thin layer of sand. Add 6 worms.
2. Tape black paper around the bottle. Cover the top with foil fastened with a rubber band. Wait 24 hours.
3. Remove the paper and foil. Observe the sand, dirt, and earthworms. Record your observations.
4. Replace the paper and the foil. Observe daily for 3 more days. Record your observations.

Earthworm Observations

Day	Sample data Observations
Day 1 (24 hours after making ecosystem)	Two tunnels are visible.
Day 2	More tunnels are visible. There are some dark castings (waste) on top of the sand.
Day 3	More tunnels are visible. There are some dark castings on top of the sand.
Day 4	More tunnels are visible. There are more dark castings on top of the sand.

26 Unit 2 Let's Investigate! Lab

Unlock the Big Question

Have students refer to the Big Question on the Unit Opener page. In pairs, have them recall what they have learned about ecosystems. Invite student pairs to share their answers to questions 5, 6, and 7 on the *Let's Investigate! Activity Card*.

Class Project: The Recycling Plan

Materials: construction paper (1 sheet per group), markers

Divide the class into small groups. Have students investigate how people harm the environment by throwing out a lot of trash. Ask them to organize a recycling plan for your school to help the environment. During the process, help them evaluate their plans' suitability and feasibility. Before presenting their projects to the class, encourage students to come up with a slogan for their plan. Have students share their ideas with the class.

Unit 2 Review

How do living organisms interact with the environment?

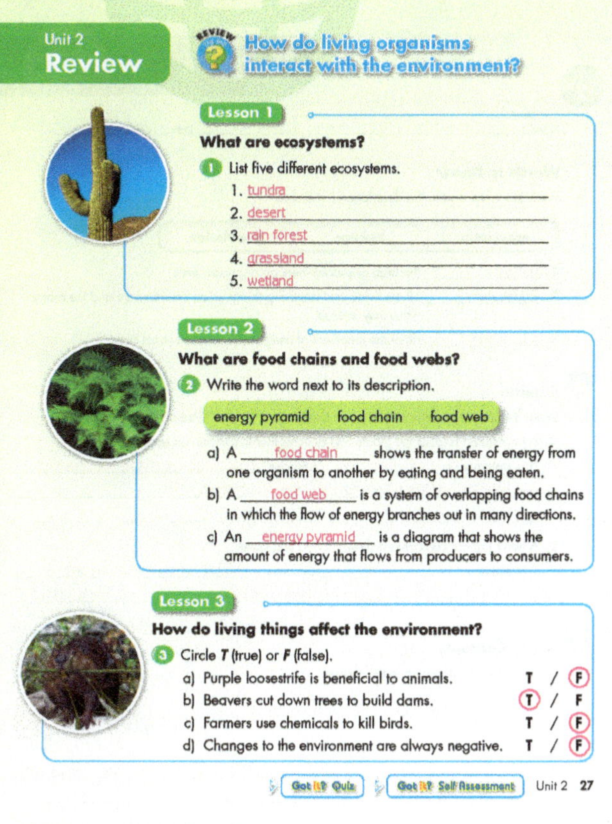
Unit 2 Review

How do living organisms interact with the environment?

Lesson 1

What are ecosystems?

1 List five different ecosystems.

1. tundra
2. desert
3. rain forest
4. grassland
5. wetland

Lesson 2

What are food chains and food webs?

2 Write the word next to its description.

energy pyramid food chain food web

a) A food chain shows the transfer of energy from one organism to another by eating and being eaten.

b) A food web is a system of overlapping food chains in which the flow of energy branches out in many directions.

c) An energy pyramid is a diagram that shows the amount of energy that flows from producers to consumers.

Lesson 3

How do living things affect the environment?

3 Circle *T* (true) or *F* (false).

a) Purple loosestrife is beneficial to animals. T / (F)

b) Beavers cut down trees to build dams. (T) / F

c) Farmers use chemicals to kill birds. T / (F)

d) Changes to the environment are always negative. T / (F)

Got it? Quiz Got it? Self Assessment Unit 2 27

Digital Resources: Print out 1 of each per student: *Got it? Self Assessment, Got it? Quiz*

ELL Language Support

Before students start working on the Review activities, have them read each question aloud along with you.

Evaluate

Strategies for Targeted Review

The following are strategies for providing targeted review for students if they encounter challenges with the content.

Lesson 1 What are ecosystems?

Question 1

If… students are having difficulty remembering the names of five ecosystems, then… direct students to page 18. Encourage them to look back at the pictures and read the texts that describe the five ecosystems.

Lesson 2 What are food chains and food webs?

Question 2

If… students are having difficulty identifying the descriptions of an energy pyramid, food chain, or food web, then… direct students to pages 20 and 22 and have them look up the definitions of these three concepts.

Lesson 3 How do living things affect the environment?

Question 3

If… students are having difficulty deciding whether the statements are true or false, then… direct students to look back over Lesson 3 and find the answers to the questions.

Got it? Self Assessment

Immediately after students have completed the Review activities, distribute a *Got it? Self Assessment* to each student. Have students complete the *Stop! Wait!* and *Go!* statements for each lesson, allowing them to look back through the lesson material if necessary.

Got it? Quiz

Distribute a Unit 2 *Got it? Quiz* to each student. Quizzes may be used for assessing students' understanding of unit concepts as well as for grading purposes.

Lesson 1 Check

Name ______________________ Date __________

Words to Know

Write the word next to the description it matches.

ecosystem	habitat	population

1. habitat ______ the area or place where an organism lives
2. ecosystem ______ all the living and nonliving things in an environment and the many ways they interact
3. population ______ all of the members of one species that live in an area.

Explain

Write if each statement is true or false. Explain your choice. Give an example.

4. A population is all of the plants and animals that live within an ecosystem.
   This statement is false because a population is all the members of one species that live within an area of an ecosystem.
5. A lion's habitat can contain shrubs, trees, and water.
   This statement is true because a habitat contains all living and nonliving things that an organism needs to survive.

Apply Concepts

6. Describe the ecosystem in the picture.

The picture shows a forest ecosystem. Some of the living things in the ecosystem are squirrels, deer, different kinds of birds, rabbits, trees, flowers, and grass.

Unit 2, *Lesson 1 Check* • What are ecosystems?

Lesson 2 Check

Name ______________________ Date __________

Words to Know

Write the word next to the description it matches.

energy pyramid	food chain	food web

1. energy pyramid ______ a diagram that shows the amount of energy that flows from producers to consumers
2. food web ______ a system of overlapping food chains in which the flow of energy branches out in many directions
3. food chain ______ a transfer of energy from one organism to another by eating and being eaten

Explain

Write if each statement is true or false. Explain your choice.

4. Producers and consumers are only eaten by one kind of organism in an ecosystem.
   This statement is false because producers and consumers may be eaten by many kinds of organisms in an ecosystem.
5. A food chain always begins with water.
   This statement is false because a food chain always begins with sunlight.

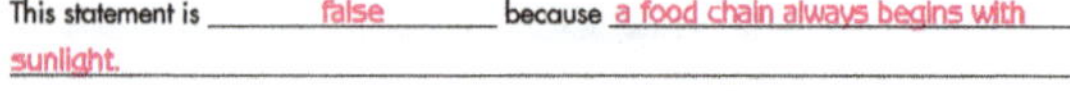

Apply Concepts

6. A food chain includes deer, which eat plants, and mountain lions, which eat deer. Describe the flow of energy through the food chain.

Energy flows from plants to deer to mountain lions. The most energy comes from the plants.

Unit 2, *Lesson 2 Check* • What are food chains and food webs?

Lesson 3 Check

Name ______________________ Date __________

Words to Know

Write the word next to the description it matches.

competition	erosion	resource

1. erosion ______ the movement of worn away rocks or soil
2. resource ______ important material that living things need
3. competition ______ what occurs when two or more living things need the same things to live

Explain

Write if each statement is true or false. Explain your choice.

4. Purple loosestrife is beneficial to animals.
   This statement is false because no animals eat purple loosestrife and it takes over space where other plants that animals might eat could grow.
5. Humans affect the environment when they build homes
   This statement is true because people cut down forests and plow up grasslands to make room for homes.

Apply Concepts

6. The United States has established national parks where humans cannot build. How can these help the environment?

Possible answer: They make sure that the habitats of plants and animals won't be destroyed by people building.

Unit 2, *Lesson 3 Check* • How do living things affect the environment?

Lesson 1 Explore My Planet! Activity Card

Name ______________________ Date __________

Let's Blog: Alligator Farm

by Emma
Middleburg, FL

Today we went to the Alligator Farm in St. Augustine. We saw American alligators. One alligator was an albino. Did you know they feed them rodents called Cavia? The worker that fed them said they only need 80 pounds of food a year. I was surprised when she walked among the alligators in the exhibit as she fed them. One of them kept hissing at her. I would have been scared.

Although we saw many crocodiles and alligators, my favorite was the albino alligator. It was all white. I learned that the parents of albino alligators don't always have to be albino themselves. We rarely see these in the wild because they can't camouflage themselves from their predators.

It was an exciting day and a great way to spend the afternoon learning.

1. Write a response to Emma's blog. Talk about an animal you have seen or read about.
   Students' responses should be about an animal that lives in the water.

Posted by: __________ Location: __________

Share your response with another student. Ask him or her to write a comment about your animal

Unit 2, Lesson 1 *Explore My Planet! Activity* • What are ecosystems?

Lesson 2 Let's Explore! Activity Card

Name _______________ Date _______________

Materials
- Food Web Cards
- yarn

How do food webs show connections?

1. Choose a card. Hold it so it can be seen. Stand in a circle with your group. Look for organisms that your organism eats or that eat it. Toss the ball of yarn to one of them, but hold onto the end of the yarn.
2. Take turns until everyone is connected. You have made a model of a food web.
3. Lay down the yarn and the cards. Using the names of the organisms, draw your food web in the space to the right.

The drawing should show each organism connected by lines to the organisms listed after its name.
kelp: kelp crab, sea urchin
kelp crab: kelp, octopus, sea otter
octopus: kelp crab
orca: sea otter
seagull: sea star, sea urchin
sea otter: kelp crab, orca, sea star, sea urchin
sea star: seagull, sea otter, sea urchin
sea urchin: kelp, seagull, sea otter, sea star

Explain Your Results

4. Look at your food web. Explain the relationships the web shows. Give examples.
Possible answer: My food web shows the relationship among producers and consumers. Kelp is a producer that is consumed by kelp crabs. My web also shows the relationship among prey and predators. Kelp crabs are prey for sea otters.

How does the food web help you understand the food chain?

Possible answer: I can see how the animals and what they eat are connected to each other.

Unit 2, Lesson 2 *Let's Explore!* Lab • What are food chains and food webs?

Lesson 3 Let's Explore! Activity Card

Name _______________ Date _______________

Materials
- Water Holes and Animals Chart
- red squares (animals)
- blue squares (water holes)
- dot cube

What happens when one part of an ecosystem is removed?

In this model of an ecosystem, you will find out what can happen to animals when a resource decreases.

1. Make a model by laying out blue paper squares. Lay two red squares of paper to overlap each blue square.
2. Roll the dot cube. Remove that number of water holes.
3. Roll the cube five times. After each roll, match animals with water holes.
4. No more than two animals can be at a water hole. Remove extra animals. Record information on the Water Holes and Animals Chart.

Explain Your Results

5. What might happen to animals when there is a long drought?
Possible answer: The animals might move to another place or they may die.

What question did you have after you did this activity that you did not have before?

Possible answer: How can humans protect water sources for animals?

Unit 2, Lesson 3 *Let's Explore!* Lab • How do living things affect the environment?

Let's Investigate! Activity Card

Name _______________ Date _______________

Analyze and Conclude

5. Explain your observations.
Possible answer: The tunnels in the soil are the same size (diameter) as the worms. I believe the worms made the tunnels and left the waste.
6. Do the earthworms get what they need from the ecosystem? Explain how you know.
Possible answer: Yes. It must be true that they get air, food, and water because they continue to move and tunnel.
7. What does the model ecosystem show about how living things interact in their environments?
Possible answer: The model shows how worms (living things) interact with the nonliving things (such as soil, water, and air) to survive in their ecosystem.

Unit 2, *Let's Investigate!* Lab • Ecosystems

Lessons 1–3 Got it? Self Assessment

Name _______________ Date _______________

Got it? Self Assessment

Complete the statements for each lesson.

Lesson 1 What are ecosystems?
- **Stop!** I need help with _______________
- **Wait!** I have a question about _______________
- **Go!** Now I know _______________

Lesson 2 What are food chains and food webs?
- **Stop!** I need help with _______________
- **Wait!** I have a question about _______________
- **Go!** Now I know _______________

Lesson 3 How do living things affect the environment?
- **Stop!** I need help with _______________
- **Wait!** I have a question about _______________
- **Go!** Now I know _______________

Unit 2, *Got it?* Self Assessment • Ecosystems

Unit 2

Got it? Quiz

Name ______________________ Date ____________

Got it? Quiz

Circle the choice you think is correct for each multiple choice question.

1. In every ecosystem ________ interact.
   - A only animals
   - B only plants
   - (C) all living and nonliving things
   - D air, water, soil, and sunlight

2. In which type of ecosystem is the ground beneath the surface frozen all year?
   - A desert
   - B wetland
   - (C) tundra
   - D rainforest

3. In which type of ecosystem do cacti grow?
   - A rainforest
   - (B) desert
   - C tundra
   - D wetland

4. ______ is the area or place where an organism lives in an ecosystem.
   - (A) A habitat
   - B A population
   - C An address
   - D An ecosystem

5. Which of the following is a producer?
   - A deer
   - (B) fern
   - C mountain lion
   - D sunlight

6. What is a decomposer?
   - A a dead animal
   - B the waste of dead animals and plants
   - C an organism that gets energy by eating only plants
   - (D) an organism that eats the remains of dead plants or animals

Unit 2, *Got it?* Quiz • Ecosystems

Unit 2

Got it? Quiz

Name ______________________ Date ____________

7. What is most likely to happen if most of a population of animals dies?
   - A The food web will stop.
   - B The decomposers will die.
   - (C) The food web will change.
   - D The food web will remain the same.

8. Which of these things are passed along in a food chain?
   - A herbivores
   - B habitats
   - C air and sunshine
   - (D) energy and matter

9. How does a beaver affect its environment? Describe how other living and nonliving things are affected.

Possible answer: A beaver modifies its environment by cutting down trees and building a dam. The dam causes a pond to form. This pond helps animals that live in still water, but some plants and animals are harmed because their habitats are flooded.

10. Rachel is looking at a wetland. She sees algae on a pond and some fish swimming. There are also lots of insects, some berries, raccoons, and cranes swooping by. Explain how the organisms Rachel sees are connected.

Possible answer: The organisms Rachel sees are connected in a food web. Algae make food using the sun's energy. Fish and insects eat the algae. Fish can also eat the insects. Cranes and raccoons eat the fish. The berries are also producers and make food using the sun's energy. The raccoons eat the berries.

Unit 2, *Got it?* Quiz • Ecosystems

Teacher's Notes

Unit 2 Study Guide

How do living things interact with their environment?

Lesson 1
What are ecosystems?

- An ecosystem is all the living and nonliving things in an environment.
- There are many ecosystems, including desert, tundra, and forest ecosystems.
- Special structures help many organisms survive in their habitats.

Lesson 2
What are food chains and food webs?

- In an ecosystem, energy flows from producers to consumers.
- Food chains show how energy is transferred.
- A food web is a system of overlapping food chains.

Lesson 3
How do living things affect the environment?

- Living things compete with each other for food and space.
- Some animals change their environment to improve their habitat.
- Changes in the environment help some living things and harm others.

Review the Big Question

How do living things interact with their environment?

Have students use what they have learned from the unit to answer the question in their own words.

How has your answer to the Big Question changed since the beginning of the unit? What are some things you learned that caused your answer to change?

Make a Concept Map

Have students make a concept map like the one shown on this page to help them organize key concepts.

Unit 2 Concept Map

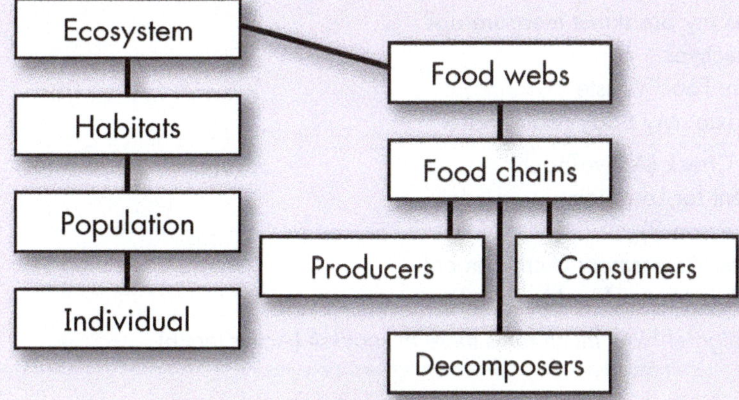

Students can make a concept map to help review the Big Question.

Body and Nutrition

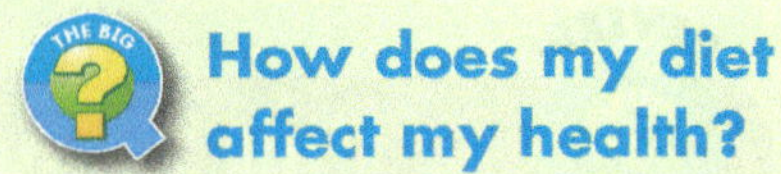

Lesson Plan

Unit Opener & Lesson 1 What are the nutrients in my food?			
	Activity	**Pages**	**Time**
Engage	• Unit Opener: Think! *What do you think food scientists study about food?* • Unit Opener: Identify different types of food. • Unit Opener: Identify why candy is a poor source of nutrients. • Think! *Is candy a source of carbohydrates?*	SB p. 28 SB p. 28 SB p. 28 TB p. 30	5 min 10 min 10 min 5 min
Explore	• Digital Lab: *Where do the minerals in your diet come from?* (ActiveTeach)	TB p. 29	10 min
Explain	• Diet and how energy content is measured in food • Carbohydrates • Fats and cholesterol • Proteins • Vitamins, minerals, and water • *Got it? 60-Second Video* (ActiveTeach)	SB p. 29 SB p. 30 SB p. 31 SB p. 32 SB p. 33 TB p. 33	20 min 20 min 20 min 20 min 20 min 10 min
Elaborate	• Science Notebook: Which bread has more fiber? • At-Home Lab: Starch Conversion • Vegetarian and Vegan Menus • Vitamin Research	TB p. 30 SB p. 31 TB p. 32 TB p. 33	15 min 15 min 20 min 20 min
Evaluate	• *Lesson 1 Check* (ActiveTeach) • Assessment for Learning • Review (Lesson 1) • *Got it? Self Assessment* (ActiveTeach) • *Got it? Quiz* (ActiveTeach)	TB p. 39a TB p. 33 SB p. 39 TB p. 39b TB p. 39b	10 min 10 min 10 min 10 min 10 min

Lesson 2 What are healthy and unhealthy diets?			
	Activity	**Pages**	**Time**
Engage	• Think! *Why might different governments make different recommendations for healthy diets?*	SB p. 35	15 min
Explore	• Digital Lab: *What are "empty" Calories?* (ActiveTeach)	TB p. 34	15 min
Explain	• Guidelines for healthy diets • Reading food labels • Malnutrition • Body mass index and obesity • *Got it? 60-Second Video* (ActiveTeach)	SB p. 34 SB p. 35 SB p. 36 SB p. 37 TB p. 37	20 min 20 min 20 min 20 min 10 min
Elaborate	• How does my breakfast measure up? • Food Detectives • Go Green: Food Waste • At-Home Lab: My Body Mass Index	TB p. 34 TB p. 35 TB p. 36 TB p. 37	15 min 15 min 15 min 10 min
Evaluate	• *Lesson 2 Check* (ActiveTeach) • Assessment for Learning • Review (Lesson 2) • *Got it? Self Assessment* (ActiveTeach) • *Got it? Quiz* (ActiveTeach)	TB p. 39a TB p. 37 SB p. 39 TB p. 39b TB p. 39b	10 min 10 min 10 min 10 min 10 min
Lab	• *Let's Investigate! How much fat is there in snacks?* (ActiveTeach)	SB p. 38	30 min

Flash Cards

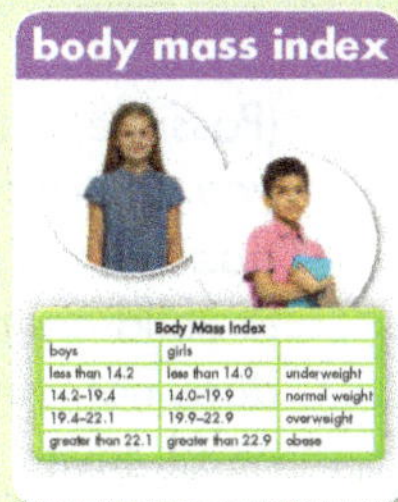

Lesson 1

Key Words	ELL Support
diet, calorie, carbohydrate, glucose, starch, fiber, cholesterol, amino acids, vegetarian	**Vocabulary:** *spinach, Calorie (kcal), kilo-, nutrients, simple carbohydrate, complex carbohydrate, saturated fats, unsaturated fats, trans fats, saliva, enzyme, essential amino acids, complete protein, vegan, calcium, iron*

Lesson 2

Key Words	ELL Support
grains, dairy, serving size, malnutrition, undernourished, anorexia nervosa, deficient, overnourished, obesity	**Vocabulary:** *balanced diet, food label, daily values, sodium, impoverished, irrational, unrealistic, anemia, type 2 diabetes, body mass index, underweight, overweight* **Prefixes Meaning *Not*:** *dis-, ir-,* and *un-*

Unit Objectives

Lesson 1: Students will learn what nutrients are in food.

Lesson 2: Students will learn what causes a diet to be healthy or unhealthy.

Vocabulary: *fish, orange, butter, spinach, rice, beans, milk, meat, producers, consumers, herbivores, carnivores, omnivores, candy, nutrients*

Introduce the Big Question

How does my diet affect my health?

Build Background Activate previous knowledge by asking students about producers, herbivores, carnivores, and omnivores in food chains and food webs. *If we looked at a food web that humans were in, what role do you think they would have?* (Possible answer: *Omnivores because they eat producers and other consumers.*) Remind students that energy and nutrients are passed through food chains and food webs. *In this unit, we are going to learn how we get energy and nutrients from the food we eat and how what we can eat can affect our health.*

Engage

Think!

What do you think food scientists study about food?

Point to the picture of the food scientist on the lower right of the page. Ask students why a scientist might study food. Encourage them to discuss the question in small groups. Circulate among the groups to monitor discussion and to provide vocabulary support as needed. Invite groups to share their ideas with the class.

1 Complete the words to label the pictures.

Point to the pictures of the food in the exercise and see if students can identify each item. Then invite them to complete the names by filling in the blanks. When they have finished, check answers and practice pronunciation by having volunteers name each item. You may also wish to reinforce prior knowledge by asking students whether each food item comes from a producer or a consumer.

2 Why can't a person live by eating only candy? Discuss as a class.

Invite volunteers to name their favorite candy. Survey the class to see how much candy they eat during a typical day. *Do you eat other food besides candy? Yes! Why do you need to eat food that is not candy?*

Unit 3 **Body and Nutrition**

How does my diet affect my health?

I will learn
- what nutrients are in my food.
- what are healthy and unhealthy diets.

1 Complete the words to label the pictures.

f_i_s_h o_r_a_ng_e b_u_tt_e_r sp_i_n_a_ch

r_i_c_e b_e_a_ns m_i_lk m_e_a_t

2 Why can't a person live by eating only candy? Discuss as a class.

3 List in your notebook what you ate for dinner last night. Compare with a partner.

Think! What do you think food scientists study about food?

28 Unit 3

(Possible answers: *Because candy is not good for you. Because there are things in other food that are good for us that you cannot get from candy.*) Explain to children that candy is okay as a treat, but that it does not contain many of the nutrients their bodies need. Invite students to look up the word *nutrient* in the Glossary and have them discuss where they have encountered the word before. (Possible answer: *In Level 4 we saw how plants get nutrients from the soil through their roots.*)

3 List in your notebook what you ate for dinner last night. Compare with a partner.

Have students form pairs and each make a list of what they had for dinner the previous night. Provide language support as necessary. Have each pair discuss whether they think the meals were healthy or not. Have them form a hypothesis as to what would make a healthy meal. Invite pairs to share their ideas with the class. Then encourage the class to come to a consensus about what makes a meal healthy.

Think! Again!

Revisit the question *What do you think food scientists study about food?* Ask *Do you think a food scientist might be able to find types of plants that might grow a lot of food? What types of areas might have difficulty growing enough food for the people who live there?* Encourage students to discuss freely. Invite them to draw on their knowledge of weather and climates as they discuss.

Lesson 1

What are the nutrients in my food?

Objective: Learn what a diet is and how the energy in food is measured in Calories (kcals).

Vocabulary: *diet, minerals, nutrients, calorie, Calorie (kcal)*

Digital Resources: Flash Card (*diet*), *Let's Explore!* Digital Lab

Unlock the Big Question

Write the following text on the board: *I will learn about the nutrients in food.*

Build Background Revisit the lists that students made the previous lesson of what they ate for dinner. Have some volunteers call out the items that they ate. Make a list of the items on the board to demonstrate the wide variety of foods that people eat. *Look at all the types of food we can eat! All of these foods could be called part of our diet.* Display the *diet* Flash Card. *We will learn what we get from our diet and what makes some diets more healthy than others.* Keep the list on the board for this lesson's exercises.

Explore

Let's Explore! Lab Where do the minerals in your diet come from?

Objective: Students will learn that the minerals in their food ultimately come from the soil.

Digital Resources: *Let's Explore!* Digital Lab, *Let's Explore! Activity Card* (1 per student)

- Invite students to recall food chains and food webs. *Remember how plants are producers and how they get energy from the sun?* Yes. *What do plants get from the soil?* Nutrients! *That's right! In this lab, we'll learn how some of those nutrients also get passed along in a food chain or food web and get from the soil to us.*
- Show the Digital Lab.
- Have students form pairs and complete the *Activity Card*.
- Review answers as a class.

Explain

1 Read. How does a cow's diet differ from a human's diet? Discuss with a partner.

Invite students to form pairs and to read the text. When they have finished, point to the list of the different kinds of food the students ate for dinner. *Look at this list again.*

Lesson 1 • What are the nutrients in my food?

Key Words
- diet
- calorie
- carbohydrate
- glucose
- starch
- fiber
- cholesterol
- amino acids
- vegetarian

1 Read. How does a cow's diet differ from a human's diet? Discuss with a partner.

Diet

All living things need energy and nutrients in order to live, grow, and repair themselves. Plants take in nutrients through their roots, and they get energy from the sun through the process of photosynthesis. Animals have to eat and drink things to get their nutrients and energy. The group of things that an animal typically eats or drinks is called that animal's **diet**.

2 Read. Circle *T* (true) or *F* (false). Correct the false statements with a partner.

Energy

Every day, you need to eat and drink to get enough energy for all the activities you do. Running, climbing mountains, and playing basketball take a lot of energy. But you need energy to do all the other things you do, too. You need energy for thinking, sleeping, and even for just sitting still. The amount of energy that your food provides is measured in calories.

One **calorie** is the amount of energy needed to raise one gram of water by one degree Celsius. Because a calorie is a very small amount of energy, the energy food provides is reported as kilocalories, or 1000 calories. A kilocalorie is often written with a capital letter "C" as a Calorie.

1. You don't need energy to sleep. T / (F)
2. Food contains energy that can be measured in calories. (T) / F
3. Five calories can raise five grams of water by five degrees Celsius. T / (F)
4. There are 100 calories in a Calorie. T / (F)

Let's Explore! Lab Unit 3 29

All of these things are part of our what? (Answer: *diet*) *What are the two things that we get from this diet?* (Answer: *energy and nutrients*) *Now let's make a list of the things that a cow might eat.* Invite students to brainstorm the items in a cow's diet. Have the pairs compare the two lists and discuss. *Why do you think a cow's diet and our diet are so different?* Have pairs discuss. Accept all logical answers and provide language support.

2 Read. Circle *T* (true) or *F* (false). Correct the false statements with a partner.

Invite students to read the text individually. Ask questions to check comprehension. *What is energy in food measured as?* Calories! *That's right. What is the difference between a calorie with a lowercase c and a Calorie with a capital C?* (Answer: *One Calorie with a capital C is 1000 calories with a lowercase c.*) Invite students to form pairs and complete the exercise. Check answers as a class.

ELL Vocabulary Support

Explain to students that a Calorie is often written as a *kilocalorie*, which can be abbreviated as *kcal*. Write *kcal* on the board. *The prefix kilo- means 1,000. So, a kilocalorie is what?* *1,000 calories!* *And what are a kilometer and a kilogram?* *1,000 meters and 1,000 grams.* *How can we write their abbreviations?* Invite two volunteers to come up and each one to write *km* or *kg*.

Lesson 1

What are the nutrients in my food?

Objective: Learn about carbohydrates and the role they play in the human diet.

Vocabulary: *nutrients, carbohydrates, simple carbohydrates, sugars, glucose, complex carbohydrates, starch, fiber, whole grain bread, flour*

Digital Resources: Flash Card (*carbohydrates*), *I Will Know…* Digital Activity

Materials: white bread, whole grain bread, hand lens (1 per small group)

Build Background Hold up one slice of white bread and one slice of whole grain bread. Ask students to identify them. Ask *Do you know what the difference between these two types of bread is?* Accept all logical answers. *They both provide you with a lot of energy, but do you think one type of bread is better for you than the other one?* Allow students to respond freely. Explain that, today, they will learn that the whole wheat bread has more of one kind of nutrient than the white bread does.

Explain

3 Read and underline three types of carbohydrates.

Read the first paragraph of the text aloud to students. Explain that some nutrients can provide energy but that other nutrients do not. *Things like vitamins, minerals, and water are nutrients, but they don't provide us energy. Why do we need them?* (Possible answer: *Because they provide other things our body needs to grow and repair itself.*) Invite students to read the text describing carbohydrates. Ask questions to check understanding. *What is the difference between a simple carbohydrate and a complex carbohydrate?* (Possible answers: *A simple carbohydrate is called a sugar, and a complex carbohydrate is a long chain of sugars.*) *Can a complex carbohydrate be broken down into simple carbohydrates? Yes!* Invite students to underline three types of carbohydrates. Check answers as a class.

4 Fill in the table below.

Review the definition of *nutrient* with the class. Then ask students to reread the paragraph describing complex carbohydrates. Invite students to form pairs and to complete the table. Invite volunteers to share their answers with the class.

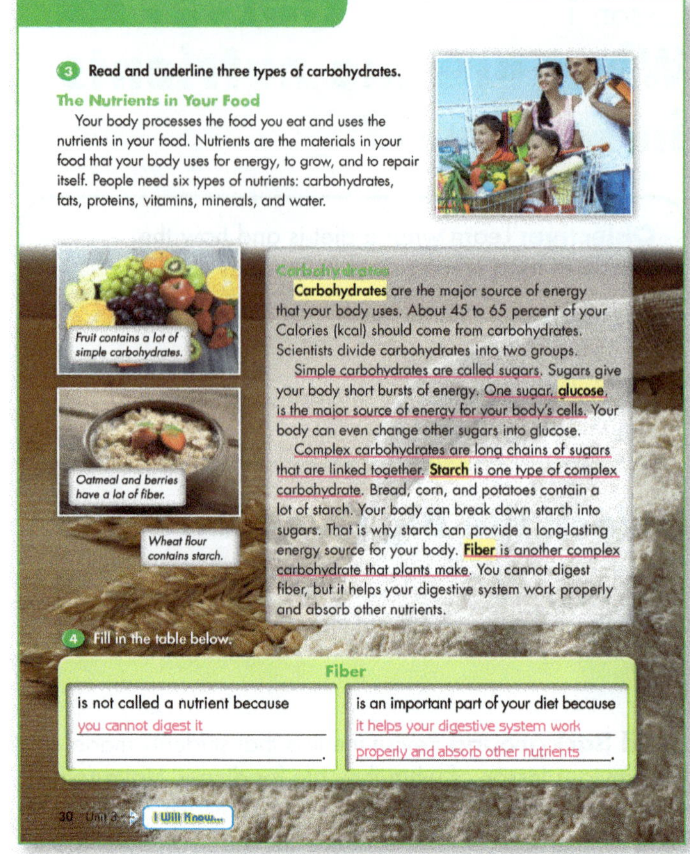

3 Read and underline three types of carbohydrates.

The Nutrients in Your Food

Your body processes the food you eat and uses the nutrients in your food. Nutrients are the materials in your food that your body uses for energy, to grow, and to repair itself. People need six types of nutrients: carbohydrates, fats, proteins, vitamins, minerals, and water.

Fruit contains a lot of simple carbohydrates.

Oatmeal and berries have a lot of fiber.

Wheat flour contains starch.

Carbohydrates

Carbohydrates are the major source of energy that your body uses. About 45 to 65 percent of your Calories (kcal) should come from carbohydrates. Scientists divide carbohydrates into two groups.

Simple carbohydrates are called sugars. Sugars give your body short bursts of energy. One sugar, **glucose**, is the major source of energy for your body's cells. Your body can even change other sugars into glucose.

Complex carbohydrates are long chains of sugars that are linked together. **Starch** is one type of complex carbohydrate. Bread, corn, and potatoes contain a lot of starch. Your body can break down starch into sugars. That is why starch can provide a long-lasting energy source for your body. **Fiber** is another complex carbohydrate that plants make. You cannot digest fiber, but it helps your digestive system work properly and absorb other nutrients.

4 Fill in the table below.

Fiber	
is not called a nutrient because you cannot digest it.	is an important part of your diet because it helps your digestive system work properly and absorb other nutrients.

30 Unit 3 I Will Know...

I Will Know...

Have students do the *I Will Know…* Digital Activity.

Think!

Is candy a source of carbohydrates?

Invite students to discuss the question as a class. Ask *What type of carbohydrates does candy have?* (Possible answer: *simple carbohydrates*) *Is candy a good source of the carbohydrates that provide long-lasting energy?* Allow students to discuss freely.

Elaborate

Science Notebook: Which bread has more fiber?

Have students form small groups. Ask *What is the main ingredient in bread?* (Answer: *flour*) Invite the groups to research on the Internet how white and whole grain flours are made. Distribute a piece of white bread, a piece of whole grain bread, and a hand lens to each group. Have the groups look at the two pieces of bread with the hand lens and see if they can find evidence of the different ways the two types of flour are made. Have them draw in their Science Notebooks a sketch of what they discover in each type of bread and label their drawings. Finally, ask, *Which type of bread has more fiber? Whole grain bread. Why is it called whole grain bread?* (Possible answers: *Because the flour used to make it is made from the whole grain of wheat. White bread is only made from the inner part of the grain.*)

Lesson 1

What are the nutrients in my food?

Objective: Learn about fats and cholesterol and the roles they play in the human diet.

Vocabulary: *fats, saturated fats, unsaturated fats, trans fats, cholesterol, waxy, saliva, enzyme*

Digital Resources: Flash Cards (*fats, carbohydrates*)

Materials: a bottle of vegetable or olive oil

Build Background Hold up the bottle of vegetable or olive oil and ask students to identify it. Have them discuss how oils like this are used in cooking. *Did you know that this oil comes from plants?* Explain that oils like this are largely composed of fats. Have them discuss what they know about fats. *Fats are essential parts of our diets. Today we will learn more about fats.*

Explain

5 Read and list one difference and one similarity between carbohydrates and fats.

Refresh students' memories about carbohydrates. *What has a lot of carbohydrates?* (Possible answers: *bread, grains, pasta, sugar,* etc.) Display the *fats* and *carbohydrates* Flash Cards. *Carbohydrates and fats are similar in some ways, but they are different in others.* Invite students to read the text. Ask questions to check for comprehension. *Which has more energy, fats or carbohydrates? Fats. What is an example of a food that has a lot of saturated fat?* (Possible answers: *butter, meat*) Hold up the bottle of vegetable or olive oil again. *What type of fat is liquid at room temperature? Unsaturated fat. Which fat is the healthiest? Unsaturated fat.* Have students complete the exercise and check answers as a class.

6 Read and, with a partner, research another food that is high in cholesterol.

Invite students to form pairs and to read the text. Ask questions to check comprehension. *Why is cholesterol important? It helps to form your cells' surfaces. How much cholesterol should you get in your diet?* (Possible answer: *You don't need cholesterol in your diet because your liver makes all the cholesterol your body needs.*) Have the pairs research other foods high in cholesterol on the Internet. Have them note what other nutrients those foods contain. Invite the groups to write their findings on the board and present them to the class. Explain that many foods that contain a lot of cholesterol contain a lot of saturated fats, which may make eating a lot of them unhealthy. *However, those same foods may contain a lot of other healthy nutrients, like protein. We'll learn about proteins later in this unit.*

5 **Read and list one difference and one similarity between carbohydrates and fats.**

Fats

Fats are like carbohydrates. They both contain a lot of energy. Your body, however, can use the fats you eat to form parts of your cells' surfaces and to make fatty tissues in your body that protect your organs and protect you from the cold. Fats provide more Calories (kcals) than carbohydrates do.

There are three kinds of fat in your food. Saturated fats are solid at room temperature and come from animals. Butter and fats in meat are saturated fats. Unsaturated fats are liquid at room temperature and come from plants. Vegetable oils are unsaturated fats. Trans fats are made when chemicals are used to change vegetable oil into a solid fat, like margarine. Unsaturated fat is healthy if you do not eat too much of it. Saturated and trans fats are not very healthy. Thirty percent or less of your daily Calories should come from fats.

Olive oil has a lot of unsaturated fat.

Beef contains a lot of saturated fat.

Margarine is made from vegetable oil and contains trans fat.

Similarity: Carbohydrates and fats are similar because they both contain a lot of energy.

Difference: Carbohydrates and fats are different because fats can be used to form parts of the cells' surfaces.

6 **Read and, with a partner, research another food that is high in cholesterol.**

Cholesterol

Cholesterol is a waxy substance that is like a fat and is also necessary to form your cells' surfaces. Many products from animals contain cholesterol. Your own liver makes all the cholesterol that your body needs, so you should avoid eating too much food that has a lot of cholesterol. Diets that are high in cholesterol can cause heart disease.

Egg yolks contain a lot of cholesterol.

At-Home Lab

Put some white bread in your mouth. Chew it for a long time. Do not swallow it. What change do you notice? What has happened to the starch in the bread?

Unit 3 31

Elaborate

ELL Content Support

Amylase is an enzyme found in the saliva of human beings and some other animals. It breaks the chemical bonds between the sugar molecules linked together in starch. Digestion of starch, therefore, begins in the mouth.

At-Home Lab

Starch Conversion

Review the meaning of *starch* and *complex carbohydrates* with students. Explain to students that there are special chemicals called enzymes in their saliva that can break down starch. Invite them to perform the lab as homework, but remind them to use white bread and to chew the bread until they notice a change in its taste. Students should notice the bread becomes sweet after they chew it for a while. *The special chemicals in your saliva broke down the complex carbohydrates in the bread. What did they break them down into?* (Possible answer: *simple carbohydrates or sugars*)

Lesson 1

What are the nutrients in my food?

Objective: Learn about proteins and the role they play in the human diet.

Vocabulary: *proteins, amino acids, essential amino acids, vegetarians, complete protein, vegan*

Digital Resources: Flash Card (*proteins*)

Build Background Remind students that a good diet provides energy and the nutrients that their bodies need to grow and repair themselves. *What do carbohydrates and fats provide your body with a lot of?* *Energy.* *That's right! Today we will learn about another nutrient in food called protein that helps your body build and repair its tissues.* Invite students to share what they know about protein and to describe some foods that contain large amounts of protein.

Explain

7 Read and fill in the blanks.

Invite students to review what they know about the structure of carbohydrates. Ask *What are simple carbohydrates?* *Sugars.* *And what is a complex carbohydrate?* (Possible answer: *A long chain of simple carbohydrates.*) Hold up the *proteins* Flash Card. *Today we will learn that proteins have a similar structure to complex carbohydrates. They are long chains of simpler molecules.*

Invite students to read the text. Allow them to underline any words that they find confusing. As a class, clarify the meanings of those words by using the context in which they appear. Then have students form pairs and fill in the blanks. Check answers as a class.

ELL Content Support

Explain to students that sources of protein that contain all the essential amino acids their bodies need are often called complete proteins. Ask *Are proteins from plants complete proteins?* *No!* *How can a vegetarian make sure he or she is getting a complete protein?* (Possible answer: *By combining different types of plant proteins.*)

8 Research your favorite form of protein. How many grams of the nutrients below does a 100 gram serving contain? Compare with a partner.

Ask volunteers to describe their favorite food that contains a lot of protein. Then have each student identify their favorite form of protein and write it in the space provided. Have students research information about that food on the Internet and enter the relevant information in the table. Invite students to compare their table with a partner's.

Then ask volunteers to share their results with the class. *Do sources of protein just contain protein, or do they contain other types of nutrients?* (Answer: *They contain other types of nutrients.*) *Protein is good for you, but what if your favorite source of protein has a lot of saturated fat, too?* (Possible answer: *It might not be as healthy as some other sources of protein.*)

Elaborate

Vegetarian and Vegan Menus

Review with the class what type of diet a vegetarian has. (Possible answer: *A diet that does not include meat or fish.*) *Did you know that some people don't eat meat, fish, eggs, or milk products? These people are called vegans.* Have students form groups and make menus for a vegetarian and a vegan breakfast, lunch, and dinner. Allow them to research recipes that include vegetarian and vegan options on the Internet. Groups should include meal options that provide complete proteins and not too much saturated fat or cholesterol. Invite the groups to write their menus on the board and to present them to the class. At the end of the exercise, have the class vote on the most interesting vegetarian and vegan breakfast, lunch, and dinner options.

---

7 Read and fill in the blanks.

Proteins

Proteins are the nutrients in your food that are the major building blocks of your body's tissues. Proteins can also provide you with energy, but only ten to 35 percent of your Calories (kcals) each day should come from protein.

Proteins are long chains of smaller molecules called **amino acids**. There are 23 amino acids that your body needs. Your body can make half of these by itself. The other half must come from your food. These amino acids are called essential amino acids.

Both animal and plant sources of food contain protein. All proteins from animal sources, including meat, fish, and eggs, contain all the essential amino acids. Proteins from plant sources, such as beans, nuts, and grains, do not contain all the essential amino acids. However, since different plants make different essential amino acids, **vegetarians**, or people who do not eat meat or fish, can combine their sources of plant protein to get all the essential amino acids in their diet.

These foods have a lot of protein.

Together, rice and beans provide all the essential amino acids.

a) Carbohydrates are made up of smaller molecules called sugars. The smaller molecules that make up proteins are __amino__ __acids__.
b) The amino acids that your body cannot make are called __essential__ amino acids. They must come from your __food__.
c) Food from __animal__ sources contain all the essential amino acids.
d) To get all the essential amino acids, __vegetarians__ must combine sources of plant protein.

8 Research your favorite form of protein. How many grams of the nutrients below does a 100 gram serving contain? Compare with a partner.

Sample answer **Nutrients in 100 gram serving of** beef		
protein = 27 g	carbohydrates = < 1 g	unsaturated fat = 5.5 g
saturated fat = 3.3 g	cholesterol = 60 mg	sodium = 65 mg

32 Unit 3

Lesson 1

What are the nutrients in my food?

Objective: Learn about vitamins, minerals, and water and the roles they play in the human diet.

Vocabulary: *vitamins, minerals, vitamin C, calcium, iron*

Digital Resources: Flash Card (*vitamins and minerals*), *Lesson 1 Check* (print out 1 per student), *Got it? 60-Second Video*

Materials: a glass of milk, bathroom scale, calculator, container with 2 liters of water

Build Background Display the glass of milk and have students identify it. *Do you know what nutrients milk has?* (Possible answers: *protein, carbohydrates, fat*) *Did you know that milk also contains something called calcium? Calcium is a mineral. Today we'll learn about minerals and vitamins, two more types of nutrients that we get from our food.* Ask students whether they take a daily vitamin and to name any other vitamins or minerals they might be familiar with.

Explain

9 Read and describe how calcium gets from the soil to your teeth. Discuss with a partner.

Invite students to read the text. Ask questions to check for understanding. *Do we need the same amount of vitamins and minerals as we do carbohydrates, fats, and proteins?* No, we need much less. *What is made by living things, vitamins or minerals?* *Vitamins.* *Where do minerals come from?* *The soil.* Invite students to form pairs to discuss the question. Check answers as a class. Have students draw on previous knowledge by asking them how the process of calcium getting from the soil to their teeth is similar to the transfer of energy in a food chain.

10 Read and calculate how many kilograms of water are in your body.

Have students read the text. Then invite them to weigh themselves on the bathroom scale and to record their weight. Have students calculate the weight of the water in their bodies. Monitor and provide assistance as needed. Once students have completed their calculations, elicit their reactions to the results.

Finally, display the container containing two liters of water on a desk or table at the front of the classroom. Ask *Do you think that you drink this much water every day?* Accept all logical answers. Then invite a volunteer to come up and lift the container.

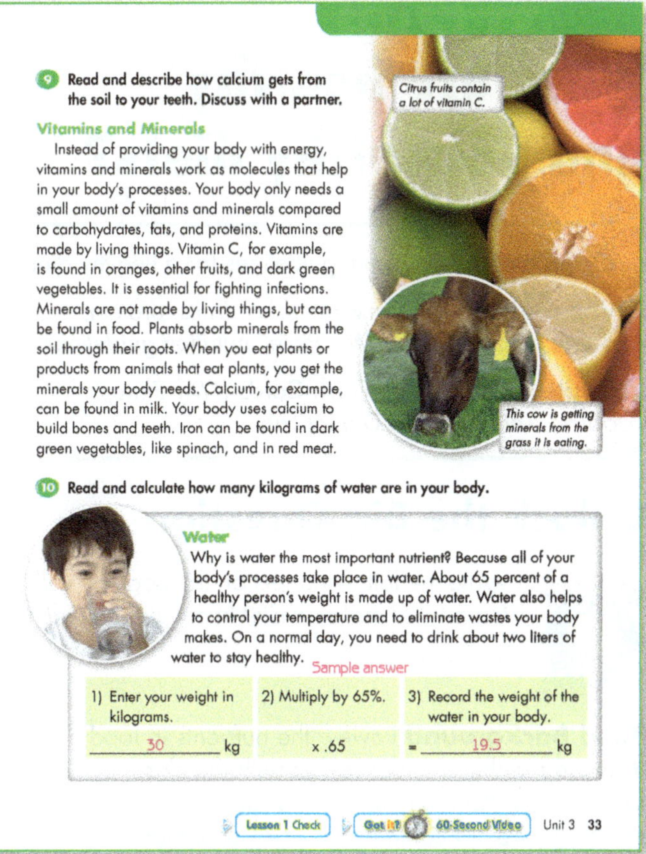

9 Read and describe how calcium gets from the soil to your teeth. Discuss with a partner.

Vitamins and Minerals

Instead of providing your body with energy, vitamins and minerals work as molecules that help in your body's processes. Your body only needs a small amount of vitamins and minerals compared to carbohydrates, fats, and proteins. Vitamins are made by living things. Vitamin C, for example, is found in oranges, other fruits, and dark green vegetables. It is essential for fighting infections. Minerals are not made by living things, but can be found in food. Plants absorb minerals from the soil through their roots. When you eat plants or products from animals that eat plants, you get the minerals your body needs. Calcium, for example, can be found in milk. Your body uses calcium to build bones and teeth. Iron can be found in dark green vegetables, like spinach, and in red meat.

Citrus fruits contain a lot of vitamin C.

This cow is getting minerals from the grass it is eating.

10 Read and calculate how many kilograms of water are in your body.

Water

Why is water the most important nutrient? Because all of your body's processes take place in water. About 65 percent of a healthy person's weight is made up of water. Water also helps to control your temperature and to eliminate wastes your body makes. On a normal day, you need to drink about two liters of water to stay healthy.

Sample answer

1) Enter your weight in kilograms.	2) Multiply by 65%.	3) Record the weight of the water in your body.
30 kg	x .65	= 19.5 kg

Lesson 1 Check   Got it? 60-Second Video   Unit 3   33

Ask whether they think it is heavy. Note that a child who weighs 30 kg has approximately ten times the amount of water in the container in his or her body.

Elaborate

Vitamin Research

Invite students to form groups and assign a vitamin, such as vitamin A, B2, B3, B12, D, E, or K, to each group. Have groups research their vitamin on the Internet. Each group should present their findings by writing on the board why humans need that vitamin, what foods are rich in it, and how much of it a person should eat daily.

Evaluate

Lesson 1 Check   Assessment for Learning

Distribute the *Lesson 1 Check* and guide students as they complete it. Check answers as a class. Then ask students to grade their progress on the topic of nutrients in food from 1 to 3: 3 = *I understand what the nutrients in food are*; 2 = *I need to study more*; 1 = *I need help!* Encourage students giving themselves a 1 or 2 to describe what they found difficult and what they need to study more.

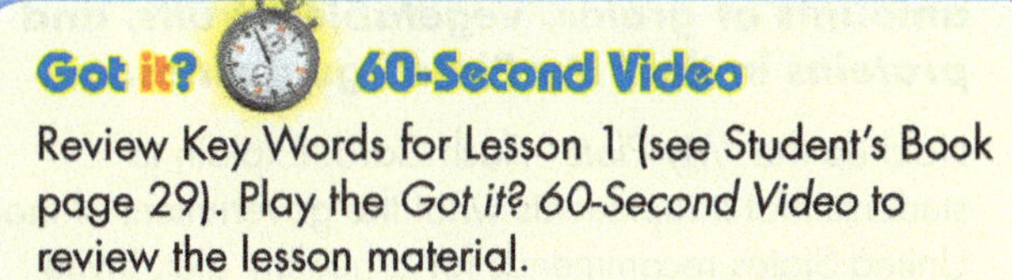

Got it? 60-Second Video

Review Key Words for Lesson 1 (see Student's Book page 29). Play the *Got it? 60-Second Video* to review the lesson material.

Lesson 2

What are healthy and unhealthy diets?

Objective: Learn about guidelines for a healthy diet.

Vocabulary: *grains, vegetables, fruits, protein, dairy, balanced diet, "My Plate" nutrition guidelines*

Digital Resources: Flash Cards (*"My Plate," dairy*), *Let's Explore!* Digital Lab

Unlock the Big Question

Write the following text on the board: *I will learn what makes a diet healthy or unhealthy.*

Build Background Review the nutrients in food that you discussed in the previous lesson with the class. *We have talked about the nutrients in food, but we have seen how the things we eat contain a mixture of nutrients. In this lesson, we will learn how we can choose food that gives us all the nutrients we need and keeps us healthy.* Invite students to share what they think a healthy diet is.

Explore

Let's Explore! Lab What are "empty" Calories?

Objective: Students will learn that some Calories are more valuable or healthy than others.

Digital Resources: *Let's Explore!* Digital Lab, *Let's Explore! Activity Card* (1 per student)

- *What are the two things that food provides? Energy and nutrients! How is the energy in food measured? In Calories! In this lab, we will find out why some foods are said to contain "empty" Calories.*
- Show the Digital Lab.
- Have students form pairs and complete the *Activity Card.*
- Review answers as a class.
- *What is an "empty" Calorie?* (Possible answer: *A Calorie that provides energy but doesn't provide many other nutrients.*)

Explain

1 Read and label the pie chart to show the amounts of *grains, vegetables, fruits,* and *proteins* in the "My Plate" guidelines.

Hold up the *"My Plate"* Flash Card. Explain to students that it represents what the government of the United States recommends for a healthy diet. Invite students to read the text and ask questions to check understanding. *What is a balanced diet?* (Possible answer: *A balanced diet does not contain too much or too little of one type of food.*) *What is an example of a grain?* (Possible answers: *wheat, rice, oats*) Hold up the *dairy* Flash Card. *What are dairy products?* (Possible answer: *Dairy products are made from milk.*) Explain to students that the diagram in the "My Plate" guidelines shows the relative amounts of grains, vegetables, protein, and fruits that should be eaten during a meal. *Look at the pie chart below. It shows the same amounts of those types of food.* Have students label the pie chart. *Grains and protein together take up how much of the pie chart?* (Answer: *One half.*) *And what combination makes up the other half?* (Answer: *Vegetables and fruits.*)

Lesson 2 • What are healthy and unhealthy diets?

1 Read and label the pie chart to show the amounts of *grains, vegetables, fruits,* and *proteins* in the "My Plate" guidelines.

Key Words
- grains
- dairy
- serving size
- malnutrition
- undernourished
- anorexia nervosa
- deficient
- overnourished
- obesity

Guidelines for Healthy Diets

A healthy diet gives your body all the energy it needs and all the nutrients necessary for it to grow and repair itself. There are some features that any healthy diet should have. First, it should provide enough Calories (kcals) and nutrients. Next, it should include a variety of different foods. Finally, it should be balanced, or not include too much or too little of one type of food.

"My Plate"

Most countries' governments issue guidelines for healthy diets. In the United States, the government uses the "My Plate" nutrition guide. It shows the amounts of different types of food that make up a healthy diet. It recommends that 30% of your diet should be **grains**, like wheat or rice; 40% vegetables; 10% fruits; and 20% proteins, like those in meat, fish, or beans. It also recommends including some **dairy** products. Dairy products, like milk and yogurt, are made from milk and contain a lot of calcium. These products are pictured in the glass. Other countries' governments may have different guidelines for healthy diets, but most agree that people should avoid eating or drinking too many things with added sugar or fats.

grains
vegetables
protein
fruits

2 With a partner, research your country's guidelines for a healthy diet. How is it the same as and different from the "My Plate" guidelines? Discuss.

34 Unit 3 Let's Explore! Lab

2 With a partner, research your country's guidelines for a healthy diet. How are they the same as and different from the "My Plate" guidelines? Discuss.

Invite students to form pairs to research their own country's guidelines. If those guidelines are not available, students may investigate the guidelines of some other country. Then have pairs present their findings to the class.

Elaborate

How does my breakfast measure up?

Invite students to write down what they had for breakfast in their notebooks and to estimate the percentages of grains, vegetables, fruits, and protein in that meal. Ask *How does your breakfast compare with the "My Plate" guidelines?*

Lesson 2

What are healthy and unhealthy diets?

Objective: Learn how to read a food nutrition label.

Vocabulary: *food label, serving size, daily values, sodium*

Digital Resources: Flash Card (*food label*), *I Will Know...* Digital Activity

Materials: nutrition labels cut out from boxes or from cans (1 per pair of students)

Build Background *We have talked about what makes a healthy meal. Today we are going to learn how to find out how many Calories and what nutrients are in some of the food we buy in a grocery store.* Hold up the *food label* Flash Card. *This is called a food label. Today we'll learn how to read it.*

Explain

3 Read and use the food label to answer the questions. Circle the correct answers.

Invite students to read the text. Encourage them to ask questions if they have any difficulty with new vocabulary. Then ask questions to check comprehension. *What does the serving size tell us?* (Possible answer: *It tells us how much of the food in the container is used to calculate all the nutrients in the table.*) *Look at the label. Does this food contain a lot of cholesterol?* (Answer: *Yes.*) *How do you know?* (Possible answer: *Because it says that one serving contains 39% of the cholesterol you should eat in a day.*) Have students complete the exercise and check answers as a class.

4 What type of food do you think this label describes? Discuss as a class.

Have students draw on previous knowledge about which foods are rich in which nutrients to discuss the questions. If the class is having trouble guessing the identity of the food, point out that this food contains a lot of saturated fat, cholesterol, and sodium. It does not contain many carbohydrates or vitamins or minerals. Accept all logical guesses.

5 With a partner, research on the Internet the nutrition facts for two fast food items.

Invite students to form pairs and to research two of their favorite fast food items on the Internet. The nutritional information for fast food can often be found on the restaurant's website. If the students cannot find the exact fast food item they are looking for, have them find the nutritional information for a generic item (for example, a hamburger or French fries) of the same kind. Have the pairs report their findings to the class.

3 Read and use the food label to answer the questions. Circle the correct answers.

Food Labels

Many food products you buy in a grocery store have food labels printed on their containers. These labels tell you some of the basic information about the food you are buying. They can help you to compare different food products and choose more healthy options.

One of the first things to look for is the **serving size**. This will tell you how much of the food in the container is used to calculate all the other information on the label. Near the top of the label, you will find the number of Calories (kcals) in one serving. Children around the age of ten should eat about 1800 Calories each day.

The label will also list the percentage of daily values for various nutrients, vitamins, and minerals. These values tell you what portion of the recommended amount one serving contains.

It is important to notice that containers often hold more than one serving. So, if you eat everything in the container at one time, you will have to multiply the numbers on the label by the number of servings!

Nutrition Facts
Serving Size 5 oz. (144g)
Servings Per Container 4

Amount Per Serving	
Calories 310	**Calories** from Fat 100
	% Daily Value*
Total Fat 15g	21%
Saturated Fat 2.6g	17%
Trans Fat 1g	
Cholesterol 118mg	39%
Sodium 560mg	28%
Total Carbohydrate 12g	4%
Dietary Fiber 1g	4%
Sugars 1g	
Protein 24g	
Vitamin A 1%	• **Vitamin C** 2%
Calcium 2%	• **Iron** 5%

*Percent Daily Values are based on a 2,000 calorie diet. Your daily values may be higher or lower depending on your calorie needs:

	Calories	2,000	2,500
Total Fat	Less Than	65g	80g
Saturated Fat	Less Than	20g	25g
Cholesterol	Less Than	300mg	300mg
Sodium	Less Than	2,400mg	2,400mg
Total Carbohydrate		300g	375g
Dietary Fiber		25g	30g

Calories per gram:
Fat 9 • Carbohydrate 4 • Protein 4

1. How many servings are there in the container? (4) / 5 / 144
2. How many Calories are there in the whole container? 100 / 310 / (1240)
3. If you ate the whole container, what percentage of your daily sodium, or salt, would you have eaten? 28% / 56% / (112%)

4 What type of food do you think this label describes? Discuss as a class.

5 With a partner, research on the Internet the nutrition facts for two fast food items.

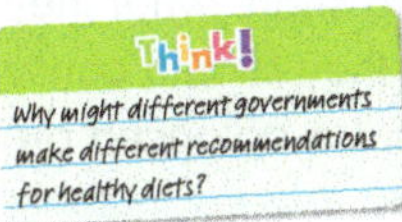

I Will Know... Unit 3 35

I Will Know...

Have students do the *I Will Know...* Digital Activity.

Think!

Why might different governments make different recommendations for healthy diets?

Invite the class to discuss the question and encourage them to consider how cultural differences, the availability of different types of food, and public health issues may contribute to the differences.

Elaborate

Food Detectives

Before class, cut out food labels from several types of food. You may want to choose items that should be easily identifiable from their nutrients, for example, rice, snack chips, luncheon meat, and candy. Keep track of what food product each food label describes. Invite students to form pairs and distribute a food label to each pair. Remind students to consider the number of Calories, the amount of fat, cholesterol, protein, carbohydrates, and sodium in a serving to guess what kind of item the food label belongs to. Have the pairs share their guesses with the class and explain what clues on the food labels they used to make their guesses. Reveal the identities of the food items at the end of the exercise and see which pairs guessed most accurately.

Lesson 2

What are healthy and unhealthy diets?

Objective: Learn about what happens when someone is either undernourished or overnourished.

Vocabulary: *malnutrition, undernourished, impoverished, anorexia nervosa, disorder, irrational, unrealistic, deficient, anemia, overnourished, type 2 diabetes*

Digital Resources: Flash Card (*"My Plate"*)

Build Background Display the *"My Plate"* Flash Card. *The "My Plate" guidelines show us what a healthy diet is. What are some of the things we learned about healthy diets?* Allow students to share what they recall about how many Calories (kcals) they should consume per day, what a varied diet is, and what a balanced diet is. *Today we will learn what can happen when someone's diet is not healthy.*

Explain

6 Read and answer *T* (true) or *F* (false). Correct the false statements with a partner.

Invite students to read the text and to underline any vocabulary they find confusing. Ask the class to help clarify the meaning of that vocabulary by using the context in which it occurs. Ask *What are the two ways that someone can be malnourished?* (Answer: *They can be undernourished or overnourished.*) *What does impoverished mean?* (Possible answers: *poor, in poverty*) *Are undernourished people always impoverished?* No! *If your diet is deficient in iron, what does that mean?* (Answer: *There is not enough iron in your diet.*) *What disease can you get if your diet is deficient in iron?* (Answer: *Anemia.*) *What disease can be caused by overnourishment?* (Answer: *Type 2 diabetes.*)

Invite students to form pairs and to complete the exercise. Check answers as a class.

7 If you had anemia, what should you eat more of? Circle the correct options.

Review what causes anemia with the class. Ask *What type of nutrient is iron?* (Answer: *A mineral.*) Then have volunteers identify the four foods pictured. Finally, have students complete the exercise. If students have difficulty recalling what nutrients each of the foods is rich in, invite them to review the previous lesson.

6 Read and answer *T* (true) or *F* (false). Correct the false statements with a partner.

Malnutrition

When someone does not have a healthy diet, they suffer from **malnutrition**. There are two ways that someone can be malnourished. They can be undernourished or overnourished.

When someone lacks Calories (kcals) or some nutrients, they are **undernourished**. Approximately 800 million people in the world suffer from undernourishment. This may mean that they do not get enough Calories (kcals). Most undernourished people in the world are impoverished and cannot get enough food. However, some people in the developed world are undernourished as well. **Anorexia nervosa** is an eating disorder that typically affects girls and young women. Its symptoms include very low body weight, an irrational fear of gaining weight, and unrealistic body image. Anorexia can lead to rapid weight loss and severe malnourishment.

Undernourishment may also occur when someone's diet lacks, or is **deficient** in, a particular nutrient. Anemia, for example, is a disease that can develop if someone does not get enough iron in their diet. When someone is anemic, they do not have enough red blood cells to carry oxygen to the rest of their body. They can become tired, weak, short of breath, and even have a hard time thinking.

When someone gets too many Calories or nutrients in his or her diet, they can become **overnourished**. This is a growing problem in the richer countries in the world. Foods like fast foods and soda pop often contain large amounts of added sugar, fat, and sodium. Overnourishment can cause people to gain weight and can also lead to other diseases. Type 2 diabetes is often caused by people having too much sugar in their diet.

1. Only people in poor countries are malnourished. T / **(F)**
2. Too much of a nutrient can cause malnourishment. **(T)** / F
3. When someone is anemic, they have too much iron in their diet. T / **(F)**
4. Type 2 diabetes can be caused by undernourishment. T / **(F)**

7 If you had anemia, what should you eat more of? Circle the correct options.

36 Unit 3

ELL Language Support

Point out to students that there are several prefixes that can be used in English to mean *not*. Invite students to circle the prefixes *dis-*, *ir-*, and *un-* that mean *not*. Ask them questions to clarify the function of these prefixes. You may also want to point out the prefix *mal-*, meaning *poor* or *bad*, in *malnutrition*.

Elaborate

Go Green

Food Waste

Explain to students that approximately half of the world's food is wasted. *Some food might be wasted when it is eaten by insects and other pests or when it dies during severe weather. More food is wasted because it isn't very pretty and the people who sell it don't think their customers will buy it. A lot of food is wasted because people don't eat it before it goes bad.* Ask students to think of ways that they can reduce the amount of food they themselves waste. Then invite students to keep a log of all the food their families throw out during one week. Have volunteers share their logs with the rest of the class at the end of the week. Have the class brainstorm some ways the food their families throw out could be saved.

Lesson 2

What are healthy and unhealthy diets?

Objective: Learn about body mass index and obesity.

Vocabulary: *body mass index, square of a number, underweight, normal weight, overweight, obese, obesity*

Digital Resources: Flash Card (*body mass index*), *Lesson 2 Check* (print out 1 per student), *Got it? 60-Second Video*

Build Background Review the meaning of *undernourished* and *overnourished* with the class. *Today we will find out how doctors and scientists make measurements to see if we are under- or overnourished.*

Explain

8 Read and calculate the body mass indexes of the boy and the girl.

Invite students to read the text. Then have them form small groups. Distribute a calculator to each group and have them practice calculating the square of some simple numbers. Have the groups calculate the body mass indexes for the boy and girl pictured. Check answers as a class. *Which one is very close to being underweight?* (Answer: *The boy.*)

9 Read and list two ways a person could reduce the number of Calories (kcals) in their diet. Discuss as a class.

Invite students to read the text. Ask questions to check for understanding. *If someone is obese, is their body mass index high or low?* (Answer: *High.*) *What are the two things an obese person should do?* (Answer: *Reduce the number of calories they eat and exercise.*) Invite students to complete the exercise and discuss their answers as a class.

Elaborate

At-Home Lab

My Body Mass Index

Invite students to calculate their own body mass index as homework. Reassure students that body mass indexes are only approximate measures of normal weight. However, if students have unusually low or high body mass indexes, they may want to discuss them with their parents.

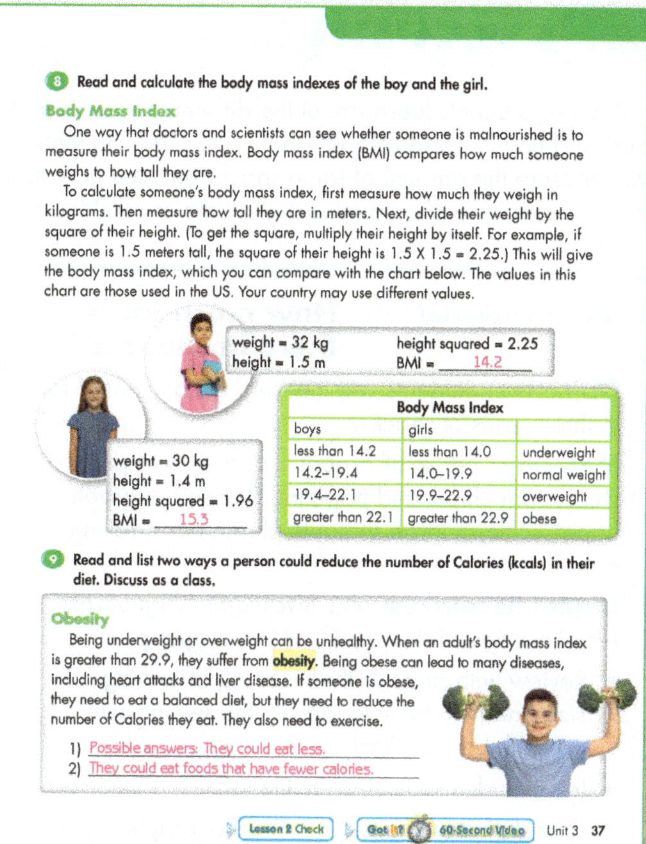

8 Read and calculate the body mass indexes of the boy and the girl.

Body Mass Index

One way that doctors and scientists can see whether someone is malnourished is to measure their body mass index. Body mass index (BMI) compares how much someone weighs to how tall they are.

To calculate someone's body mass index, first measure how much they weigh in kilograms. Then measure how tall they are in meters. Next, divide their weight by the square of their height. (To get the square, multiply their height by itself. For example, if someone is 1.5 meters tall, the square of their height is 1.5 X 1.5 = 2.25.) This will give the body mass index, which you can compare with the chart below. The values in this chart are those used in the US. Your country may use different values.

weight = 32 kg
height = 1.5 m
height squared = 2.25
BMI = 14.2

weight = 30 kg
height = 1.4 m
height squared = 1.96
BMI = 15.3

Body Mass Index		
boys	girls	
less than 14.2	less than 14.0	underweight
14.2–19.4	14.0–19.9	normal weight
19.4–22.1	19.9–22.9	overweight
greater than 22.1	greater than 22.9	obese

9 Read and list two ways a person could reduce the number of Calories (kcals) in their diet. Discuss as a class.

Obesity

Being underweight or overweight can be unhealthy. When an adult's body mass index is greater than 29.9, they suffer from **obesity**. Being obese can lead to many diseases, including heart attacks and liver disease. If someone is obese, they need to eat a balanced diet, but they need to reduce the number of Calories they eat. They also need to exercise.

1) Possible answers: They could eat less.
2) They could eat foods that have fewer calories.

Lesson 2 Check | Got it? 60-Second Video

Evaluate

Lesson 2 Check **Assessment for Learning**

Distribute the *Lesson 2 Check* and guide students as they complete it. Check answers as a class. Then ask students to grade their progress on the topic of healthy and unhealthy diets from 1 to 3: 3 = *I understand what are healthy and unhealthy diets*; 2 = *I need to study more*; 1 = *I need help!* Encourage students giving themselves a 1 or 2 to describe what they found difficult and what they need to study more.

Review Key Words for Lesson 2 (see Student's Book page 34). Play the *Got it? 60-Second Video* to review the lesson material.

Let's Investigate!

In this unit, students learn about the different nutrients in food and how to read a food label. In this lab, students will measure the amount of fat in snacks and see how their results compare with amounts listed on the snacks' food labels.

Let's Investigate! Lab How much fat is there in snacks?

Objective: Students will measure the amount of fat in three types of snacks.

Materials: per small group: 1 bag each of potato chips, corn chips, and popcorn, 3 sheets of graph paper, scale, 3 large sealable plastic bags, rolling pin

Digital Resources: *Let's Investigate!* Digital Lab, *Let's Investigate! Activity Card* (1 per group)

- Review with students the meaning of *serving size* and help them to identify the serving size on each bag of snacks.
- Invite students to complete the lab.
- If most of the squares on a sheet of graph paper are stained by grease, suggest to students that they count the number of unstained squares and subtract them from the total number of squares on the graph paper.
- Circulate among the groups to make sure that they fill in the bar graph correctly.
- Ask students to complete the *Activity Card* and share their results with the class.

Teacher Time-Saving Option: Show the *Let's Investigate!* Digital Lab as an alternative to the hands-on lab activity.

Let's Investigate!

How much fat is there in snacks?

1. Cut the sheets of graph paper so they fit inside the plastic bags. Put one in each bag.
2. Use the serving size on the food labels to weigh out one serving of each snack. Put each in a separate bag. Seal the bags, but leave a small portion open for air to escape.
3. Use the rolling pin to crush the two kinds of chips and popcorn in their bags. Let the crushed chips and popcorn sit on the graph paper in the bags for 15 minutes.
4. Remove the sheets of graph paper and brush off all crumbs.
5. Hold up the sheets of paper to a bright light and count the number of squares that are stained by grease. Record your results.
6. Make a bar graph and compare your results with the amount of fat on the nutritional labels.

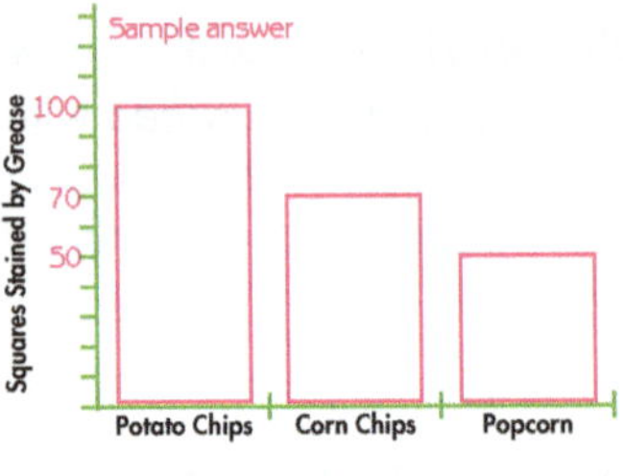

38 Unit 3 Let's Investigate! Lab

Unlock the Big Question

Have students refer to the Big Question on the Unit Opener page. In pairs, have them recall what they have learned about the nutrients in food and healthy and unhealthy diets. Have pairs complete questions 7 and 8 on the *Activity Card.*

Class Project: Healthier Options

Materials: large sheet of construction paper (1 per group), art supplies

Invite students to form small groups. Have them investigate a favorite food item that may not be all that healthy. They should then investigate a healthier option and draw the two items on their piece of construction paper. Have students list the relevant nutritional information underneath each food item. Invite groups to present their posters to the class and to explain what makes the healthier option healthier. Encourage students to ask their parents to try the healthier option at home.

Unit 3 Review

How does my diet affect my health?

Digital Resources: Print out 1 of each per student: *Got it? Self Assessment*, *Got it? Quiz*

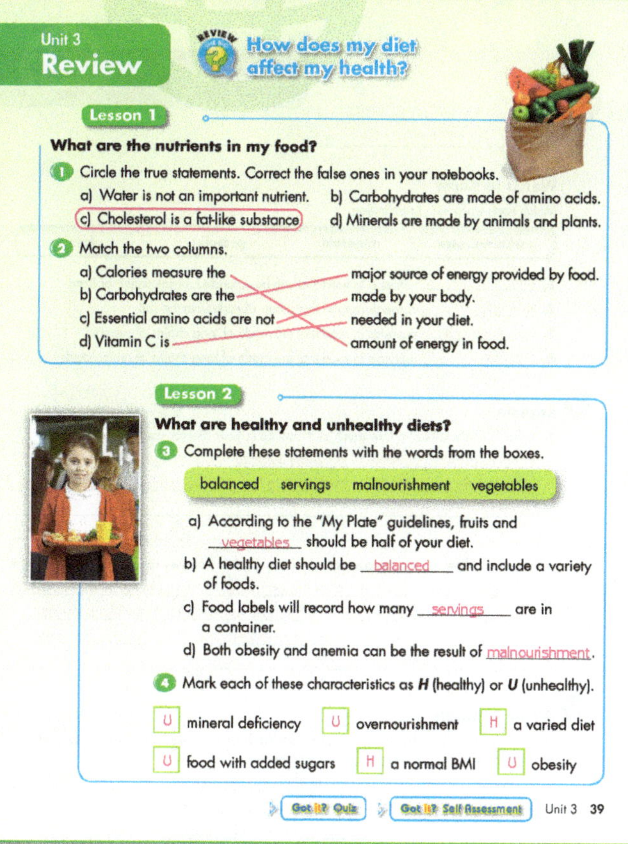
Unit 3 Review

How does my diet affect my health?

Lesson 1

What are the nutrients in my food?

1 Circle the true statements. Correct the false ones in your notebooks.

a) Water is not an important nutrient.
b) Carbohydrates are made of amino acids.
c) Cholesterol is a fat-like substance.
d) Minerals are made by animals and plants.

2 Match the two columns.

a) Calories measure the — amount of energy in food.
b) Carbohydrates are the — major source of energy provided by food.
c) Essential amino acids are not — made by your body.
d) Vitamin C is — needed in your diet.

Lesson 2

What are healthy and unhealthy diets?

3 Complete these statements with the words from the boxes.

balanced servings malnourishment vegetables

a) According to the "My Plate" guidelines, fruits and vegetables should be half of your diet.
b) A healthy diet should be balanced and include a variety of foods.
c) Food labels will record how many servings are in a container.
d) Both obesity and anemia can be the result of malnourishment.

4 Mark each of these characteristics as *H* (healthy) or *U* (unhealthy).

U mineral deficiency U overnourishment H a varied diet
U food with added sugars H a normal BMI U obesity

Got it? Quiz Got it? Self Assessment Unit 3 39

Evaluate

Strategies for Targeted Review

The following are strategies for providing targeted review for students if they encounter challenges with the content.

Lesson 1 What are the nutrients in my food?

Question 1

If… students are having difficulty circling the correct answer and correcting the false answers, then… direct students to Lesson 1 and have them reread the relevant sections.

Question 2

If… students are having difficulty matching the two columns, then… review the meanings of the key words on page 29.

Lesson 2 What are healthy and unhealthy diets?

Question 3

If… students are having difficulty choosing the correct words to complete the sentences, then… direct students to Lesson 2 and have them reread the relevant sections.

Question 4

If… students are having difficulty distinguishing between the healthy and unhealthy characteristics, then… remind students that unhealthy diets can be the result of too much or too little of a particular nutrient.

ELL Language Support

Before students start working on the Review activities, have them read each question aloud along with you.

Got it? Self Assessment

Immediately after students have completed the Review activities, distribute a *Got it? Self Assessment* to each student. Have students complete the *Stop! Wait!* and *Go!* statements for each lesson, allowing them to look back through the lesson material if necessary.

Got it? Quiz

Distribute a Unit 3 *Got it? Quiz* to each student. Quizzes may be used for assessing students' understanding of unit concepts as well as for grading purposes.

Lesson 1 Check

Name _______________ Date _______________

Words to Know

Match each word with its definition.

carbohydrates	minerals	proteins	fats

1. fats nutrients in food that can be saturated, unsaturated, or trans
2. minerals nutrients in food that plants absorb from the soil
3. carbohydrates nutrients in food that are made of long chains of sugars
4. proteins nutrients in food that are made of long chains of amino acids

Explain

5. What is cholesterol and how much of it must be in your diet for you to stay healthy?
Possible answer: Cholesterol is a waxy substance that is like fat. It helps form part of your cells' surfaces. You do not have to eat any cholesterol because your liver makes all that you need.

6. Starch and fiber are two complex carbohydrates. Explain how they are the same and different.
Possible answer: Starch and fiber are both long chains of sugars that are made by plants. Starch can be digested, but fiber cannot. Starch provides energy. Fiber does not provide energy, but it does help your digestive system work properly and absorb other nutrients.

Apply Concepts

7. You are stuck on a deserted island that does not have any fresh water for you to drink. Explain why this can be very bad for your health.
Possible answer: Water makes up about 65 percent of your body's weight. All of your body's processes take place in water. Water also helps to control your body's temperature and to eliminate the wastes your body makes.

Unit 3, *Lesson 1 Check* • What are the nutreints in my food?

Lesson 2 Check

Name _______________ Date _______________

Words to Know

Match each word with its definition.

dairy	fruits	grains	complete

1. fruits what covers 10% of the plate in the "My Plate" guidelines
2. dairy foods that are made from milk
3. complete a protein that provides all of the essential amino acids
4. grains foods like wheat or rice that contain a lot of carbohydrates

Explain

5. What can a person's body mass index tell you about their diet?
Possible answer: A person's body mass index can tell you if they are undernourished or overnourished. If their body mass index is very low, they may not be eating enough Calories. If their body mass index is very high, they may be eating too many Calories.

6. Only people from impoverished countries are malnourished. Is this statement true or false? Explain your answer.
This statement is false because people in the developed world can be undernourished as well. They can have anorexia nervosa or a deficiency. They can also be overnourished.

Apply Concepts

7. Charlie is watching television and eating chips. When his show is over, he realizes that he ate the whole bag of chips. What can he do to find out how many Calories he just ate?
Possible answer: He can look at the food label and multiply the number of Calories in a serving by the number of servings in the bag.

Unit 3, *Lesson 2 Check* • What are healthy and unhealthy diets?

Lesson 1 Let's Explore! Activity Card

Name _______________ Date _______________

Materials
- *Let's Investigate!* Digital Lab

Where do the minerals in your diet come from?

1. Watch the Digital Lab again with a partner. Discuss.
2. Draw a diagram to show the two ways you can get iron in your diet.

Students' drawings should show the iron in the soil moving to the grass and the spinach. Then the iron in the spinach moves from the spinach to the dinner plate. The iron in the grass moves from the grass to the cow, which becomes the steak on the dinner plate.

Explain Your Results

3. How are the movements of iron and energy in a food chain alike and different?
Possible answer: The movements of iron and energy are similar because they go from producers to consumers in food chains. Their movements are different because energy comes from the sun and passes to the producers and iron comes from the soil and passes to producers.

Magnesium and iodine are two other minerals you need in your diet. Research on the Internet where the magnesium and iodine in your diet come from.

Possible answers: Magnesium is found in the soil and in foods like seeds, nuts, and bananas. Iodine is mainly found in seawater and soil. Foods that contain a lot of iodine include seafood, dairy products, and iodized salt.

Unit 3, Lesson 1 *Let's Explore!* Lab • What are the nutrients in my food?

Lesson 2 Let's Explore! Activity Card

Name _______________ Date _______________

Materials
- *Let's Explore!* Digital Lab

What are "empty" Calories?

1. Watch the Digital Lab again with a partner.
2. Share your answers to the questions.
3. Make a list of all the foods you can think of that contain "empty" Calories.
Possible answers: candy, cake, donuts, ice cream, butter

Explain Your Results

4. What could happen to you if you ate too many "empty" calories?
Possible answer: I might gain a lot of weight. My diet might not be very healthy because I would not get enough of the nutrients I need.

5. If foods with "empty" Calories have few nutrients compared to the number of Calories they provide, what do healthier foods have?
Possible answer: Healthier foods have a lot of nutrients compared to the number of Calories they provide.

Are "empty" Calories really empty?

Possible answer: No, because the foods do contain other nutrients, but they do not contain as many nutrients as other types of food do.

Unit 3, Lesson 2 *Let's Explore!* Lab • What are healthy and unhealthy diets?

Unit 3 — Let's Investigate! Activity Card

Name ______________________ Date __________

Analyze and Conclude

7. What was the relationship between the number of squares stained by grease and the amount of fat in the snacks.

Possible answer: The potato chips had the most amount of fat per serving and stained the greatest number of squares on the graph paper with grease. The popcorn had the least amount of fat per serving and stained the least number of squares on the graph paper with grease.

8. Which of these snacks would be the healthiest to eat?

Possible answer: The popcorn would be the healthiest because it has the least amount of fat.

Unit 3, *Let's Investigate! Lab* • Body and Nutrition

Unit 3 — Lessons 1 & 2 Got it? Self Assessment

Name ______________________ Date __________

Got it? Self Assessment

Complete the statements for each lesson.

Lesson 1 What are the nutrients in my food?

Stop! I need help with ______________________

Wait! I have a question about ______________________

Go! Now I know ______________________

Lesson 2 What are healthy and unhealthy diets?

Stop! I need help with ______________________

Wait! I have a question about ______________________

Go! Now I know ______________________

Unit 3, *Got it? Self Assessment* • Body and Nutrition

Unit 3 — Got it? Quiz

Name ______________________ Date __________

Got it? Quiz

Circle the choice you think is correct for each multiple choice question.

1. Complex carbohydrates are long chains of __________.
- **A** amino acids
- **B** cholesterol
- **(C)** sugars
- **D** fats

2. The main building blocks of your body's tissues are __________.
- **A** calories
- **(B)** proteins
- **C** saturated fats
- **D** vitamins

3. The amount of energy your food provides is measured in __________.
- **A** degrees
- **B** calcium
- **C** vitamins
- **(D)** Calories

4. You need smaller amounts of __________ in your diet than the amounts of carbohydrates, fats, and proteins you need.
- **A** energy
- **B** Calories
- **(C)** vitamins
- **D** water

5. An essential amino acid is one that your body __________.
- **(A)** does not make
- **B** always makes
- **C** can make
- **D** does not need

6. A healthy diet is __________.
- **A** full of saturated fat
- **(B)** balanced and varied
- **C** contains enough cholesterol
- **D** does not contain iron

Unit 3, *Got it? Quiz* • Body and Nutrition

Unit 3 — Got it? Quiz

Name ______________________ Date __________

7. If someone is obese, they have __________.
- **(A)** a high body mass index
- **B** a vitamin deficiency
- **C** a mineral deficiency
- **D** anorexia nervosa

8. Anemia is a type of __________.
- **A** overnourishment
- **B** obesity
- **(C)** mineral deficiency
- **D** diabetes

9. Why do vegetarians need to eat multiple sources of protein?

Possible answer: Vegetarians do not eat meat or fish, which contain all of the essential amino acids. Individual plants do not contain all of the essential amino acids, so vegetarians need to combine multiple sources of plant protein to make sure they get all the essential amino acids in their diet.

10. Malnutrition can mean many things. Describe three types of malnutrition.

Possible answers: Malnutrition can mean not getting enough Calories in your diet. Poverty prevents some people from getting enough food, and anorexia nervosa is an eating disorder that prevents some people from getting adequate Calories and a healthy diet. Malnutrition can also mean having a diet that lacks a nutrient. Anemia is a disease that can occur when someone's diet lacks, or is deficient in, iron. Malnutrition can also mean being overnourished. Obese people are malnourished because they have too many Calories in their diet.

Unit 3, *Got it? Quiz* • Body and Nutrition

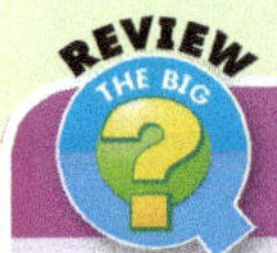

Unit 3 Study Guide

How does my diet affect my health?

Lesson 1
What are the nutrients in my food?

- A healthy diet provides enough energy, which is measured in Calories.
- A healthy diet provides enough nutrients.
- The nutrients in food include carbohydrates, fats, proteins, vitamins, and minerals.

Lesson 2
What are healthy and unhealthy diets?

- A healthy diet is balanced and varied.
- People can become malnourished if they get too many or too few nutrients in their diets.

Review the Big Question

How does my diet affect my health?

Have students use what they have learned from the unit to answer the question in their own words.

How has your answer to the Big Question changed since the beginning of the unit? What are some things you learned that caused your answer to change?

Make a Concept Map

Have students make a concept map like the one shown on this page to help them organize key concepts.

Unit 3 Concept Map

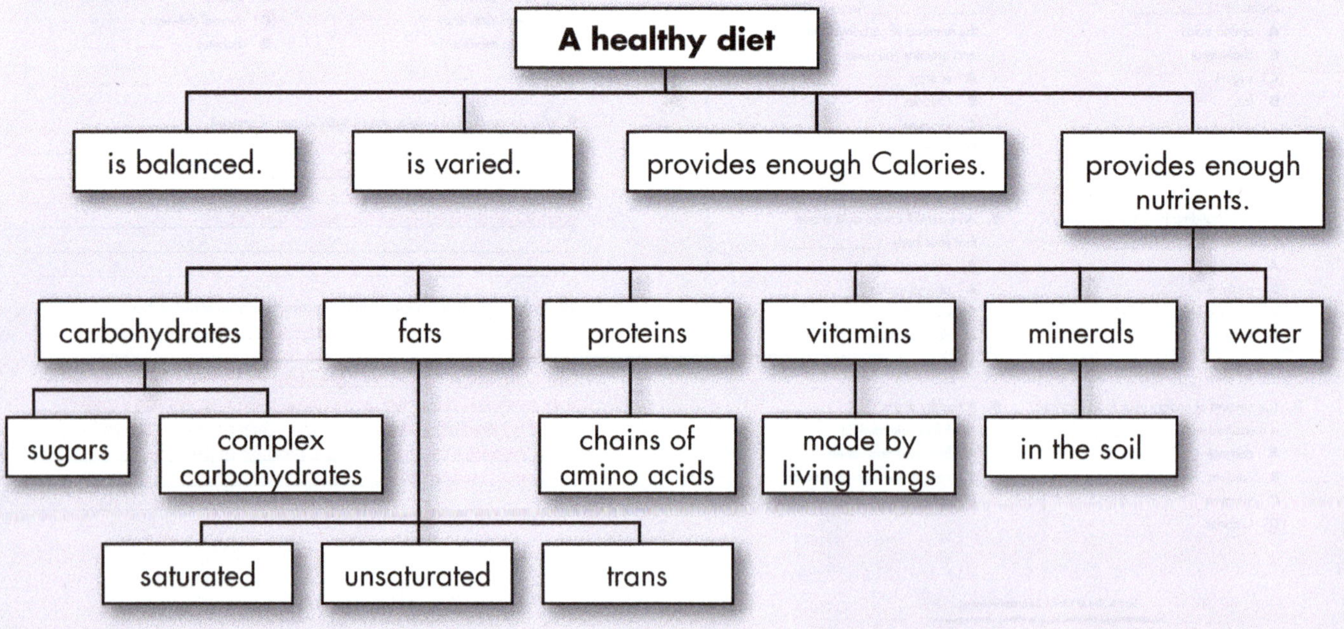

A healthy diet

- is balanced.
- is varied.
- provides enough Calories.
- provides enough nutrients.
  - carbohydrates
    - sugars
    - complex carbohydrates
  - fats
    - saturated
    - unsaturated
    - trans
  - proteins
    - chains of amino acids
  - vitamins
    - made by living things
  - minerals
    - in the soil
  - water

Students can make a concept map to help review the Big Question.

Teacher's Notes

Lesson Plan

Unit Opener & Lesson 1 How can Earth's surface change rapidly?			
	Activity	Pages	Time
Engage	• Think! *What forces shaped this formation?* • Unit Opener: List how a volcano's eruption changes its surroundings. • Unit Opener: Discuss the dangers of living on a mountainside or near a river. • Think! *How might ash and other particles change Earth's temperature?* • Think! *Why is it important to locate the epicenter of an earthquake?*	SB p. 40 SB p. 40 SB p. 40 SB p. 41 SB p. 42	5 min 10 min 10 min 5 min 5 min
Explore	• Digital Activity: *Science Stats: Earthquakes* (ActiveTeach)	TB p. 41	15 min
Explain	• Earth's plates and volcanoes • Earthquakes • Changes caused by landslides, floods, and droughts • *Got it? 60-Second Video* (ActiveTeach)	SB p. 41 SB p. 42 SB p. 43 TB p. 43	15 min 15 min 15 min 5 min
Elaborate	• A Volcano in My Country • Three Largest Earthquakes • Too much rain! Too little rain! • Flash Lab: Earthquake Model	TB p. 41 TB p. 42 TB p. 43 SB p. 43	20 min 20 min 20 min 15 min
Evaluate	• *Lesson 1 Check* (ActiveTeach) • Assessment for Learning • Review (Lesson 1) • *Got it? Self Assessment* (ActiveTeach) • *Got it? Quiz* (ActiveTeach)	TB p. 51a TB p. 43 SB p. 51 TB p. 51b TB p. 51c	10 min 10 min 10 min 10 min 10 min

Lesson 2 Where is Earth's water?			
	Activity	Pages	Time
Engage	• Think! *Why do people dig wells to reach groundwater?*	SB p. 46	10 min
Explore	• Digital Lab: *Where is Earth's water?* (ActiveTeach)	TB p. 44	15 min
Explain	• Where Earth's water is • Bodies of water • Importance of groundwater and water treatment plants • *Got it? 60-Second Video* (ActiveTeach)	SB p. 44 SB p. 45 SB p. 46 TB p. 46	15 min 15 min 15 min 5 min
Elaborate	• Science Notebook: Water on Earth • Water on Earth Graphs • How can animals live in very salty water? • Bodies of Water in the World • Science Notebook: Groundwater • Saving Water	TB p. 44 TB p. 44 TB p. 45 TB p. 45 TB p. 46 TB p. 46	20 min 20 min 20 min 20 min 15 min 10 min
Evaluate	• *Lesson 2 Check* (ActiveTeach) • Assessment for Learning • Review (Lesson 2) • *Got it? Self Assessment* (ActiveTeach) • *Got it? Quiz* (ActiveTeach)	TB p. 51a TB p. 46 SB p. 51 TB p. 51b TB p. 51c	10 min 10 min 10 min 10 min 10 min

Lesson 3 What is the water cycle?			
	Activity	**Pages**	**Time**
Engage	• Think! *Could the water falling as rain fall again in the future as snow?* • Think! *What is the difference between evaporation and condensation?*	SB p. 47 TB p. 48	5 min 5 min
Explore	• Digital Activity: *Did You Know: Wetlands* (ActiveTeach)	TB p. 47	20 min
Explain	• Three steps in the recycling of Earth's water • The water cycle • Relationship between the water cycle and weather • *Got it? 60-Second Video* (ActiveTeach)	SB p. 47 SB p. 48 SB p. 48 TB p. 48	15 min 15 min 15 min 5 min
Elaborate	• Water Droplets • Water Cycle Phases • Story of a Raindrop • Water Cycle in Action	TB p. 48 TB p. 48 TB p. 48 TB p. 49	20 min 20 min 20 min 20 min
Evaluate	• *Lesson 3 Check* (ActiveTeach) • Assessment for Learning • Review (Lesson 3) • *Got it? Self Assessment* (ActiveTeach) • *Got it? Quiz* (ActiveTeach)	TB p. 51a TB p. 46 SB p. 51 TB p. 51b TB p. 51c	10 min 10 min 10 min 10 min 10 min
Lab	• *Let's Investigate! How does the steepness of a stream affect how water flows?* (ActiveTeach)	SB p. 50	30 min

Flash Cards

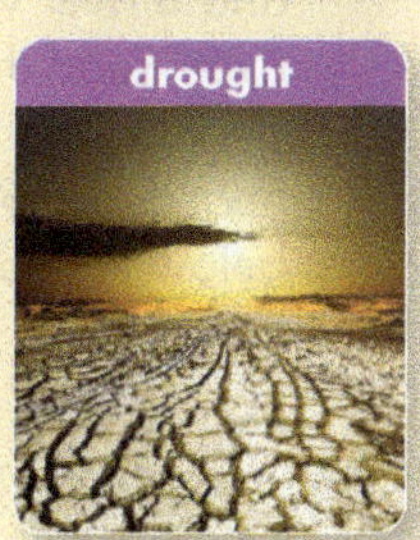

Lesson 1

Key Words	ELL Support
plates, volcano, fault, earthquake, focus, epicenter	**Vocabulary:** *lava, landslide, river, pond, eruption, Earth's crust, layer, mantle, magma, gases, ash, vents, tsunamis, heavy rains, steep slope, gravity, downhill, soil, flash flood, uproot, carry away, mud, flood* (v/n), *sand*

Lesson 2

Key Words	ELL Support
fresh water, glacier, ice cap, groundwater	**Vocabulary:** *polar ice caps, bodies of water, gas, water vapor, surface water, salt water, ocean, seas, land, lakes, ponds, rivers, flow* (v), *snow* (n), *melt, cracks, underground, spring* (n), *dig* (v), *wells, germs, chemicals, wash* (v), *filter* (v), *remove, dirt, water treatment plants, cleaning process, pipes*

Lesson 3

Key Words	ELL Support
water vapor, evaporation, condensation, precipitation, water cycle	**Vocabulary:** *atmosphere, phases, heat* (v), *rise* (v), *cool* (v), *droplets, ice crystals, sleet, hail, storage, soak into the ground, run off, climate, weather pattern, rainfall* **Suffix *-ion*:** *precipitation, condensation, evaporation*

Unit Objectives

Lesson 1: Students will describe how some processes can cause rapid changes to Earth's surface.

Lesson 2: Students will explain where water collects on Earth.

Lesson 3: Students will learn about the water cycle.

Vocabulary: *lava, landslide, river, pond, ash, flood, volcano's eruption, surroundings, dangerous, mountainside, river, formation, stone pillars, magma, flowed, crystallization*

Materials: pictures of different landforms (mountain, hill, coastline, valley, sand dune, island, cave, volcano, etc.) and bodies of water (ocean, lake, river, pond, glacier, etc.)

Introduce the Big Question

How do Earth's resources change?

Build Background Display the pictures of different landforms and bodies of water and have students name them. Draw a two-column chart on the board with the headings: *Landforms* and *Bodies of Water.* Have one volunteer at a time come to the front and write the word you call out in the corresponding column.

Engage

Think!

What forces shaped this formation?

Point to the photo on the bottom right and have students identify what is shown. *What are they? Rocks! This rock formation is known as Giant's Causeway. It consists of about 40,000 stone pillars in Northern Ireland. The tops of the pillars form a path of stepping-stones to the sea. Humans did not build these stone pillars. What forces do you think might have shaped these stone pillars?* Divide the class into small groups and have them discuss their ideas.

1 Look and match. Label.

Use the photos to elicit vocabulary and teach new words. Have students label the photos. Review the answers by pointing to the pictures for students to say the words.

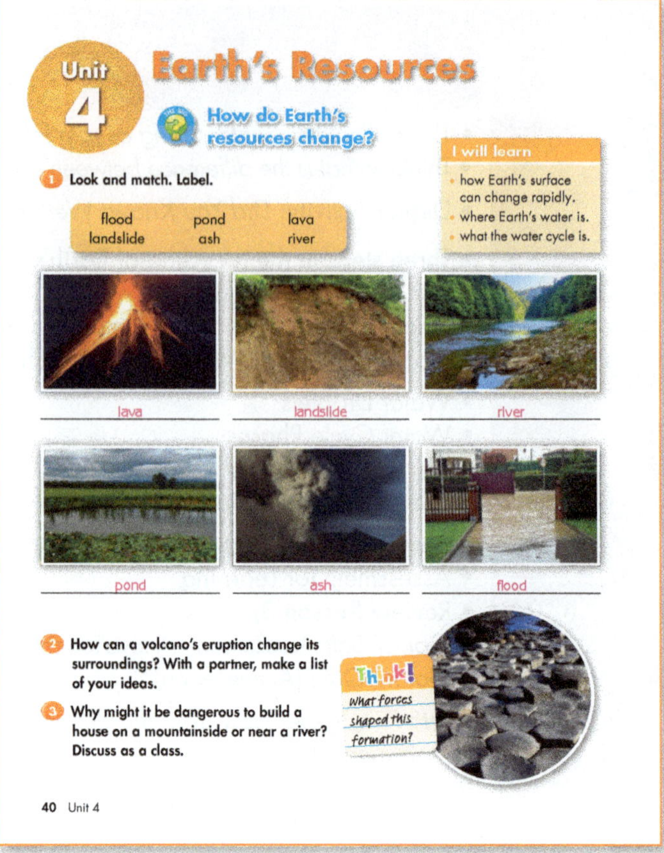

Unit 4 Earth's Resources

How do Earth's resources change?

I will learn
- how Earth's surface can change rapidly.
- where Earth's water is.
- what the water cycle is.

1 Look and match. Label.

flood pond lava
landslide ash river

lava landslide river
pond ash flood

2 How can a volcano's eruption change its surroundings? With a partner, make a list of your ideas.

3 Why might it be dangerous to build a house on a mountainside or near a river? Discuss as a class.

Think! What forces shaped this formation?

40 Unit 4

2 How can a volcano's eruption change its surroundings? With a partner, make a list of your ideas.

Display the picture of a volcano. Have students identify the two pictures on this page that are associated with volcanoes (*lava* and *ash*). Elicit explanations of what *lava* and *ash* are. Ask students what kind of surroundings a volcano can have. (Possible answers: *forests, towns, the sea,* etc.) Have pairs discuss and make a list of how a volcano's eruption can change its surroundings.

3 Why might it be dangerous to build a house on a mountainside or near a river? Discuss as a class.

On the board, draw some houses on a mountainside and a river with some houses on both sides of the river. Read the question out loud. *Why might it be dangerous to build a house on a mountainside or near a river?* Write students' ideas on the board, below the corresponding drawings.

Think! Again!

Point to the photo on the bottom right and revisit the question. *What forces shaped this formation?* Ask students to share their ideas along with their reasonings with the class. Ask students to think about ways that lava could have formed such formations.

Lesson 1

How can Earth's surface change rapidly?

Objective: Learn about the effects of Earth's moving plates and volcanoes.

Vocabulary: *Earth's crust, layer, mantle, plates, Earth's surface, volcanoes, earthquakes, landform, magma, erupt, lava, gases, ash, burst* (v), *vents, spread over, reshaped, floods, landslides, tsunamis, ocean waves*

Digital Resources: Flash Cards (*earthquake, plates, volcano*), *Explore My Planet!* Digital Activity

Materials: globe, poster board, art supplies

Unlock the Big Question

Write the following text on the board:
I will know how some processes can make rapid changes to Earth's surface.

Build Background Display the *earthquake* Flash Card. On the board, draw a KWL chart with the following headings: K (What I *know*), W (What I *want* to know), L (What I *learned*). *What do you know about earthquakes? What do you want to know?* Complete the first two columns of the chart with students' responses.

Explore

Explore My Planet! Science Stats: Earthquakes

Objective: Students will learn about the Wabash Valley Fault and discuss how people might prepare for earthquakes.

Digital Resources: *Explore My Planet!* Digital Activity, *Explore My Planet! Activity Card* (1 per student)

- Use a globe to show students where the Wabash Valley Fault is.
- Show the *Explore My Planet!* Guide students in a discussion about what happens during an earthquake or other natural disasters.
- Have students complete the *Activity Card* and check their answers in small groups or pairs.

Explain

1 Read and complete the statements. Then compare your answers with a partner.

Display the *plates* Flash Card and have students describe it. Have them read to find out the name of the plates in the picture. Use board drawings for students to visualize the concepts of *mantle, crust,* and *plates*. Ask

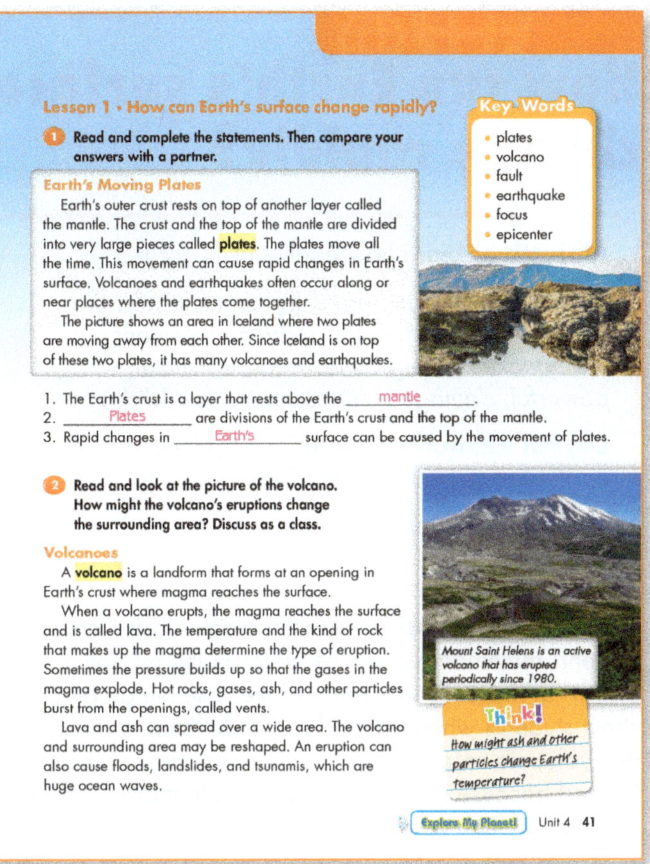
Lesson 1 • How can Earth's surface change rapidly?

Key Words
- plates
- volcano
- fault
- earthquake
- focus
- epicenter

1 Read and complete the statements. Then compare your answers with a partner.

Earth's Moving Plates

Earth's outer crust rests on top of another layer called the mantle. The crust and the top of the mantle are divided into very large pieces called **plates**. The plates move all the time. This movement can cause rapid changes in Earth's surface. Volcanoes and earthquakes often occur along or near places where the plates come together.

The picture shows an area in Iceland where two plates are moving away from each other. Since Iceland is on top of these two plates, it has many volcanoes and earthquakes.

1. The Earth's crust is a layer that rests above the ___mantle___.
2. ___Plates___ are divisions of the Earth's crust and the top of the mantle.
3. Rapid changes in ___Earth's___ surface can be caused by the movement of plates.

2 Read and look at the picture of the volcano. How might the volcano's eruptions change the surrounding area? Discuss as a class.

Volcanoes

A **volcano** is a landform that forms at an opening in Earth's crust where magma reaches the surface.

When a volcano erupts, the magma reaches the surface and is called lava. The temperature and the kind of rock that makes up the magma determine the type of eruption. Sometimes the pressure builds up so that the gases in the magma explode. Hot rocks, gases, ash, and other particles burst from the openings, called vents.

Lava and ash can spread over a wide area. The volcano and surrounding area may be reshaped. An eruption can also cause floods, landslides, and tsunamis, which are huge ocean waves.

Mount Saint Helens is an active volcano that has erupted periodically since 1980.

Think!
How might ash and other particles change Earth's temperature?

Explore My Planet! Unit 4 41

students to read again and complete the statements. Then have pairs compare their answers.

2 Read and look at the picture of the volcano. How might the volcano's eruptions change the surrounding area? Discuss as a class.

Display the *volcano* Flash Card. Have students explain where volcanoes often occur. (Answer: *Along or near places where plates come together.*) Ask students to read and underline what things burst from a volcano's openings. Discuss with the class how the surroundings of the volcano in the picture changed because of its eruption.

Elaborate

A Volcano in My Country

Place students in groups. Ask each group to find a volcano in their country on their continent and research it using the Internet. Students' research should uncover any available data about the volcano including height, when it was formed, and its location. Have each group use this data as well as images of the volcano to make a poster. Each group should present its poster to the class.

Think!

Discuss with the class how ash and other particles might change Earth's temperature. (Possible answer: *Temperatures may become cooler because they may block sunlight.*)

Lesson 1

How can Earth's surface change rapidly?

Objective: Learn how earthquakes occur.

Vocabulary: *fault, break* (n), *crack* (n), *Earth's crust, get stuck, stress* (n), *sudden movement, earthquake, underground, focus, epicenter, powerful, damage* (v/n), *tsunamis, landslides*

Digital Resources: Flash Cards (*plates, earthquake*), *I Will Know...* Digital Activity

Materials: globe

Build Background Display the *plates* Flash Card. Help students remember where the plates in the picture are and what this area in Iceland is like. Use the globe to show students where Iceland is. Guide them to conclude that volcanic eruptions and earthquakes are common in Iceland because it sits atop two plates that are moving away from each other.

ELL Content Support

Iceland consists of a number of islands and lies across two tectonic plates: the Eurasian plate and the North American plate. The islands have been formed by lava from the volcanoes on the plate borders. The last addition of an island to Iceland ocurred in 1963. This island is called Surtsey and is located 30 km off the main island. Iceland is one of the few places in the world where the effects of two major tectonic plates drifting apart can be easily observed above sea level.

Explain

3 Read and circle the area in the picture where the most damage would occur from the earthquake.

Display the *earthquake* Flash Card. Have students say if they have ever experienced an earthquake and, if they have, ask them to describe what happened. If they have not, ask them to say what they think experiencing an earthquake may be like. Have students read the text and circle the area in the picture where the most damage would occur from the earthquake.

ELL Vocabulary Support

Write the following words on the board for students to investigate and write down their definitions: *fault, focus, earthquake, epicenter.*

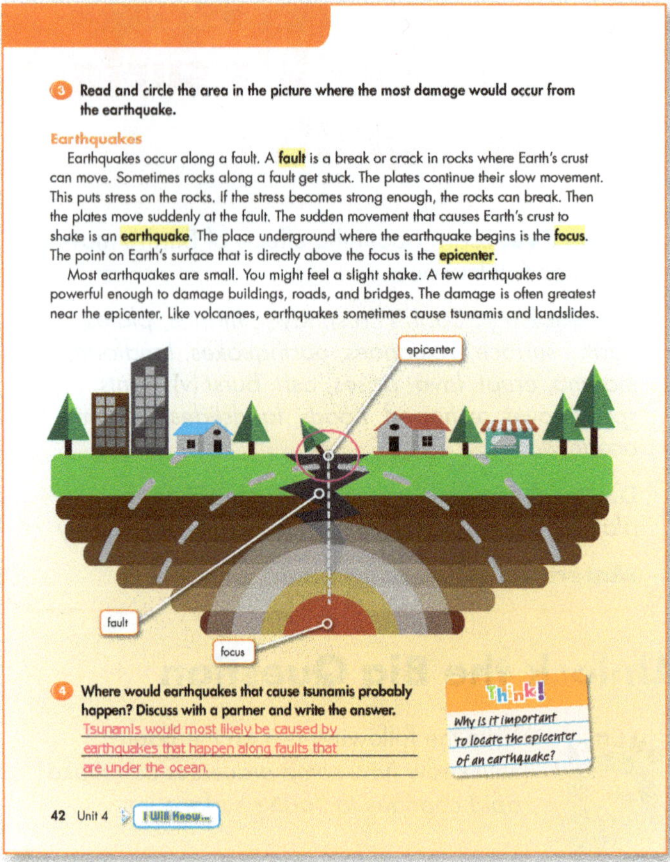

3 Read and circle the area in the picture where the most damage would occur from the earthquake.

Earthquakes

Earthquakes occur along a fault. A **fault** is a break or crack in rocks where Earth's crust can move. Sometimes rocks along a fault get stuck. The plates continue their slow movement. This puts stress on the rocks. If the stress becomes strong enough, the rocks can break. Then the plates move suddenly at the fault. The sudden movement that causes Earth's crust to shake is an **earthquake**. The place underground where the earthquake begins is the **focus**. The point on Earth's surface that is directly above the focus is the **epicenter**.

Most earthquakes are small. You might feel a slight shake. A few earthquakes are powerful enough to damage buildings, roads, and bridges. The damage is often greatest near the epicenter. Like volcanoes, earthquakes sometimes cause tsunamis and landslides.

4 Where would earthquakes that cause tsunamis probably happen? Discuss with a partner and write the answer.

Tsunamis would most likely be caused by earthquakes that happen along faults that are under the ocean.

42 Unit 4 I Will Know...

4 Where would earthquakes that cause tsunamis probably happen? Discuss with a partner and write the answer.

Elicit from students what tsunamis are. Have pairs discuss where tsunamis probably happen and what usually causes them. (Possible answer: *Tsunamis are huge waves of water that are usually caused by earthquakes or volcanic eruptions under the ocean.*)

Elaborate

Three Largest Earthquakes

Have students find out about the three largest recorded earthquakes by researching on the Internet. Divide the class into small groups. Have students compare their information and make a table that ranks the earthquakes and show the earthquakes' locations, dates, and magnitudes.

Ask *Why is it important to locate the epicenter of an earthquake?* Divide the class into small groups and have them discuss the answer to the question. (Possible answer: *The epicenter is directly above the earthquake's focus. The largest amount of damage may occur at the earthquake's epicenter.*)

I Will Know...

Have students do the *I Will Know...* Digital Activity.

Lesson 1

How can Earth's surface change rapidly?

Objective: Learn about the effects of landslides, floods, and droughts.

Vocabulary: *heavy rains, earthquakes, loosen, steep slope, gravity, loosened, downward, downhill, soil, landslide, sliding, flash flood, uproot, carry away, mud, flood* (v/n), *sand*

Digital Resources: *Lesson 1 Check* (print out 1 per student), *Got it? 60-Second Video*

5 Read and label the photos.

Landslides and Floods

Heavy rains or earthquakes may loosen material on a steep slope. Gravity then pulls the loosened material downward. Bits of rock and soil may travel slowly downhill a little at a time. But sometimes they travel rapidly. The rapid downhill movement of a large amount of rock and soil is a landslide. Buildings, cars, trees, and other objects are sometimes carried along with the sliding soil.

Floods also can produce rapid changes. Fast moving water in flash floods can uproot trees, carry away big rocks, and change where rivers flow. Floods can also cover land with sand and mud. In the United States in 1993, huge amounts of rain caused the Mississippi and Missouri Rivers to overflow. Some areas were flooded for almost 200 days. The floods covered thousands of acres with sand and mud.

6 Read again and underline two sentences that describe how the effects of landslides and floods are similar.

7 Read. Why does drought increase the possibility of soil erosion? Discuss as a class and write the answer.

Droughts

Sometimes an area experiences a period of weather that is much drier than usual. Periods of unusually low rain or snowfall are called droughts. A severe drought can change the surface of Earth by reducing the amount of water in rivers and lakes. Droughts can destroy farmland and kill trees and other plants. Droughts increase soil erosion, too. In the 1930s, the United States experienced a severe drought. As crops died, high winds blew away the dry soil. Winds sometimes carried great, dark clouds of dust thousands of miles away.

Drought increases the possibility of soil erosion because the dry soil caused by a drought can be easily blown away by the wind.

Flash Lab

Earthquake Model Create a hill of soil or sand in a pan. Give the pan a quick shake. How does the movement affect the hill?

Lesson 1 Check | Got it? 60-Second Video | Unit 4 43

Build Background On the board, write *Extreme Weather Conditions*. Elicit examples and write students' ideas on the board. Have students discuss how people may be affected by those extreme conditions.

Explain

5 Read and label the photos.

Have some volunteers describe the two pictures at the top of the page and say what they think happened before the photographs were taken. Ask students to read and label the pictures.

6 Read again and underline two sentences that describe how the effects of landslides and floods are similar.

Have pairs read again and underline two sentences that describe how the effects of landslides and floods are similar. Ask *How are landslides' and floods' effects different?* Accept all logical answers.

7 Read. Why does drought increase the possibility of soil erosion? Discuss as a class and write the answer.

Point to the picture at the bottom of the page and have students discuss why they think the soil is so dry. Ask students to read and confirm their predictions. Read the question and have students explain what *soil erosion* means. Then have them read the text again to find out why drought increases soil erosion. Discuss the answer with the class and ask students to write the answer using their own words.

Elaborate

Too much rain! Too little rain!

Have pairs draw a two-column chart with the headings *Too Much Rain* and *Too Little Rain*. Then have them discuss the effects of these two extreme weather conditions and write them in the corresponding columns.

Flash Lab

Earthquake Model

Materials: sand or soil, pan

Divide the class into small groups and distribute materials. Have groups discuss how the movement affects the hill. Guide students to realize how this experiment resembles what can happen to hills when there is an earthquake.

Evaluate

Lesson 1 Check **Assessment for Learning**

Distribute the *Lesson 1 Check* and guide students as they complete it. Check answers as a class. Then ask students to grade their progress on the topic of how Earth's surface can change rapidly from 1 to 3: 3 = *I understand how Earth's surface can change rapidly*; 2 = *I need to study more*; 1 = *I need help!* Encourage students giving themselves a 1 or 2 to describe what they found difficult and what they need to study more.

Got it? 60-Second Video

Review Key Words for Lesson 1 (see Student's Book page 41). Play the *Got it? 60-Second Video* to review the lesson material.

Lesson 2

Where is Earth's water?

Objective: Learn where water on Earth is and the difference between salt water and fresh water.

Vocabulary: *glaciers, polar ice caps, bodies of water, gas, water vapor, surface water, salt water, fresh water, ocean water, salty, healthy, crops, underground*

Digital Resources: *Let's Explore!* Digital Lab

Materials: glass of water, pictures of a lake, a river, a stream, groundwater, ice caps, glaciers, an ocean, a sea, and the atmosphere, a globe, flip-chart paper, markers, pencils

Unlock the Big Question

Write the following text on the board:
I will learn where water collects on Earth.

Build Background Show students a glass of water. Guide a discussion about where they think people get drinking water. Have students reflect on where liquid water can be found on Earth.

Explore

Let's Explore! Lab Where is Earth's water?

Objective: Students will make and observe a model of all of Earth's water.

Digital Resources: *Let's Explore!* Digital Lab, *Let's Explore! Activity Card* (1 per student) (*Optional*: Do the lab in class; refer to the *Activity Card* for materials and steps.)

- Review or pre-teach key vocabulary by using board drawings or pictures and/or by writing the following words on the board: *lake, river, stream, groundwater, ice caps, glaciers, ocean, sea, atmosphere.*
- Show the Digital Lab. Have students complete the *Activity Card* and check their answers in pairs or small groups. Provide support as needed.

Explain

1 Read and look at the picture. Why do you see more blue water than green land? Underline the answer.

Have students look at the picture at the top of the page and describe it. Ask them to name the countries and oceans that can be seen in the picture. Read the question out loud. Invite students to read the text and underline the answer.

Lesson 2 • Where is Earth's water?

Key Words
- fresh water
- glacier
- ice cap
- groundwater

1 Read and look at the picture. Why do you see more blue water than green land? Underline the answer.

Water on Earth

When you look at our planet, you see much more blue water than green land. That is because almost three-fourths of Earth's surface is covered with water.

Water exists as a solid, liquid, and gas. Ice is solid water. Some of Earth's water is frozen in glaciers and polar ice caps. You see liquid water in rivers, lakes, the ocean, and other bodies of water. When water gets hot enough, it turns into an invisible gas called water vapor. Some of the water near Earth's surface is water vapor in the air. The water vapor rises from water on Earth's surface and becomes part of the atmosphere.

2 Read. Why is fresh water an important resource? Discuss as a class.

Surface Water

Surface water is any water that is above the ground on Earth. You can classify water bodies by the type of water they contain: salt water or fresh water.

Salt Water

If you have ever tasted ocean water, you know it tastes very salty. However, taste is not the main problem. Ocean water is not healthy for drinking. More than ninety-seven percent of Earth's water is salty water in the ocean and seas. Why is the ocean salty? Ocean water is a mixture of water and dissolved salts. These salts come mostly from rocks on land. As rivers flow over land, they dissolve salts from rocks. They carry the salts to the ocean.

Fresh Water

Only three percent of Earth's water is **fresh water**. People need fresh water for drinking, cooking, growing crops, and many other activities. Most of Earth's fresh water is frozen in glaciers and ice caps. People cannot use that water. People depend on the small amount of fresh water available in rivers and lakes. People also get drinking water from underground.

44 Unit 4 Let's Explore! Lab

2 Read. Why is fresh water an important resource? Discuss as a class.

Write the following words on the board for students to read and underline their definitions: *surface water, salt water, fresh water.* Have students discuss the differences among the types of water. Then have them discuss why fresh water is an important resource.

Elaborate

Science Notebook: Water on Earth

Have students draw a graphic organizer in their Science Notebooks and write examples of the different states in which water can be found on Earth.

Water on Earth Graphs

Divide the class into small groups. Have students research on the Internet the percentages of fresh and salt water on Earth and the percentages of fresh water found in glaciers, ice caps, and groundwater. Ask groups to make two bar graphs that show both groups of percentages on flip-chart paper with the headings *Surface Water* and *Fresh Water.*

Lesson 2

Where is Earth's water?

Objective: Learn about the kinds of water different bodies of water contain.

Vocabulary: *ocean, seas, land, lakes, ponds, fresh water, salt water, rivers, downhill, flow* (v), *glaciers, ice caps, snow* (n), *melt*

Digital Resources: *I Will Know...* Digital Activity

Materials: map of the world

Build Background On the board, draw a two-column chart with the headings *Salt Water* and *Fresh Water*. Have one volunteer at a time write the name of the body of water you call out in the corresponding column. Tell students that they will check their answers after reading the texts.

Explain

3 Read and match.

Have students read and match the texts with the pictures.

4 Circle the names of bodies of water that can contain salt water. Then, circle the picture of the body of water that contains most of Earth's fresh water.

Invite pairs of students to read the texts again and circle the names of bodies of water that contain salt water. Then have them circle the picture of the body of water that contains most of Earth's fresh water. Go back to the chart on the board and check students' answers.

ELL Content Support

The reason salt water is unhealthy to drink is that salt water causes cells in the body to lose water. Since most of Earth's water is salt water and there is not as much naturally occurring fresh water, humans have invented ways to remove the salt from salt water to create fresh water. This process is called desalination. One method of desalination involves boiling salt water so that the water changes into water vapor. The water vapor is quickly cooled, causing condensation. Condensation produces fresh water, and the salt is left behind.

I Will Know...

Have students do the *I Will Know...* Digital Activity.

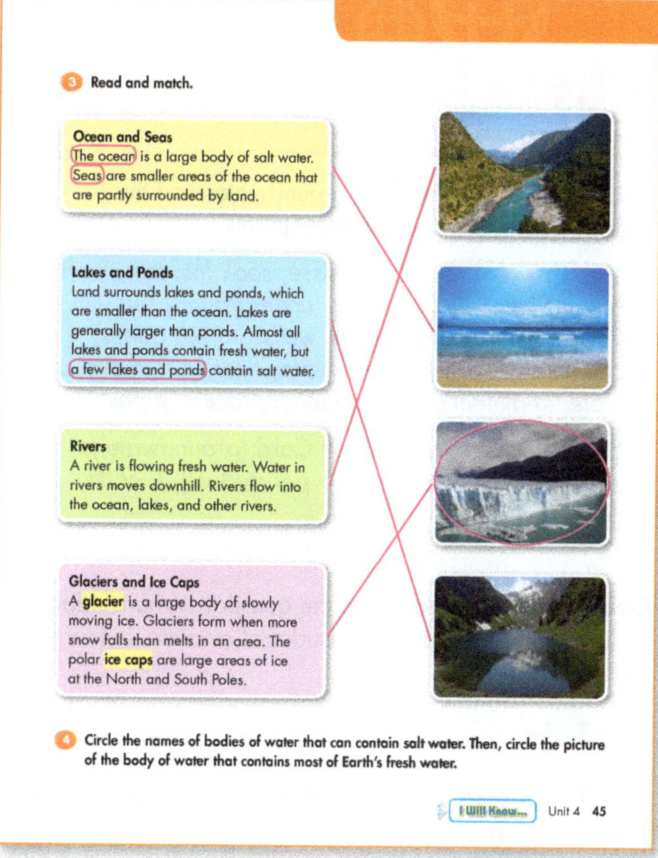

3 Read and match.

Ocean and Seas
The ocean is a large body of salt water. Seas are smaller areas of the ocean that are partly surrounded by land.

Lakes and Ponds
Land surrounds lakes and ponds, which are smaller than the ocean. Lakes are generally larger than ponds. Almost all lakes and ponds contain fresh water, but a few lakes and ponds contain salt water.

Rivers
A river is flowing fresh water. Water in rivers moves downhill. Rivers flow into the ocean, lakes, and other rivers.

Glaciers and Ice Caps
A **glacier** is a large body of slowly moving ice. Glaciers form when more snow falls than melts in an area. The polar **ice caps** are large areas of ice at the North and South Poles.

4 **Circle the names of bodies of water that can contain salt water. Then, circle the picture of the body of water that contains most of Earth's fresh water.**

I Will Know... Unit 4 45

Elaborate

How can animals live in very salty water?

Ask students *If salt water is unhealthy to drink, how can animals like fish live in very salty water? There are several lakes around the world that contain water that is much saltier than the ocean.* Invite students to form groups and research Lake Van in Turkey, The Great Salt Lake in the United States, and Qinghai Lake in China. They should investigate why these lakes are so salty, what animals live in them, and what adaptations allow these animals to live in such salty water. Have groups share their findings with the class.

Bodies of Water in the World

Show students a map of the world, pointing out an ocean. With your finger, trace a path to show how the ocean is actually one huge body of water. Read aloud the names people have given to the various areas of the ocean. Divide the class into small groups. Have students choose one country and research on the Internet the names of the main bodies of water that can be found in the country they chose.

Lesson 2

Where is Earth's water?

Objective: Learn about groundwater and the importance of water treatment plants.

Vocabulary: *bodies of water, soak into the ground, groundwater, fill* (v), *cracks, underground, soil, flow out, spring* (n), *dig* (v), *wells, stored, germs, chemicals, wash* (v), *fresh water, filter* (v), *remove, dirt, water treatment plants, cleaning process, pipes*

Digital Resources: Flash Card (*groundwater*), *Lesson 2 Check* (print out 1 per student), *Got it? 60-Second Video*

Build Background Write the question *Where's Earth's water?* on the board. Write students' answers below the question.

Explain

5 Read and complete the captions with words from the box.

Invite students to look at the picture at the top of the page and describe it. Ask students to read and complete the captions. Display the *groundwater* Flash Card and have students explain in their own words what it is.

6 Read and list three things that are removed from water at a water treatment plant.

Why is fresh water an important resource? Because we need fresh water to live. Where do people get drinking water? People get drinking water from the small amount of fresh water available in rivers, lakes, and underground. Should we drink fresh water straight from a river or lake? We shouldn't because the water is not clean. Have students look at the picture of the water treatment plant and say what it is for. Ask students to read and list three things that are removed from water at a water treatment plant. (Answers: *germs, chemicals, dirt*)

Elaborate

Science Notebook: Groundwater

Have students review the information on groundwater and write a short paragraph that summarizes the main concepts about groundwater. Students' paragraphs should mention how groundwater collects underground and how it becomes surface water again.

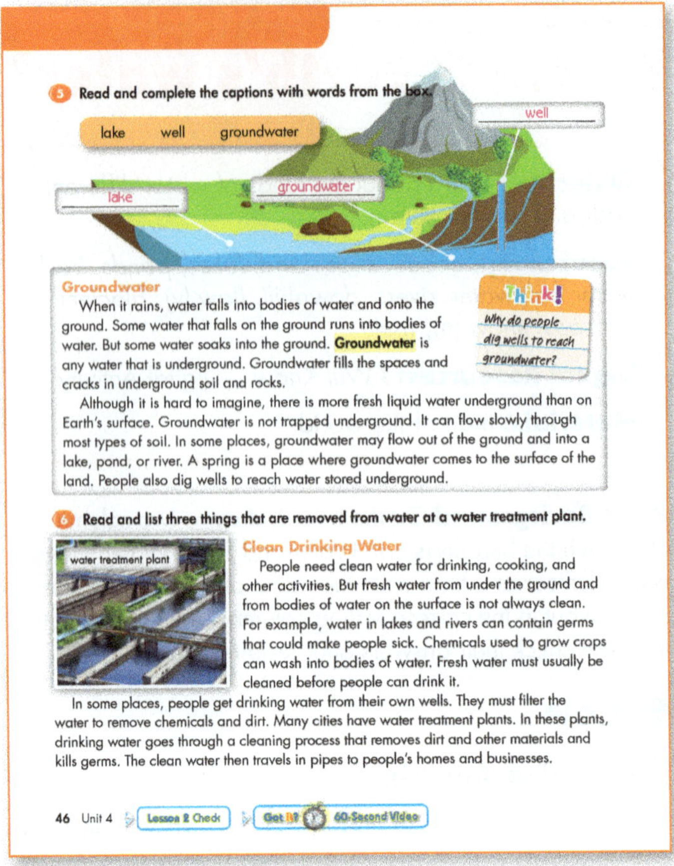

5 Read and complete the captions with words from the box.

lake well groundwater

Groundwater

When it rains, water falls into bodies of water and onto the ground. Some water that falls on the ground runs into bodies of water. But some water soaks into the ground. **Groundwater** is any water that is underground. Groundwater fills the spaces and cracks in underground soil and rocks.

Although it is hard to imagine, there is more fresh liquid water underground than on Earth's surface. Groundwater is not trapped underground. It can flow slowly through most types of soil. In some places, groundwater may flow out of the ground and into a lake, pond, or river. A spring is a place where groundwater comes to the surface of the land. People also dig wells to reach water stored underground.

Think! Why do people dig wells to reach groundwater?

6 Read and list three things that are removed from water at a water treatment plant.

water treatment plant

Clean Drinking Water

People need clean water for drinking, cooking, and other activities. But fresh water from under the ground and from bodies of water on the surface is not always clean. For example, water in lakes and rivers can contain germs that could make people sick. Chemicals used to grow crops can wash into bodies of water. Fresh water must usually be cleaned before people can drink it.

In some places, people get drinking water from their own wells. They must filter the water to remove chemicals and dirt. Many cities have water treatment plants. In these plants, drinking water goes through a cleaning process that removes dirt and other materials and kills germs. The clean water then travels in pipes to people's homes and businesses.

46 Unit 4 Lesson 2 Check Got it? 60-Second Video

Have pairs discuss the answer to the question *Why do people dig wells to reach groundwater?* (Possible answer: *They need fresh water for drinking or for watering crops.*)

Saving Water

Divide the class into small groups. Have students list ways they use water in their school. Allow them to go out of the classroom to make their research easier. Then have them discuss and list ways they can use less water. Ask students to read their lists to the class. Students' responses should include all the ways that water is used, such as for water fountains and in science labs.

Evaluate

Lesson 2 Check Assessment for Learning

Distribute the *Lesson 2 Check* and guide students as they complete it. Check answers as a class. Then ask students to grade their progress on the topic of where Earth's water is from 1 to 3: 3 = *I understand where Earth's water is;* 2 = *I need to study more;* 1 = *I need help!* Encourage students giving themselves a 1 or 2 to describe what they found difficult and what they need to study more.

Review Key Words for Lesson 2 (see Student's Book page 44). Play the *Got it? 60-Second Video* to review the lesson material.

Lesson 3

What is the water cycle?

Objective: Learn how Earth's water is constantly being recycled.

Vocabulary: *lake, sun's energy, Earth's surface, atmosphere, fall back, solid, liquid, gas, phases, heat* (v), *water vapor, evaporation, rise* (v), clouds, *cool* (v), *droplets, condensation, ice crystals, precipitation, rain, snow, sleet, hail*

Digital Resources: *Explore My Planet!* Digital Activity

Materials: pictures of bodies of water (*river, lake, sea, ocean, pond, spring, swamp, glaciers,* etc.)

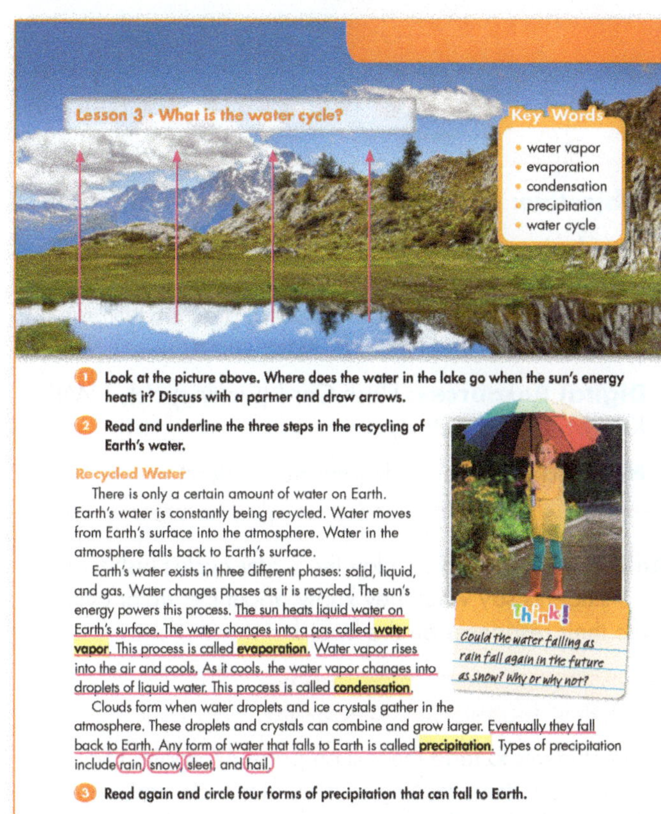

Lesson 3 • What is the water cycle?

Key Words
- water vapor
- evaporation
- condensation
- precipitation
- water cycle

1. **Look at the picture above. Where does the water in the lake go when the sun's energy heats it? Discuss with a partner and draw arrows.**
2. **Read and underline the three steps in the recycling of Earth's water.**

Recycled Water

There is only a certain amount of water on Earth. Earth's water is constantly being recycled. Water moves from Earth's surface into the atmosphere. Water in the atmosphere falls back to Earth's surface.

Earth's water exists in three different phases: solid, liquid, and gas. Water changes phases as it is recycled. The sun's energy powers this process. The sun heats liquid water on Earth's surface. The water changes into a gas called **water vapor**. This process is called **evaporation**. Water vapor rises into the air and cools. As it cools, the water vapor changes into droplets of liquid water. This process is called **condensation**.

Clouds form when water droplets and ice crystals gather in the atmosphere. These droplets and crystals can combine and grow larger. Eventually they fall back to Earth. Any form of water that falls to Earth is called **precipitation**. Types of precipitation include rain, snow, sleet, and hail.

Think!

Could the water falling as rain fall again in the future as snow? Why or why not?

3. **Read again and circle four forms of precipitation that can fall to Earth.**

Let's Explore! Lab Unit 4 47

Unlock the Big Question

Write the following text on the board: *I will know how water moves through the water cycle.*

Build Background Write the phrase *Bodies of Water* on the board. Use pictures of bodies of water to elicit vocabulary. Ask students in which ecosystems those bodies of water can be found.

Explore

Explore My Planet! Did You Know: Wetlands

Objective: Students will learn about the importance of wetlands as cleaners of polluted water.

Digital Resources: *Explore My Planet!* Digital Activity, *Explore My Planet! Activity Card* (1 per student)

- Review or pre-teach key vocabulary using board drawings.
- Write the title of the activity on the board. Ask *What are wetlands? Why do you think wetlands are important for water preservation?* Accept all logical answers.
- Show the *Explore My Planet!* Ask students to work independently or in pairs to complete the *Activity Card*.
- Provide support as needed. Check answers as a class.

Explain

1 Look at the picture above. Where does the water in the lake go when the sun's energy heats it? Discuss with a partner and draw arrows.

Look at the picture. What body of water is it? A lake! What's the weather like? It's sunny! Where does the water in the lake go when the sun's energy heats it? Have pairs answer the question and draw arrows that show how the water moves.

2 Read and underline the three steps in the recycling of Earth's water.

Have students discuss what happens to some of the water in bodies of water like ponds, rivers, lakes, and oceans when the sun's energy heats it. Explain that water changes phases as it is recycled. Ask students to read and underline the three steps in the recycling of water.

3 Read again and circle four forms of precipitation that can fall to Earth.

Have students describe in their own words the precipitation phase of the water cycle. Then have them circle four forms of precipitation.

ELL Content Support

Have students say the word *precipitation*. Explain that the suffix *-ion* is used to end a word that describes a process or something happening. Ask students to say and describe the other *-ion* words from the lesson.

Think!

Ask *Could the water falling as rain fall again in the future as snow? Why or why not?* Elicit answers from the class. (Possible answer: *Yes. This water may move from Earth's surface to the atmosphere. In very cold weather, it may fall as snow.*)

Lesson 3

What is the water cycle?

Objective: Learn the steps in the water cycle.

Vocabulary: *atmosphere, water cycle, steps, condensation, droplets, ice crystals, clouds, precipitation, gravity, storage, soak into the ground, groundwater, run off, evaporation, heat* (v), *gas, water vapor*

Digital Resources: Flash Card (*water cycle*), *I Will Know…* Digital Activity

Materials: white cards (one per student)

Build Background Display the *water cycle* Flash Card. Have students say what they know about the water cycle. Write their responses on the board.

ELL Vocabulary Support

Ask students to read the text on page 47 again. Elicit vocabulary items relating to the water cycle and write them on the board. Write the following words on the board to ensure students know their meanings: *droplets, ice crystals, clouds, groundwater,* and *water vapor.*

Explain

4 Read and complete the captions using information from the previous page.

Before reading, have students explain what happens at each step in the water cycle. Ask students to read and complete the captions using the words written on the board. Check answers with the whole class.

Think!

On the board write the question *What is the difference between evaporation and condensation?* Ask pairs to discuss the answer. (Possible answer: *During evaporation, water becomes water vapor. During condensation, water vapor becomes water.*)

ELL Content Support

Tell students that all water is part of the water cycle, including the water they drink. Show students a cold glass of water with ice. Explain that, when warm air comes into contact with a cool glass, water vapor in the air turns to liquid and condenses on the glass. Have students make a list in their notebooks of other places where they have seen condensation. For example, they may have seen dew on the grass or fogged-up windows and mirrors.

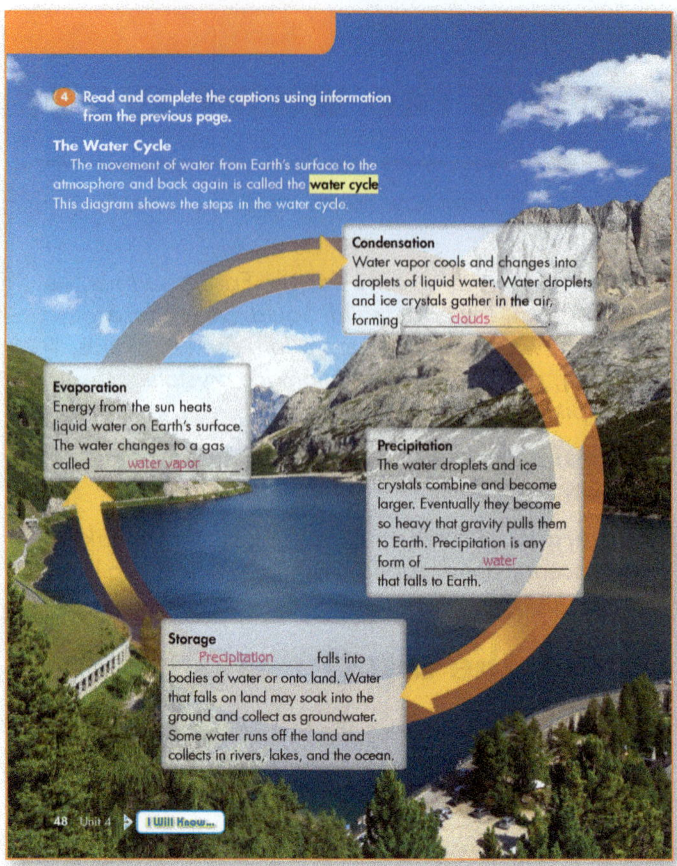

4 Read and complete the captions using information from the previous page.

The Water Cycle

The movement of water from Earth's surface to the atmosphere and back again is called the **water cycle**. This diagram shows the steps in the water cycle.

Condensation
Water vapor cools and changes into droplets of liquid water. Water droplets and ice crystals gather in the air, forming ___clouds___.

Evaporation
Energy from the sun heats liquid water on Earth's surface. The water changes to a gas called ___water vapor___.

Precipitation
The water droplets and ice crystals combine and become larger. Eventually they become so heavy that gravity pulls them to Earth. Precipitation is any form of ___water___ that falls to Earth.

Storage
___Precipitation___ falls into bodies of water or onto land. Water that falls on land may soak into the ground and collect as groundwater. Some water runs off the land and collects in rivers, lakes, and the ocean.

48 Unit 4 I Will Know...

Elaborate

Water Droplets

Suppose that you just took a hot shower at home. The mirror in your bathroom would probably be foggy. Have students work in pairs to explain why they see droplets of water on the mirror. Students should describe how, as the liquid water from the shower heats up, it changes to water vapor. Then, as the water vapor cools, it condenses on the mirror, which makes the mirror look foggy.

Water Cycle Phases

Give each student a white card. Divide the class into groups of four. Within each group, have one student draw a picture showing evaporation, another student draw a picture showing condensation, another student draw a picture showing groundwater, and the last student draw a picture showing precipitation. Have the students discuss and arrange their pictures to show the water cycle.

Story of a Raindrop

Have students write a comic strip about a raindrop. Ask students to describe how the raindrop passes through the water cycle in their comic strips. Students should write about different parts of the water cycle, showing how water can be recycled in many different ways.

I Will Know...

Have students do the *I Will Know…* Digital Activity.

Lesson 3

What is the water cycle?

Objective: Learn how the water cycle affects weather.

Vocabulary: *weather report, cloudy, rain* (n), *water cycle, weather conditions, clouds, precipitation, temperature, freezing, sleet* (n), *snow* (n), *water vapor, ice crystals, hail* (n), *drops, freeze, climate, weather pattern, evaporate, wind, rainfall, body of water*

Digital Resources: Flash Cards (*thunderstorm, flood, drought*), *Lesson 3 Check* (print out 1 per student), *Got it? 60-Second Video*

Materials: weather map for local area or country

Build Background Ask *What's the weather like?* Write students' responses on the board. Use board drawings to elicit words to describe weather and write them on the board, too. (Possible answers: *rainy, sunny, cloudy, cold, hot, mild, humid,* etc.)

Explain

5 **Read. Why does the weather depend partly on the water cycle? Discuss as a class.**

Ask *Why does the weather depend partly on the water cycle?* Elicit some ideas. Once students have read the text, guide a discussion to answer the question.

6 **Read the captions and underline the names of two natural disasters related to the water cycle.**

Display the *thunderstorm, flood,* and *drought* Flash Cards, covering their names. Have students describe the pictures and say what specific weather conditions they associate with each picture. Ask students to read the captions and underline the names of two natural disasters that relate to the water cycle.

7 **Read. Why do you think rain forests receive so much rain? Discuss as a class.**

Point to the picture and encourage students to describe what rain forests' climates are like. Ask *Why do you think rain forests receive so much rain?* Once students have read the text, have them discuss the answer to the question.

Elaborate

Water Cycle in Action

Bring a weather map for the local area or the entire country to class. Discuss the map and its symbols with the class. Then have students identify where the evaporation, condensation, and precipitation steps of the water cycle are probably taking place.

5 **Read. Why does the weather depend partly on the water cycle? Discuss as a class.**

Water Cycle and Weather

You watch the local weather report. Will it be cloudy? Will there be rain? The answer to these questions depends partly on the water cycle. The water cycle is a main factor that causes daily weather conditions such as clouds and precipitation.

Weather is the result of how water, air, and temperature interact. For example, the temperature of air in the atmosphere affects the type of precipitation that falls. Water falls as rain when air in the atmosphere is above freezing. If the air near Earth's surface is colder than the air higher up, rain may freeze as it falls. Sleet is rain that has frozen on its way to Earth.

Snow may form when the air in the atmosphere is very cold. Snow forms when water vapor changes directly into ice crystals. The crystals eventually grow large enough to fall out of the clouds. Hail can also develop at freezing temperatures. Hail forms when drops of water in clouds freeze and are coated with multiple layers of ice.

The pictures show other types of weather that relate to the water cycle.

A thunderstorm can happen when warm, moist air rises quickly into much cooler air.

Long periods of rainfall or heavy storms can cause flooding.

A drought is a long period of dryness with little or no precipitation.

6 **Read the captions and underline the names of two natural disasters related to the water cycle.**

7 **Read. Why do you think rain forests receive so much rain? Discuss as a class.**

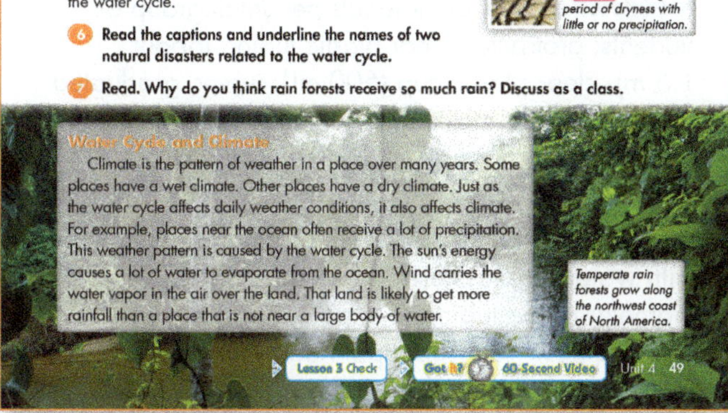
Water Cycle and Climate

Climate is the pattern of weather in a place over many years. Some places have a wet climate. Other places have a dry climate. Just as the water cycle affects daily weather conditions, it also affects climate. For example, places near the ocean often receive a lot of precipitation. This weather pattern is caused by the water cycle. The sun's energy causes a lot of water to evaporate from the ocean. Wind carries the water vapor in the air over the land. That land is likely to get more rainfall than a place that is not near a large body of water.

Temperate rain forests grow along the northwest coast of North America.

Lesson 3 Check Got it? 60-Second Video Unit 4 49

ELL Content Support

Are weather and climate the same? Students may find it difficult to understand the difference between weather and climate. Explain that climate is the pattern of weather of an area over a long period of time and that weather is the day-to-day changes in temperature, wind speed and direction, and precipitation. *When planning a trip, you might consult a travel brochure months in advance to get an idea of the region's climate. While packing for the trip, you might review weather reports to get an idea of the what the weather is like at the moment.*

Evaluate

Lesson 3 Check **Assessment for Learning**

Distribute the *Lesson 3 Check* and guide students as they complete it. Check answers as a class. Then ask students to grade their progress on the topic of the water cycle from 1 to 3: 3 = *I understand the water cycle*; 2= I need to study more; 1 = *I need help!* Encourage students giving themselves a 1 or 2 to describe what they found difficult and what they need to study more.

Got it?  **60-Second Video**

Review Key Words for Lesson 3 (see Student's Book page 47). Play the *Got it? 60-Second Video* to review the lesson material.

Let's Investigate!

In this unit, students learn how Earth's surface can change rapidly. In this lab, students will observe that the steeper a stream's angle is, the faster the water flows in it.

Let's Investigate! Lab **How does the steepness of a stream affect how water flows?**

Objective: Measure the time it takes water to flow through a tube set at different angles.

Materials: 1 set of materials per small group of students: protractor, clear plastic tubing (about 1.5 m), clear plastic cup (500 mL), water, graduated cylinder, funnel, masking tape, timer, pouring container (teacher use), duct tape (*optional*), clothespin (*optional*)

Advance Preparation: Fill a pouring container with water for class use. Make sure the inside diameter of the tubing is large enough to allow water to flow freely through it and that the funnels fit inside the tubing. If they do not, use duct tape to seal the connection between the funnels and the tubing.

- Divide students into groups of four and distribute materials.
- Have students make a model of a stream using the tubing. Have one student set the stream angle at 10° and another student place a cup at the low end of the stream.
- Have a third student measure 50 mL of water into a graduated cylinder and attach the funnel to the top of the tubing.
- Have a fourth student start a timer as the third student pours the water into the tubing, stop the timer when all the water has flowed into the cup, and record the time.
- Explain to students that they should repeat the same process but change the stream's angle to 20°, 40°, and 55°.

Teacher Time-Saving Option: Show the *Let's Investigate!* Digital Lab as an alternative to the hands-on lab activity.

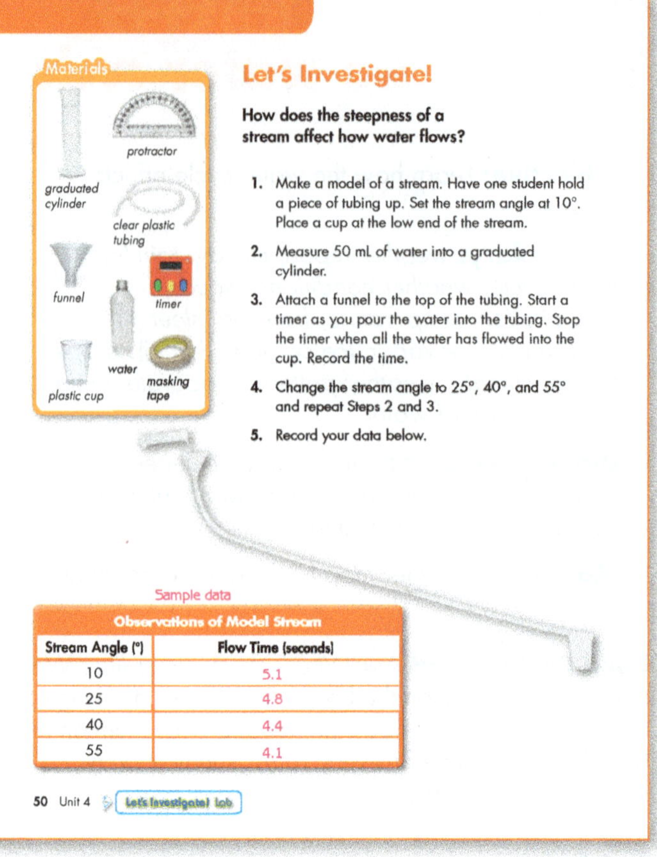

Materials

Let's Investigate!

How does the steepness of a stream affect how water flows?

1. Make a model of a stream. Have one student hold a piece of tubing up. Set the stream angle at 10°. Place a cup at the low end of the stream.
2. Measure 50 mL of water into a graduated cylinder.
3. Attach a funnel to the top of the tubing. Start a timer as you pour the water into the tubing. Stop the timer when all the water has flowed into the cup. Record the time.
4. Change the stream angle to 25°, 40°, and 55° and repeat Steps 2 and 3.
5. Record your data below.

Sample data

Observations of Model Stream

Stream Angle (°)	Flow Time (seconds)
10	5.1
25	4.8
40	4.4
55	4.1

50 Unit 4 Let's Investigate! Lab

Class Project: Climate Maps

Materials: construction paper (2 sheets per group), markers

Display a climate map. Tell students that climate maps use color coding to show areas that have similar climates. Divide the class into small groups. Ask students to think about three places in their country that they have visited or learned about. Have students draw a climate map and write a profile of the three locations' climates. Ask students to make a chart that identifies the locations and includes descriptions of their general temperature, precipitation, and other climate features. Have students present their maps and charts to the class.

Unlock the Big Question

Have students refer to the Big Question on the Unit Opener page. In pairs, have them recall what they have learned about how the steepness of a stream affects water flow. Invite pairs to share their answers to questions 6 and 7 on the *Let's Investigate! Activity Card*.

Unit 4 Review

How do Earth's resources change?

Digital Resources: Print out 1 of each per student: *Got it? Self Assessment, Got it? Quiz*

Evaluate

Strategies for Targeted Review

The following are strategies for providing targeted review for students if they encounter challenges with the content.

Lesson 1 How can Earth's surface change rapidly?

Question 1

If… students are having difficulty identifying the two causes of the rapid changes to Earth's surface, then… direct students to pages 41 and 42. Encourage them to look again at the pictures and say what changes each of the things in the pictures can cause.

Lesson 2 Where is Earth's water?

Question 2

If… students are having difficulty deciding whether the statements are true or false, then… direct students to page 44 and have them find the answers to the questions.

Lesson 3 What is the water cycle?

Question 3

If… students are having difficulty matching the sentences, then… direct students to review Lesson 3 to find the answers to the questions.

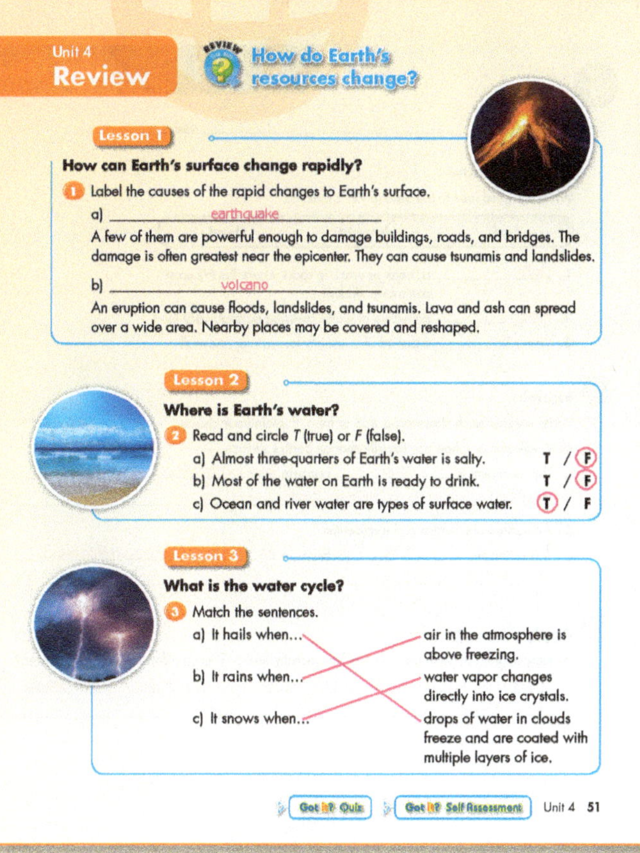

Unit 4 Review — How do Earth's resources change?

Lesson 1

How can Earth's surface change rapidly?

1 Label the causes of the rapid changes to Earth's surface.

a) earthquake

A few of them are powerful enough to damage buildings, roads, and bridges. The damage is often greatest near the epicenter. They can cause tsunamis and landslides.

b) volcano

An eruption can cause floods, landslides, and tsunamis. Lava and ash can spread over a wide area. Nearby places may be covered and reshaped.

Lesson 2

Where is Earth's water?

2 Read and circle *T* (true) or *F* (false).

a) Almost three-quarters of Earth's water is salty. T / (F)

b) Most of the water on Earth is ready to drink. T / (F)

c) Ocean and river water are types of surface water. (T) / F

Lesson 3

What is the water cycle?

3 Match the sentences.

a) It hails when… — drops of water in clouds freeze and are coated with multiple layers of ice.

b) It rains when… — air in the atmosphere is above freezing.

c) It snows when… — water vapor changes directly into ice crystals.

Got it? Quiz · Got it? Self Assessment · Unit 4 51

ELL Language Support

Before students start working on the Review activities, have them read each question aloud along with you.

Got it? Self Assessment

Immediately after students have completed the Review activities, distribute a *Got it? Self Assessment* to each student. Have students complete the *Stop! Wait!* and *Go!* statements for each lesson, allowing them to look back through the lesson material if necessary.

Got it? Quiz

Distribute a Unit 4 *Got it? Quiz* to each student. Quizzes may be used for assessing students' understanding of unit concepts as well as for grading purposes.

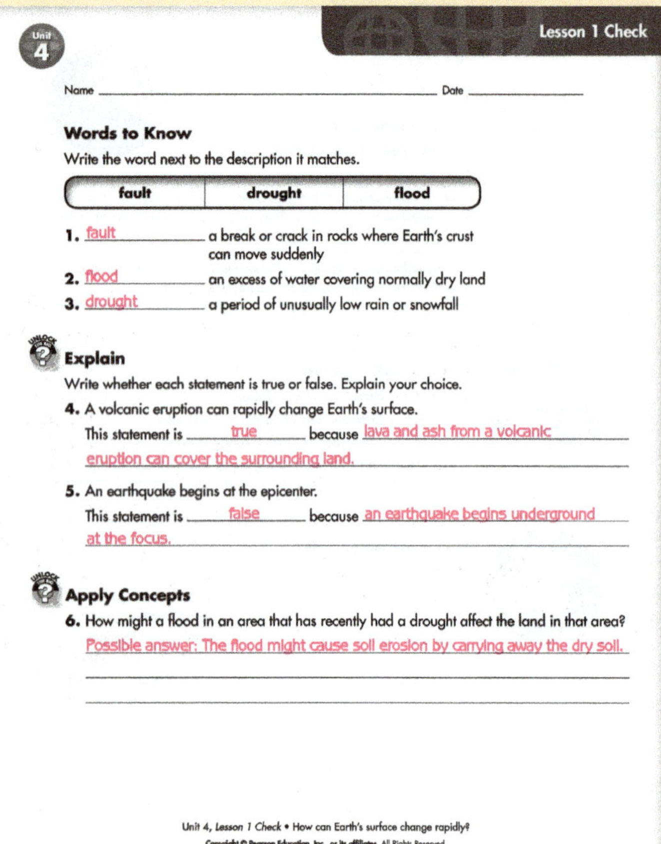

Unit 4

Lesson 1 Check

Name ______________________ Date ____________

Words to Know

Write the word next to the description it matches.

fault	drought	flood

1. fault ________ a break or crack in rocks where Earth's crust can move suddenly
2. flood ________ an excess of water covering normally dry land
3. drought ________ a period of unusually low rain or snowfall

Explain

Write whether each statement is true or false. Explain your choice.

4. A volcanic eruption can rapidly change Earth's surface.
   This statement is ___true___ because lava and ash from a volcanic eruption can cover the surrounding land.
5. An earthquake begins at the epicenter.
   This statement is ___false___ because an earthquake begins underground at the focus.

Apply Concepts

6. How might a flood in an area that has recently had a drought affect the land in that area?
   Possible answer: The flood might cause soil erosion by carrying away the dry soil.

Unit 4, *Lesson 1 Check* • How can Earth's surface change rapidly?

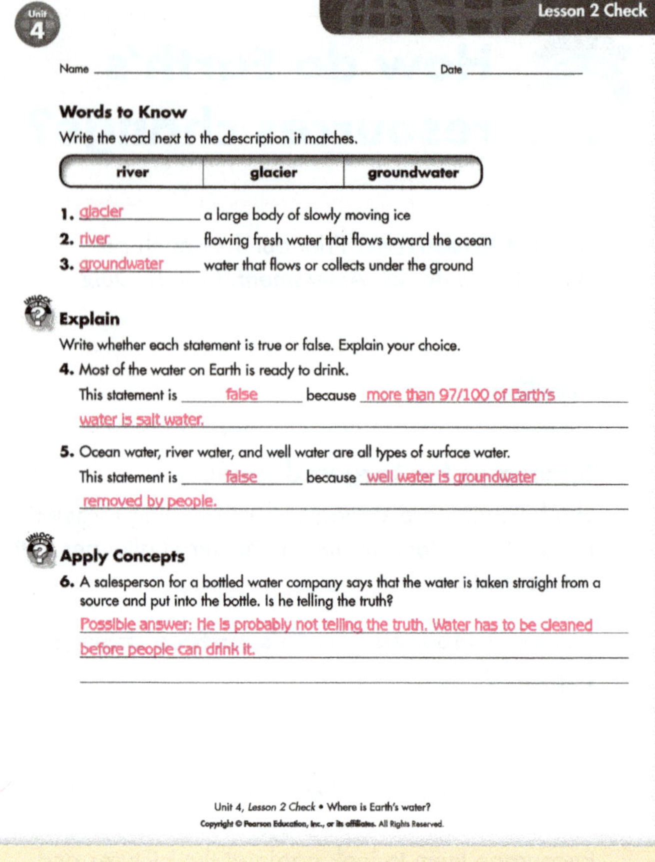

Unit 4

Lesson 2 Check

Name ______________________ Date ____________

Words to Know

Write the word next to the description it matches.

river	glacier	groundwater

1. glacier ________ a large body of slowly moving ice
2. river ________ flowing fresh water that flows toward the ocean
3. groundwater ________ water that flows or collects under the ground

Explain

Write whether each statement is true or false. Explain your choice.

4. Most of the water on Earth is ready to drink.
   This statement is ___false___ because more than 97/100 of Earth's water is salt water.
5. Ocean water, river water, and well water are all types of surface water.
   This statement is ___false___ because well water is groundwater removed by people.

Apply Concepts

6. A salesperson for a bottled water company says that the water is taken straight from a source and put into the bottle. Is he telling the truth?
   Possible answer: He is probably not telling the truth. Water has to be cleaned before people can drink it.

Unit 4, *Lesson 2 Check* • Where is Earth's water?

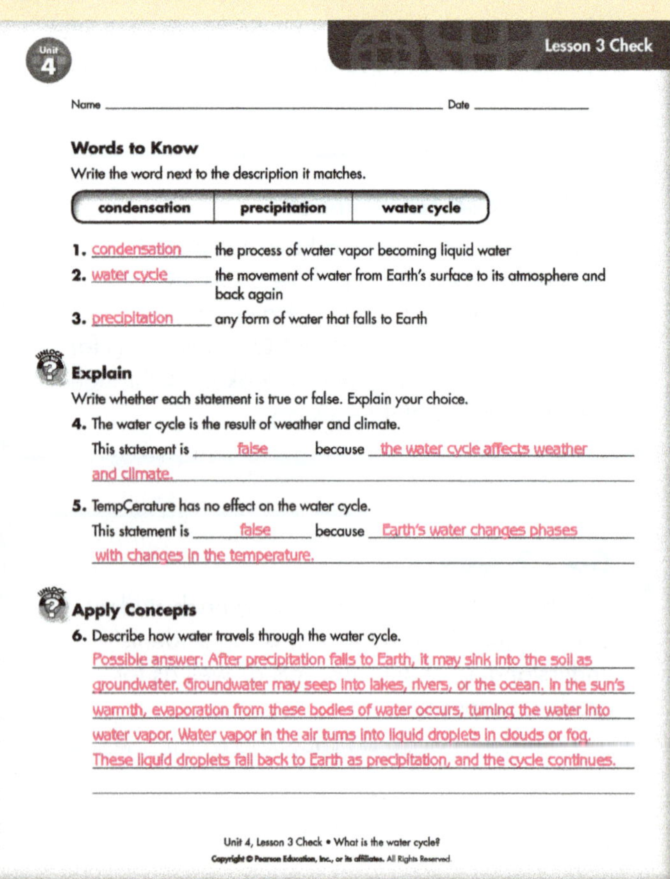

Unit 4

Lesson 3 Check

Name ______________________ Date ____________

Words to Know

Write the word next to the description it matches.

condensation	precipitation	water cycle

1. condensation ________ the process of water vapor becoming liquid water
2. water cycle ________ the movement of water from Earth's surface to its atmosphere and back again
3. precipitation ________ any form of water that falls to Earth

Explain

Write whether each statement is true or false. Explain your choice.

4. The water cycle is the result of weather and climate.
   This statement is ___false___ because the water cycle affects weather and climate.
5. Temperature has no effect on the water cycle.
   This statement is ___false___ because Earth's water changes phases with changes in the temperature.

Apply Concepts

6. Describe how water travels through the water cycle.
   Possible answer: After precipitation falls to Earth, it may sink into the soil as groundwater. Groundwater may seep into lakes, rivers, or the ocean. In the sun's warmth, evaporation from these bodies of water occurs, turning the water into water vapor. Water vapor in the air turns into liquid droplets in clouds or fog. These liquid droplets fall back to Earth as precipitation, and the cycle continues.

Unit 4, Lesson 3 Check • What is the water cycle?

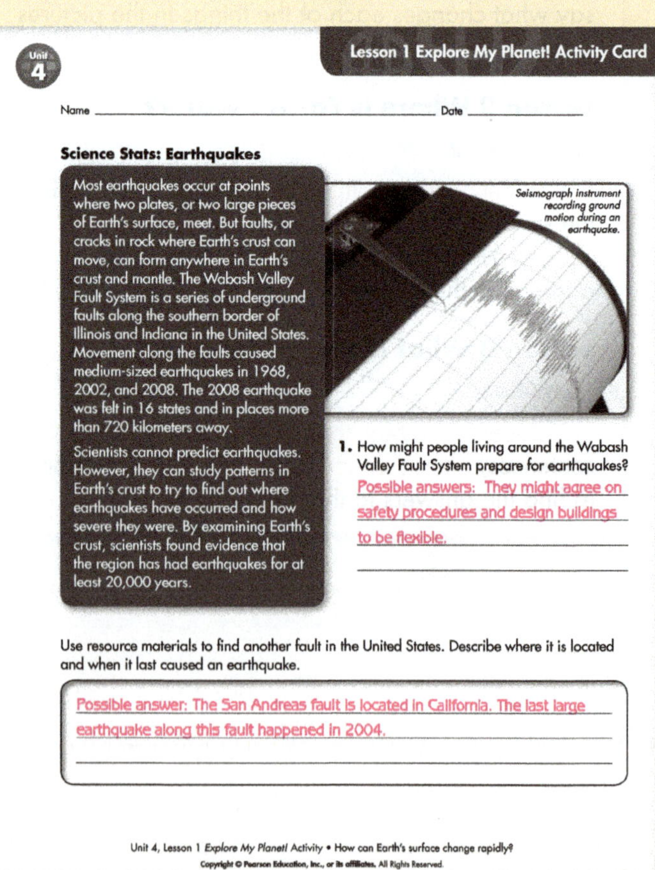

Unit 4

Lesson 1 Explore My Planet! Activity Card

Name ______________________ Date ____________

Science Stats: Earthquakes

Most earthquakes occur at points where two plates, or two large pieces of Earth's surface, meet. But faults, or cracks in rock where Earth's crust can move, can form anywhere in Earth's crust and mantle. The Wabash Valley Fault System is a series of underground faults along the southern border of Illinois and Indiana in the United States. Movement along the faults caused medium-sized earthquakes in 1968, 2002, and 2008. The 2008 earthquake was felt in 16 states and in places more than 720 kilometers away.

Scientists cannot predict earthquakes. However, they can study patterns in Earth's crust to try to find out where earthquakes have occurred and how severe they were. By examining Earth's crust, scientists found evidence that the region has had earthquakes for at least 20,000 years.

Seismograph instrument recording ground motion during an earthquake.

1. How might people living around the Wabash Valley Fault System prepare for earthquakes?
   Possible answers: They might agree on safety procedures and design buildings to be flexible.

Use resource materials to find another fault in the United States. Describe where it is located and when it last caused an earthquake.

Possible answer: The San Andreas fault is located in California. The last large earthquake along this fault happened in 2004.

Unit 4, Lesson 1 *Explore My Planet! Activity* • How can Earth's surface change rapidly?

Unit 4

Lesson 2 Let's Explore! Activity Card

Name ______________________ Date ______________

Materials

- plastic bottle (2 L with cap) filled with water
- 4 clear plastic cups
- masking tape
- graduated cylinder
- funnel
- plastic dropper

Where is Earth's water?

1. Label 4 plastic cups: *Ice caps and glaciers; Lakes, rivers, and streams; Groundwater; Atmosphere*
2. Look at the chart. Find the amount of water shown for the atmosphere. Put that much water in the labeled cup.

Earth's Water	Amount
Atmosphere (fresh water)	about 12 drop
Lakes, rivers, streams (fresh water)	about 4 drops
Groundwater (fresh water)	13 mL
Icecaps and glaciers (fresh water)	47 mL
Oceans and seas (salt water)	2,139 mL

3. Repeat for the next three places on the chart. Measure the water.
4. Label the 2 L bottle *Oceans and seas.*

Explain Your Results

5. Use your model. Make an inference about the amount of fresh water available for human use.

Possible answer:
I infer that there is a small amount of fresh water on Earth that people can easily use.

Why might you want to show your model to other people?

Possible answer: The model shows clearly why we should not waste fresh water resources. When you look at it, you really understand how little fresh water there is for all living things to use.

Unit 4, Lesson 2 *Let's Explore!* Lab • Where is Earth's water?

Unit 4

Lesson 3 Explore My Planet! Activity Card

Name ______________________ Date ______________

Did You Know: Wetlands

What percentage of the water on Earth do you think is fresh water that we can drink? Most water, ninety-seven percent of the water found in oceans, rivers, and lakes is salty or undrinkable. Another two percent of Earth's water is trapped in glaciers or in the polar ice caps. This leaves only one percent of Earth's water that is possible for us to drink. Because of this, water pollution can be a very serious problem.

You already know that wetlands are a type of habitat. But wetlands also play an important role in cleaning water that is polluted. When large storms take place, the water that runs off land used for farming can contain high amounts of fertilizer, pesticides, and dangerous bacteria. When this runoff water finds its way to a wetland, it is trapped and cleaned. Wetlands act like huge sponges, absorbing extra water. They can help prevent flooding in the area, and their plants and soil can help to trap pollutants in water. The water that flows out of a wetland is much cleaner than the water that flows into it.

Unfortunately, over half of Earth's wetlands have been lost since 1990. Some wetlands have been drained to create farmland or to build on. Other wetlands have been destroyed by invasive species and by global warming. However, many people are recognizing the importance of wetlands and are taking steps to protect them.

1. Why are wetlands important?

Why is water pollution a problem?

Unit 4, Lesson 3 *Explore My Planet!* Activity • What is the water cycle?

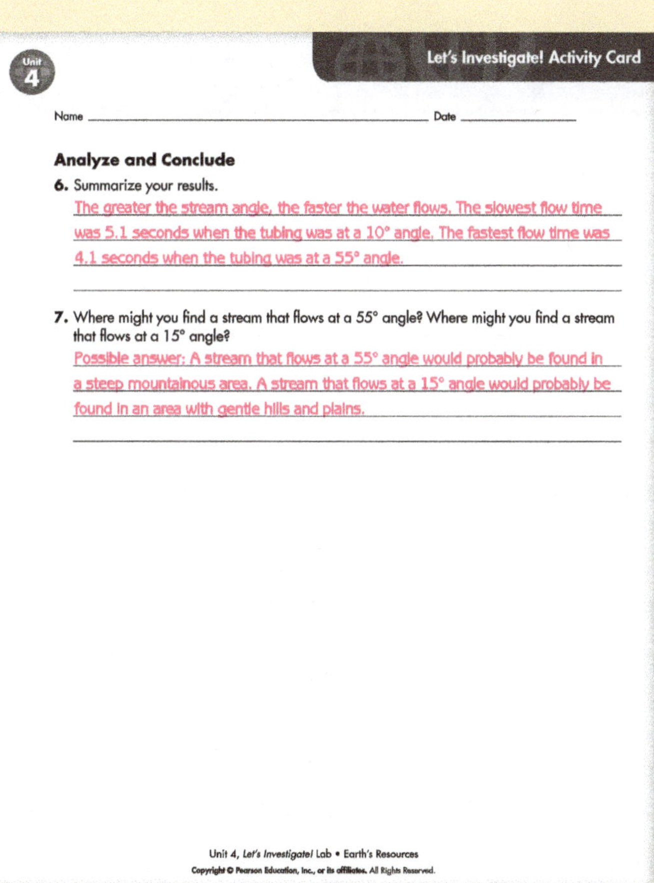

Unit 4

Let's Investigate! Activity Card

Name ______________________ Date ______________

Analyze and Conclude

6. Summarize your results.

The greater the stream angle, the faster the water flows. The slowest flow time was 5.1 seconds when the tubing was at a 10° angle. The fastest flow time was 4.1 seconds when the tubing was at a 55° angle.

7. Where might you find a stream that flows at a 55° angle? Where might you find a stream that flows at a 15° angle?

Possible answer: A stream that flows at a 55° angle would probably be found in a steep mountainous area. A stream that flows at a 15° angle would probably be found in an area with gentle hills and plains.

Unit 4, *Let's Investigate!* Lab • Earth's Resources

Unit 4

Lessons 1–3 Got it? Self Assessment

Name ______________________ Date ______________

Got it? Self Assessment

Complete the statements for each lesson.

Lesson 1 How can Earth's surface change rapidly?

Stop! I need help with ______________

Wait! I have a question about ______________

Go! Now I know ______________

Lesson 2 Where is Earth's water?

Stop! I need help with ______________

Wait! I have a question about ______________

Go! Now I know ______________

Lesson 3 What is the water cycle?

Stop! I need help with ______________

Wait! I have a question about ______________

Go! Now I know ______________

Unit 4, *Got it? Self Assessment* • Earth's Resources

Unit 4 | Got it? Quiz

Name ______________________ Date ____________

Got it? Quiz

Circle the choice you think is correct for each multiple choice question.

1. Earth's outer crust rests on top of another layer called __________.
   - (A) the mantle
   - B the plate
   - C volcano
   - D earthquake

2. When a volcano erupts, the magma that reaches the surface is called __________.
   - A gas
   - B ash
   - C vent
   - (D) lava

3. The ________ is the point on Earth's surface that is directly above the focus.
   - (A) epicenter
   - B fault
   - C earthquake
   - D plate

4. Which of the following can rapidly change Earth's surface?
   - A glaciers
   - (B) earthquakes
   - C rivers
   - D wells

5. Where do earthquakes always occur?
   - (A) along a fault
   - B in the mountains
   - C in the ocean
   - D in big cities

6. ____________ can also cause rapid changes on Earth's surface.
   - (A) Landslides and floods
   - B Droughts
   - C Lakes and ponds
   - D Glaciers and ice caps

Unit 4, *Got it? Quiz* • Earth's Resources

Unit 4 | Got it? Quiz

Name ______________________ Date ____________

7. What is an example of liquid water on Earth's surface?
   - A a glacier
   - (B) a pond
   - C an ice cap
   - D water vapor

8. What is the sun's role in the water cycle?
   - A The sun's warmth changes water from gas to liquid.
   - B The sun makes water particles disappear.
   - C Water condenses when it is warmed by the sun.
   - (D) Water evaporates in the sun's warmth.

9. Explain how a volcano's eruption can cause Earth's surface to change rapidly.

Possible answers: Lava and ash from a volcano can spread over a wide area. The volcano and its surrounding area can be reshaped. A volcano can also cause landslides, floods, and tsunamis.

10. Draw and label a diagram that shows the steps of the water cycle.

Unit 4, *Got it? Quiz* • Earth's Resources

Teacher's Notes

Unit 4 Study Guide

How do Earth's resources change?

Lesson 1

How can Earth's surface change rapidly?

- Volcanic eruptions, earthquakes, landslides, and floods can cause Earth's surface to change rapidly.
- Earthquakes occur when Earth's plates move suddenly along faults.

Lesson 2

Where is Earth's water?

- Earth's fresh water exists as lakes, ponds, rivers, glaciers, and groundwater.
- Most of Earth's water is salt water in the ocean.

Lesson 3

What is the water cycle?

- The water cycle is the constant movement of water from Earth's surface to the atmosphere and back again.
- Evaporation, condensation, and precipitation are parts of the cycle.

Review the Big Question

How do Earth's resources change?

Have students use what they have learned from the unit to answer the question in their own words.

How has your answer to the Big Question changed since the beginning of the unit? What are some things you learned that caused your answer to change?

Make a Concept Map

Have students make a concept map like the one shown on this page to help them organize key concepts.

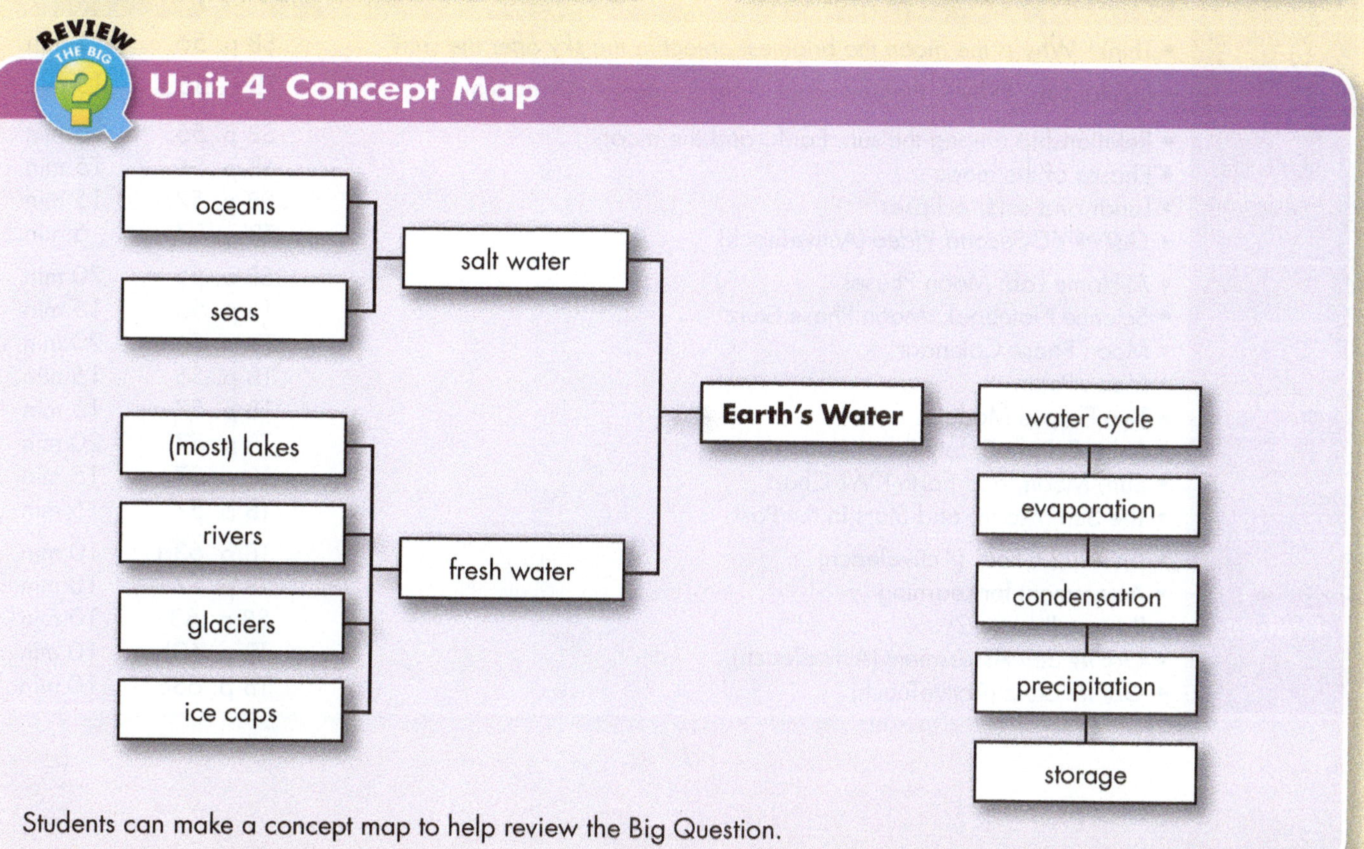

Students can make a concept map to help review the Big Question.

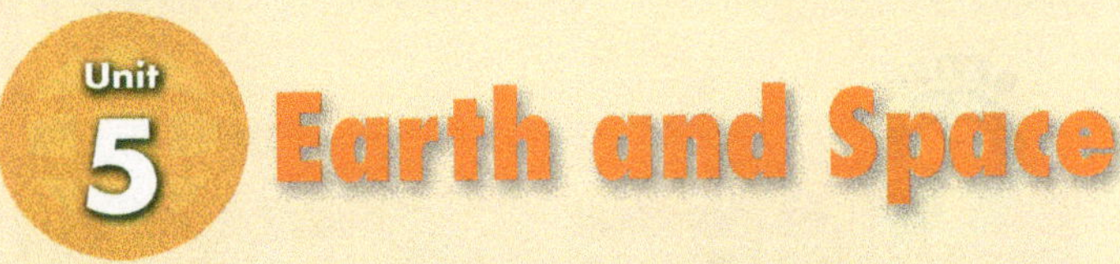

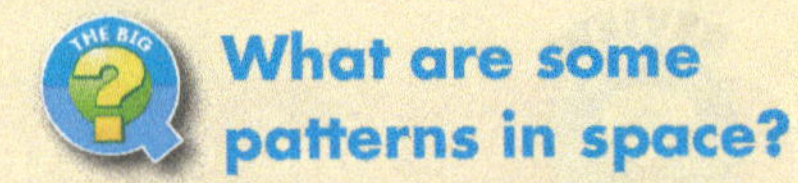

THE BIG
What are some patterns in space?

Lesson Plan

Unit Opener & Lesson 1 How do star patterns change?			
	Activity	**Pages**	**Time**
Engage	• Unit Opener: Think! *What makes the moon glow?* • Unit Opener: List characteristics of the sun, Earth, and the moon. • Unit Opener: Discuss how the sun, Earth, and the moon move in space. • Think! *How can we learn more about the stars in the Milky Way?*	SB p. 52 SB p. 52 SB p. 52 SB p. 53	5 min 10 min 10 min 5 min
Explore	• Digital Lab: *What star patterns can you see?* (ActiveTeach)	TB p. 53	15 min
Explain	• Star patterns • Star positions and the North Star • *Got it? 60-Second Video* (ActiveTeach)	SB p. 53 SB p. 54 TB p. 54	15 min 15 min 5 min
Elaborate	• At-Home Lab: Pictures in the Sky	SB p. 54	15 min
Evaluate	• *Lesson 1 Check* (ActiveTeach) • Assessment for Learning • Review (Lesson 1) • *Got it? Self Assessment* (ActiveTeach) • *Got it? Quiz* (ActiveTeach)	TB p. 63a TB p. 54 SB p. 63 TB p. 63b TB p. 63c	10 min 10 min 10 min 10 min 10 min

Lesson 2 What are the phases of the moon?			
	Activity	**Pages**	**Time**
Engage	• Think! *Why is the moon the brightest object in the sky after the sun?*	SB p. 56	5 min
Explore	• Digital Lab: *Why is the new moon hard to see?* (ActiveTeach)	TB p. 55	15 min
Explain	• Relationship among the sun, Earth, and the moon • Phases of the moon • Lunar and solar eclipses • *Got it? 60-Second Video* (ActiveTeach)	SB p. 55 SB p. 56 SB p. 57 TB p. 57	15 min 15 min 15 min 5 min
Elaborate	• At-Home Lab: Moon Phases • Science Notebook: Moon Phase Quiz • Moon Phase Calendar • Moon Poem • Two Eclipse Models • Solar Eclipse Safety Precautions • Sun, Moon, and Earth KWL Chart • The Sun, Moon, and Stars in the Past	SB p. 55 TB p. 56 TB p. 56 TB p. 56 TB p. 57 TB p. 57 TB p. 57 TB p. 57	20 min 15 min 20 min 15 min 15 min 20 min 15 min 15 min
Evaluate	• *Lesson 2 Check* (ActiveTeach) • Assessment for Learning • Review (Lesson 2) • *Got it? Self Assessment* (ActiveTeach) • *Got it? Quiz* (ActiveTeach)	TB p. 63a TB p. 57 SB p. 63 TB p. 63b TB p. 63c	10 min 10 min 10 min 10 min 10 min

Lesson 3 What is the solar system?			
	Activity	**Pages**	**Time**
Engage	• Think! *What effect does the sun's gravity have on the planets in the solar system?* • Think! *What natural resources help support life on Earth?* • Think! *How are the inner planets alike and different?*	TB p. 58 SB p. 59 TB p. 60	5 min 5 min 5 min
Explore	• Digital Activity: *Fun Fact: Pluto* (ActiveTeach)	TB p. 58	15 min
Explain	• Objects in the solar system • Planets and moons in the solar systems • Inner planets • Outer planets • *Got it? 60-Second Video* (ActiveTeach)	SB p. 58 SB p. 59 SB p. 60 SB p. 61 TB p. 61	15 min 15 min 15 min 15 min 5 min
Elaborate	• Science Notebook: Our Solar System • The Sun, Moon, and Stars in Greek and Roman Mythology • Go Green: Solar Energy • Science Notebook: Inner Planets • Inner Planet Chart • Outer Planet Chart	TB p. 58 TB p. 59 SB p. 59 TB p. 60 TB p. 60 TB p. 61	20 min 20 min 15 min 15 min 20 min 15 min
Evaluate	• *Lesson 3 Check* (ActiveTeach) • Assessment for Learning • Review (Lesson 3) • *Got it? Self Assessment* (ActiveTeach) • *Got it? Quiz* (ActiveTeach)	TB p. 63a TB p. 61 SB p. 63 TB p. 63b TB p. 63c	10 min 10 min 10 min 10 min 10 min
Lab	• *Let's Investigate! What is the shape of a planet's path?* (ActiveTeach)	SB p. 62	20 min

Flash Cards

asteroid

solar eclipse

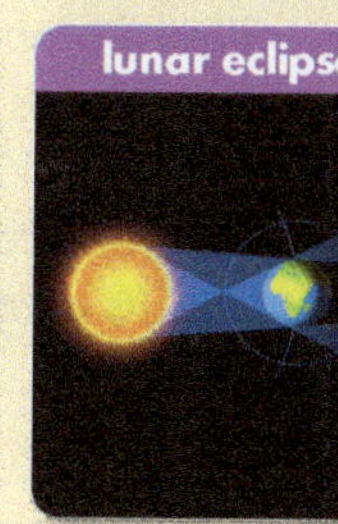

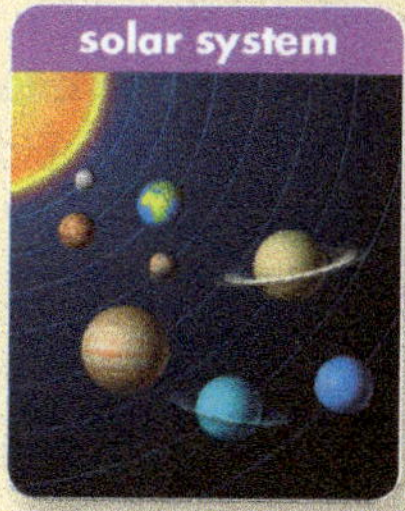

Lesson 1

Key Words	ELL Support
star, constellation, astronomer	**Vocabulary:** *sextillion, axis, move across, orbit (v), seasons, Orion, Northern Hemisphere, Southern Hemisphere, Polaris, North Star, rise (v), set (v), revolve, Ursa Major (the Great Bear), Ursa Minor (the Little Bear), Big Dipper, dipper's bowl, Antarctica*

Lesson 2

Key Words	ELL Support
eclipse, lunar eclipse, solar eclipse	**Vocabulary:** *reflect, moon's surface, gravity, orbit (v/n), spin (v/n), half, shape (n), lighted, full moon, waxing moon, waning moon, new moon, crescent moon, first quarter moon, full moon, last quarter moon, cast (v), shadow, partial eclipse, total lunar eclipse, block (v)*

Lesson 3

Key Words	ELL Support
solar system, gravity, ellipse, planet, asteroid, comet, inner planets, outer planets	**Vocabulary:** *revolve, pull toward, orbit (n), dwarf planet, dust, diameter, average surface temperature, craters, dents, swirling clouds, burning hot, poisonous, dust storms, rings, cloudy atmosphere, hydrogen, helium, windy, stormy*

Unit Objectives

Lesson 1: Students will understand that constellations stay the same but appear to change nightly and throughout the year.

Lesson 2: Students will describe the phases of the moon.

Lesson 3: Students will demonstrate an understanding that the sun, the planets and their moons, and other objects are part of the solar system.

Vocabulary: *sun, Earth, moon, space, glow* (v)

Materials: pictures of different landscapes with daytime and nighttime skies, large sheet of paper

What are some patterns in space?

Build Background Display pictures of different landscapes with daytime and nighttime skies where the sun, the moon, and stars can be seen. Ask students why the sky can look so different.

Think!

What makes the moon glow?

Point to the photo on the bottom right and have students describe it. Divide the class into small groups and have them discuss the answer.

1 Look and label.

Use the photos to elicit vocabulary. Have students label the photos. Review the answers by pointing to the pictures for students to say the words.

2 What do you know about the three objects in space above? With a partner, make a list of your ideas.

Have pairs answer the question and make a list. On a large sheet of paper, draw a three-column KWL chart with the following headings: K (What I *know*), W (What I *want* to know), L (What I *learned*). Complete the first two columns of the chart with students' responses about the three objects in space. Keep the chart on a classroom wall for further use.

Unit 5 Earth and Space

What are some patterns in space?

I will learn
- how star patterns change.
- about the phases of the moon.
- about the solar system.

1 Look and label.

sun | Earth | moon

2 What do you know about the three objects in space above? With a partner, make a list of your ideas.

3 Do these objects in space move? Discuss as a class.

Think! What makes the moon glow?

52 Unit 5

3 Do these objects in space move? Discuss as a class.

Do you think these objects in space move? Why or why not? Ask volunteers to share their answers with the class along with the reasons for their answers. (Possible answers: *Yes, but it's hard to tell. The moon, sun, and stars appear to move across the sky because of the way Earth moves.*)

Think! Again!

Revisit the question *What makes the moon glow?* (Possible answer: *The moon reflects light from the sun.*)

ELL Content Support

Explain to students that the sun shines with its own energy. *The light you see (and feel as heat) is given off by the sun's surface, called the photosphere.* Some of the energy the sun produces must escape, or it would build up and explode. *What we call moonlight is actually sunshine reflected by the moon. The moon gives off no light of its own but reflects sunlight from its rocky surface.*

Lesson 1

How do star patterns change?

Objective: Learn about the sun, stars, and constellations.

Vocabulary: *scientists, sextillion, stars, universe, hot gas, sky, bigger, brighter, hotter, smaller, dimmer, cooler, faint* (adj)

Digital Resources: Flash Card (*constellation*), *Let's Explore!* Digital Lab, (*Optional*: Star Finder Pattern Resource (1 copy per student), Star Wheel Resource (1 copy per student)), *I Will Know...* Digital Activity

Materials: picture of a telescope

Unlock the Big Question

Write the following text on the board: *I will understand that constellations stay the same but appear to change nightly and throughout the year.*

Build Background Display a picture of a telescope. Have students say what telescopes are used for and if they have ever used one. Explain to students that, apart from the fact that there are different types of telescopes, all of them are used to see objects that are far away, mostly the stars and planets.

Explore

Let's Explore! Lab What star patterns can you see?

Objective: Students will use Star Finders to determine constellations visible at different times on a given night.

Digital Resources: *Let's Explore!* Digital Lab, *Let's Explore! Activity Card* (1 per student)

- Display the *constellation* Flash Card and have students describe it. Ask *How many stars do you think someone can see in the clear night sky?* Explain that a person may be able to see 2,000–2,500 stars at any one time. *What can astronomers do to study so many stars? They group them into patterns called constellations.*
- Show the *Let's Explore!* Digital Lab. Guide students in a discussion about why different constellations can be seen at different times of the night and year.
- Have students complete the *Activity Card* and check their answers in small groups or pairs. Provide support as needed. * You may wish to provide materials for students to make their own Star Finders and Star Wheels for use at home.

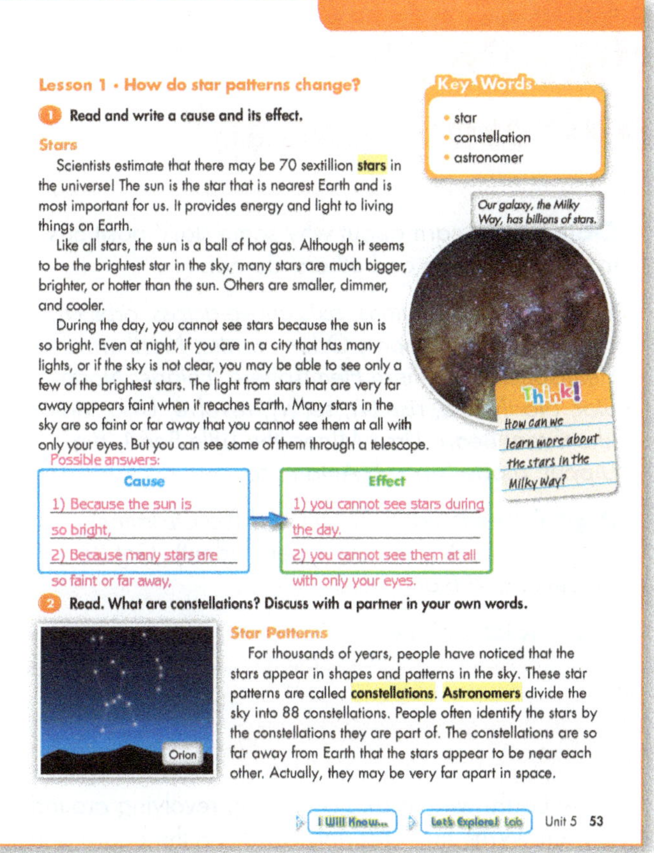

Lesson 1 • How do star patterns change?

Key Words
- star
- constellation
- astronomer

1 Read and write a cause and its effect.

Stars

Scientists estimate that there may be 70 sextillion **stars** in the universe! The sun is the star that is nearest Earth and is most important for us. It provides energy and light to living things on Earth.

Like all stars, the sun is a ball of hot gas. Although it seems to be the brightest star in the sky, many stars are much bigger, brighter, or hotter than the sun. Others are smaller, dimmer, and cooler.

During the day, you cannot see stars because the sun is so bright. Even at night, if you are in a city that has many lights, or if the sky is not clear, you may be able to see only a few of the brightest stars. The light from stars that are very far away appears faint when it reaches Earth. Many stars in the sky are so faint or far away that you cannot see them at all with only your eyes. But you can see some of them through a telescope.

Our galaxy, the Milky Way, has billions of stars.

Think! How can we learn more about the stars in the Milky Way?

Possible answers:

Cause	Effect
1) Because the sun is so bright,	1) you cannot see stars during the day.
2) Because many stars are so faint or far away,	2) you cannot see them at all with only your eyes.

2 Read. What are constellations? Discuss with a partner in your own words.

Star Patterns

For thousands of years, people have noticed that the stars appear in shapes and patterns in the sky. These star patterns are called **constellations**. **Astronomers** divide the sky into 88 constellations. People often identify the stars by the constellations they are part of. The constellations are so far away from Earth that the stars appear to be near each other. Actually, they may be very far apart in space.

Orion

I Will Know... Let's Explore! Lab Unit 5 53

Explain

1 Read and write a cause and its effect.

How many stars do scientists estimate there might be in the universe? Write on the board the full number, 70 followed by 21 zeros, for students to appreciate such a large amount. *When something happens, it is usually caused by something else. What happens is called an effect. The effect answers the question: What happened? The cause answers the question: What made it happen? Look for a cause and its effect.* Read the remaining paragraphs for students and have them write a cause and its effect.

2 Read. What are constellations? Discuss with a partner in your own words.

Display the *constellation* Flash Card. Have students read the text to find out what constellations are. Then ask students to close their books and explain to a partner what constellations are, using their own words.

Think!

Ask *How can we learn more about the stars in the Milky Way?* (Possible answer: *We can use telescopes to study the stars in our galaxy.*)

I Will Know...

Have students do the *I Will Know...* Digital Activity.

Lesson 1

How do star patterns change?

Objective: Learn about why some stars' positions in the sky change and others' do not.

Vocabulary: *positions, axis, move across, orbit* (v), *constellations, seasons, Orion, winter, Northern Hemisphere, Southern Hemisphere, Polaris, North Star, North Pole, rise* (v), *set* (v), *revolve, Ursa Major (the Great Bear), Ursa Minor (the Little Bear), Big Dipper, dipper's bowl, Antarctica*

Digital Resources: Flash Card (*constellation*), *Lesson 1 Check* (print out 1 per student), *Got it? 60-Second Video*

Materials: a globe

ELL Vocabulary Support

Draw Earth, with its axis and orbit, revolving around the sun. Write the following words on the board and ask eight volunteers to label the picture on the board: *Earth, axis, orbit, Northern Hemisphere, Southern Hemisphere, North Pole, South Pole,* and *Antarctica.*

Explain

3 Read. Why do constellations seem to move across the sky? Discuss as a class.

Display the *constellation* Flash Card. *What are constellations? Star patterns! How many constellations are there? 88!* Point to the *constellation* Flash Card. *What is the name of this constellation? Orion! Can people in the Northern and Southern Hemispheres see Orion at the same time? Why or why not? No. People in the Northern and Southern Hemispheres see different parts of the sky, and some constellations are only visible during certain seasons.* Then discuss with the class why constellations seem to move across the sky. (Possible answer: *Because Earth turns on its axis and orbits the sun.*)

4 Read and complete the sentences with a partner.

The picture shows a constellation that appears above the North Pole. What makes it different from other constellations? Ask students to read the text and find its two names. Then have pairs complete the sentences. Check answers on the board.

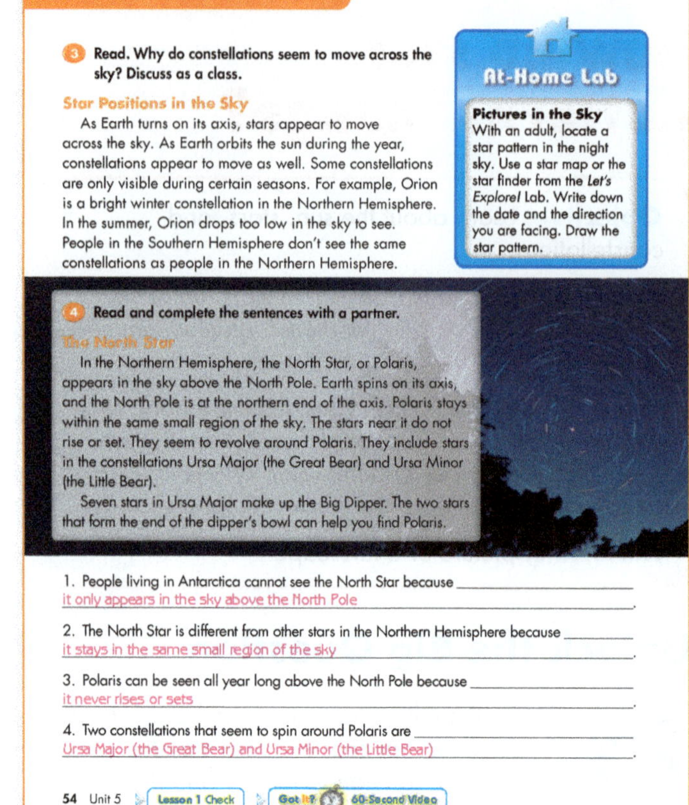
3 Read. Why do constellations seem to move across the sky? Discuss as a class.

Star Positions in the Sky

As Earth turns on its axis, stars appear to move across the sky. As Earth orbits the sun during the year, constellations appear to move as well. Some constellations are only visible during certain seasons. For example, Orion is a bright winter constellation in the Northern Hemisphere. In the summer, Orion drops too low in the sky to see. People in the Southern Hemisphere don't see the same constellations as people in the Northern Hemisphere.

At-Home Lab

Pictures in the Sky
With an adult, locate a star pattern in the night sky. Use a star map or the star finder from the *Let's Explore!* Lab. Write down the date and the direction you are facing. Draw the star pattern.

4 Read and complete the sentences with a partner.

The North Star

In the Northern Hemisphere, the North Star, or Polaris, appears in the sky above the North Pole. Earth spins on its axis, and the North Pole is at the northern end of the axis. Polaris stays within the same small region of the sky. The stars near it do not rise or set. They seem to revolve around Polaris. They include stars in the constellations Ursa Major (the Great Bear) and Ursa Minor (the Little Bear).

Seven stars in Ursa Major make up the Big Dipper. The two stars that form the end of the dipper's bowl can help you find Polaris.

1. People living in Antarctica cannot see the North Star because it only appears in the sky above the North Pole.
2. The North Star is different from other stars in the Northern Hemisphere because it stays in the same small region of the sky.
3. Polaris can be seen all year long above the North Pole because it never rises or sets.
4. Two constellations that seem to spin around Polaris are Ursa Major (the Great Bear) and Ursa Minor (the Little Bear).

54 Unit 5 | Lesson 1 Check | Got it? 60-Second Video

Elaborate

Pictures in the Sky

Materials: drawing supplies, Star Wheel and Star Finder Resources

Have students locate a constellation in the night sky using the resources from the *Let's Explore!* Activity. Ask them to draw the star pattern and write down the date and the direction they are facing.

Evaluate

Lesson 1 Check Assessment for Learning

Distribute the *Lesson 1 Check* and guide students as they complete it. Check answers as a class. Then ask students to grade their progress on the topic of how star patterns change from 1 to 3: 3 = *I understand how star patterns change;* 2 = *I need to study more;* 1 = *I need help!* Encourage students giving themselves a 1 or 2 to describe what they found difficult and what they need to study more.

Review Key Words for Lesson 1 (see Student's Book page 53). Play the *Got it? 60-Second Video* to review the lesson material.

Lesson 2

What are the phases of the moon?

Objective: Learn about the relationship among the sun, the moon, and Earth.

Vocabulary: *sunlight, reflect, moon's surface, gravity, orbit* (v/n), *spin* (v/n), *axis, shape* (n), *half, lighted, full moon, phases, waxing, waning*

Digital Resources: Flash Cards (*full moon, new moon*), *Let's Explore!* Digital Lab

Unlock the Big Question

Write the following text on the board: *I will be able to describe the phases of the moon.*

Build Background Ask students to think about what the moon looks like on different nights and at different times. Guide students to conclude that the shape, color, size, and position seem to change.

Explore

Let's Explore! Lab Why is the new moon hard to see?

Objective: Students will use models to observe how the phases of the moon are visible from Earth.

Digital Resources: *Let's Explore!* Digital Lab, *Let's Explore! Activity Card* (1 per student)

- Display the *full moon* and *new moon* Flash Cards to review their names. *Does the moon produce its own light? No. We can see the moon because sunlight reflects off the moon's surface.*
- Show the Digital Lab and have students complete the *Activity Card* in pairs.
- Have students check their answers in small groups or pairs. Provide support as needed.

Explain

1 Read. Why is the same side of the moon always facing Earth? Discuss as a class.

Look at the picture. What do the red arrows represent? That the moon spins around its axis. And what do the blue arrows represent? That the moon revolves around Earth. Is the same side of the moon always facing Earth? Have students read to find the answer to the question. Then have the class discuss why this happens. (Possible answer: *The moon rotates once on its axis in the same amount of time that it takes to revolve once around Earth.*)

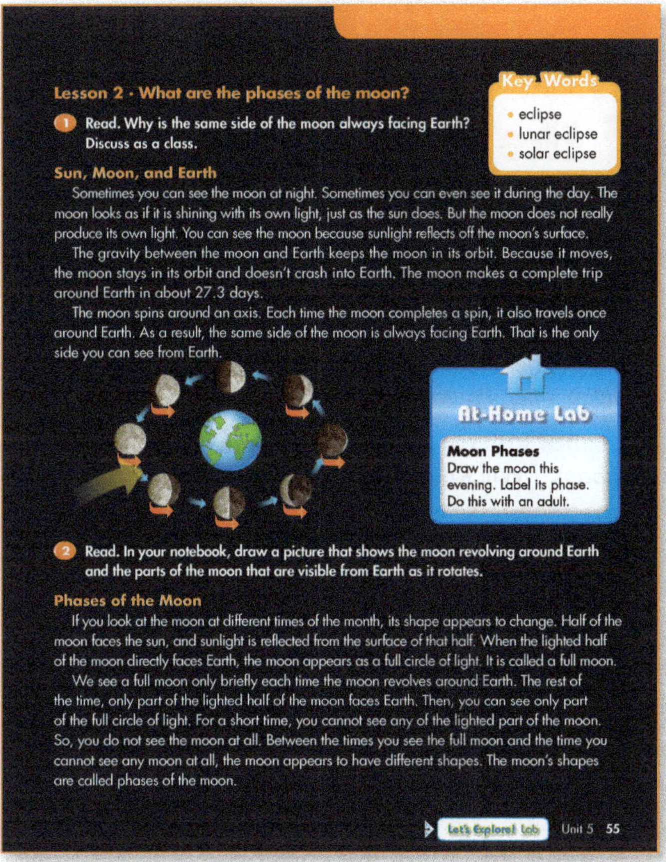

Lesson 2 · What are the phases of the moon?

Key Words
- eclipse
- lunar eclipse
- solar eclipse

1 Read. Why is the same side of the moon always facing Earth? Discuss as a class.

Sun, Moon, and Earth

Sometimes you can see the moon at night. Sometimes you can even see it during the day. The moon looks as if it is shining with its own light, just as the sun does. But the moon does not really produce its own light. You can see the moon because sunlight reflects off the moon's surface.

The gravity between the moon and Earth keeps the moon in its orbit. Because it moves, the moon stays in its orbit and doesn't crash into Earth. The moon makes a complete trip around Earth in about 27.3 days.

The moon spins around an axis. Each time the moon completes a spin, it also travels once around Earth. As a result, the same side of the moon is always facing Earth. That is the only side you can see from Earth.

At-Home Lab

Moon Phases
Draw the moon this evening. Label its phase. Do this with an adult.

2 Read. In your notebook, draw a picture that shows the moon revolving around Earth and the parts of the moon that are visible from Earth as it rotates.

Phases of the Moon

If you look at the moon at different times of the month, its shape appears to change. Half of the moon faces the sun, and sunlight is reflected from the surface of that half. When the lighted half of the moon directly faces Earth, the moon appears as a full circle of light. It is called a full moon.

We see a full moon only briefly each time the moon revolves around Earth. The rest of the time, only part of the lighted half of the moon faces Earth. Then, you can see only part of the full circle of light. For a short time, you cannot see any of the lighted part of the moon. So, you do not see the moon at all. Between the times you see the full moon and the time you cannot see any moon at all, the moon appears to have different shapes. The moon's shapes are called phases of the moon.

Let's Explore! Lab Unit 5 55

2 Read. In your notebook, draw a picture that shows the moon revolving around Earth and the parts of the moon that are visible from Earth as it rotates.

Read the text with students. *How would you define the phases of the moon? The different shapes the moon appears to have.* Have students identify the picture of the full moon. *When is it called a full moon? When the moon appears as a full circle of light.* Ask students to draw a picture that shows the moon revolving around Earth and its different phases.

Elaborate

Moon Phases

Materials: drawing compass (*optional*), markers

Ask students to observe the moon this evening and draw it. Have them investigate what the name of the phase is. Answers will vary with the time of the month. Students should identify which of the following phases the moon they saw most resembles: new, waxing crescent, first quarter, full, last quarter, waning crescent.

Lesson 2

What are the phases of the moon?

Objective: Learn about the phases of the moon.

Vocabulary: *waxing moon, waning moon, new moon, crescent moon, first quarter moon, full moon, last quarter moon, lighted, half, sliver, dark, unlighted, entire, infer, figurative language*

Digital Resources: Flash Cards (*full moon, new moon*), *I Will Know...* Digital Activity

Build Background Display the *new moon* Flash Card. *When the moon is in its new moon phase, what can you infer about the opposite side of the moon?* *That it is completely lighted.* *And when the moon is in its full moon phase, what can you infer about the opposite side of the moon?* *That it is completely dark.*

Explain

3 Read and label the captions.

Have students identify the full moon and the new moon phases in the diagram. Have them explain in their own words what happens to the moon phases between the new moon and the full moon phase. Before asking students to read and label the captions, have them describe the eight moon phases. Check answers with the class.

4 What causes the changing appearance of the moon over the course of a month? Discuss as a class.

Draw the sun, Earth, and the moon on the board. Have students say the ways in which they are related. Ask *What causes the changing appearance of the moon over the course of a month?* Guide students to conclude that the moon phases depend on the different amounts of the lighted half of the moon facing Earth.

Elaborate

Science Notebook: Moon Phase Quiz

Have students draw the eight moon phases in their Science Notebooks. Before having them label the eight phases, ask them to investigate what the missing three names are. On the board, write *Which phase of the moon follows a full moon?* Elicit the answer from the whole class. Have students write a five-question quiz using the sentence frame on the board as a basis. Then have pairs ask and answer questions.

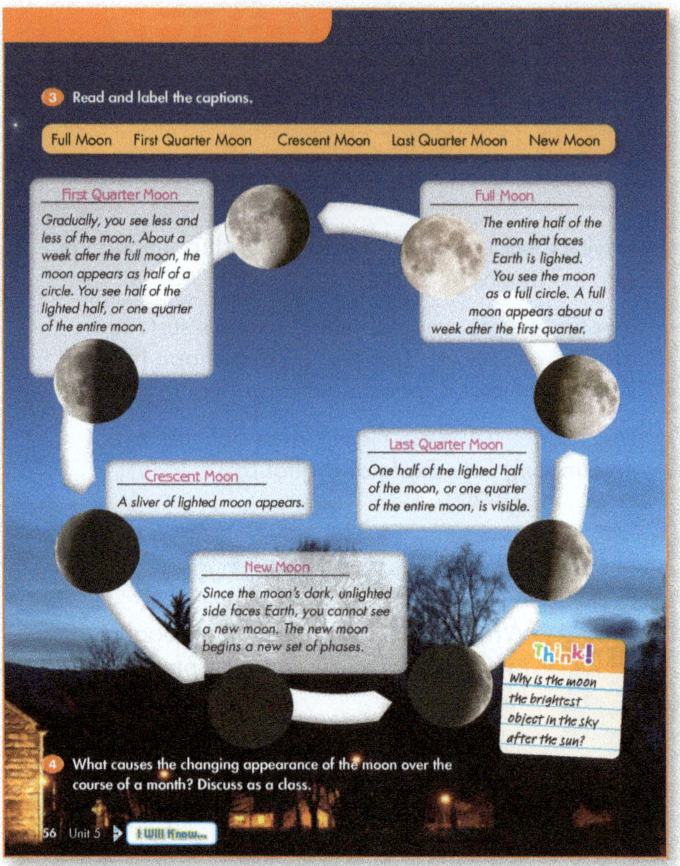

3 Read and label the captions.

Full Moon First Quarter Moon Crescent Moon Last Quarter Moon New Moon

4 What causes the changing appearance of the moon over the course of a month? Discuss as a class.

56 Unit 5 I Will Know...

Moon Phase Calendar

Provide students with the moon phase information shown in the local newspaper or posted on the Internet. Divide the class into pairs. Ask students to make calendars of the current month on poster board. Have students draw a picture of each moon phase on the day it occurs. Display the calendars on the classroom walls.

Moon Poem

Have students write and illustrate a poem about the phases of the moon. Ask students to use figurative language to describe the moon in different phases. On the board, draw the crescent moon and write *The crescent moon points its horns like an angry bull.*

Think!

Ask *Why is the moon the brightest object in the sky after the sun?* (Possible answer: *Because it reflects the sun's light and is close to Earth.*)

I Will Know...

Have students do the *I Will Know...* Digital Activity.

Lesson 2

What are the phases of the moon?

Objective: Learn about the lunar and solar eclipses.

Vocabulary: *cast* (v), *shadow, eclipse, full moon, lunar eclipse, partial eclipse, total lunar eclipse, solar eclipse, last* (v), *block* (v)

Digital Resources: Flash Cards (*lunar eclipse, solar eclipse*), *Lesson 2 Check* (print out 1 per student), *Got it? 60-Second Video*

Materials: a flashlight, an orange, a table tennis ball

Build Background Draw the sun, Earth orbiting around it, and the moon orbiting around Earth. Have students discuss what happens if the moon gets between the sun and Earth. Write students' ideas on the board.

Explain

5 Read and match the columns.

Have students read and underline the definition of *eclipse*. Explain that there are two types of eclipses: lunar and solar. Ask students to read and underline the descriptions of each type of eclipse. Display the *lunar eclipse* and *solar eclipse* Flash Cards. Guide two volunteers to describe what happens in each picture using their own words. Have pairs read the texts again and match the columns.

Elaborate

Two Eclipse Models

Have students model the two types of eclipses with a flashlight (the sun), an orange (Earth), and a table tennis ball (the moon). Have them shine the flashlight on the orange. To simulate a solar eclipse, students can place the ball between the flashlight and the orange. For a lunar eclipse, have students place the orange between the ball and the flashlight.

Solar Eclipse Safety Precautions

Explain to students that it is dangerous to look directly at a solar eclipse. However, it is safe to look at a lunar eclipse, since it is merely sunlight reflected off the moon. Have students research the safety precautions needed for people to view a solar eclipse.

Sun, Moon, and Earth KWL Chart

Display the sun, moon, and Earth KWL chart that you used in the first class of this unit. Go through the questions from the second column and elicit answers from the class. Finally elicit what students learned in the last two lessons. Complete the third column with students' ideas.

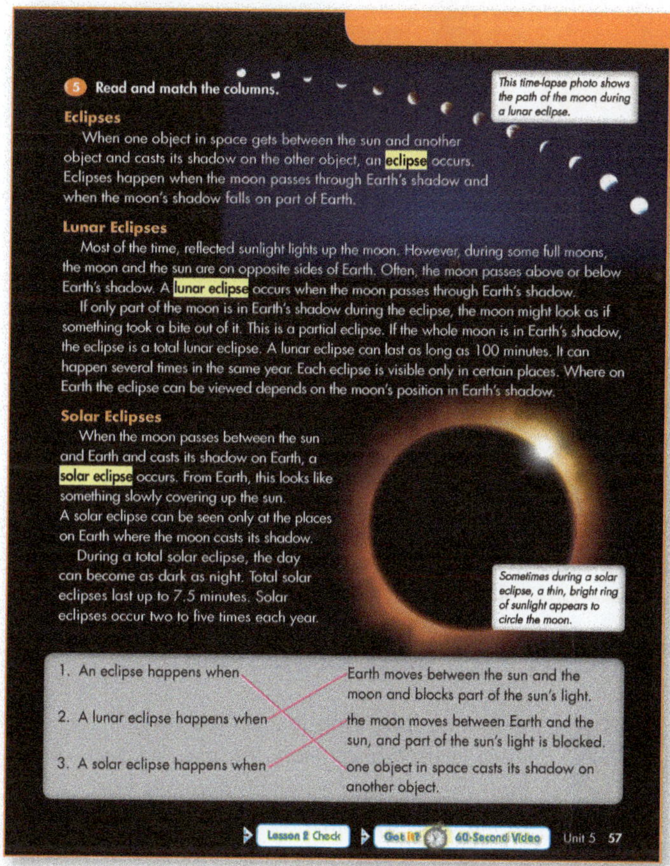

5 Read and match the columns.

This time-lapse photo shows the path of the moon during a lunar eclipse.

Eclipses

When one object in space gets between the sun and another object and casts its shadow on the other object, an eclipse occurs. Eclipses happen when the moon passes through Earth's shadow and when the moon's shadow falls on part of Earth.

Lunar Eclipses

Most of the time, reflected sunlight lights up the moon. However, during some full moons, the moon and the sun are on opposite sides of Earth. Often, the moon passes above or below Earth's shadow. A lunar eclipse occurs when the moon passes through Earth's shadow.

If only part of the moon is in Earth's shadow during the eclipse, the moon might look as if something took a bite out of it. This is a partial eclipse. If the whole moon is in Earth's shadow, the eclipse is a total lunar eclipse. A lunar eclipse can last as long as 100 minutes. It can happen several times in the same year. Each eclipse is visible only in certain places. Where on Earth the eclipse can be viewed depends on the moon's position in Earth's shadow.

Solar Eclipses

When the moon passes between the sun and Earth and casts its shadow on Earth, a solar eclipse occurs. From Earth, this looks like something slowly covering up the sun. A solar eclipse can be seen only at the places on Earth where the moon casts its shadow.

During a total solar eclipse, the day can become as dark as night. Total solar eclipses last up to 7.5 minutes. Solar eclipses occur two to five times each year.

Sometimes during a solar eclipse, a thin, bright ring of sunlight appears to circle the moon.

1. An eclipse happens when	Earth moves between the sun and the moon and blocks part of the sun's light.
2. A lunar eclipse happens when	the moon moves between Earth and the sun, and part of the sun's light is blocked.
3. A solar eclipse happens when	one object in space casts its shadow on another object.

Lesson 2 Check Got it? 60-Second Video Unit 5 57

The Sun, Moon, and Stars in the Past

Divide the class into small groups. Have students use the Internet or other sources to find out about early civilizations, such as the Maya, Aztec, Inca, Egyptian, and Greek, whose rituals and customs were shaped by their ideas about the sun, the stars, and the moon. Ask students to write a report explaining what they learned.

Evaluate

Lesson 2 Check Assessment for Learning

Distribute the *Lesson 2 Check* and guide students as they complete it. Check answers as a class. Then ask students to grade their progress on the topic of the moon's phases from 1 to 3: 3 = *I can describe the moon's phases*; 2 = *I need to study more*; 1 = *I need help!* Encourage students giving themselves a 1 or 2 to describe what they found difficult and what they need to study more.

Review Key Words for Lesson 2 (see Student's Book page 55). Play the *Got it? 60-Second Video* to review the lesson material.

Lesson 3

What is the solar system?

Objective: Learn that the sun, the planets and their moons, and other objects are part of the solar system.

Vocabulary: *solar system, planets, moons, medium-sized, star, revolve, path, gravity, force, pull toward, massive, curved, ellipse, stretched-out circle, elliptical, orbit* (n), *dwarf planet, Pluto, asteroid, rocky, asteroid belt, Mars, Jupiter, comet, frozen, ice, gases, dust*

Digital Resources: Flash Card (*solar system*), *Explore My Planet!* Digital Activity

Unlock the Big Question

Write the following text on the board:
I will learn that the sun, the planets and their moons, and other objects are part of the solar system.

Build Background Display the *solar system* Flash Card. Explain to students that *solar* has the root word *sol* meaning *sun*. One definition for *system* is *an orderly arrangement of things that make up a whole*. Based on this information, ask students to explain what a solar system is.

Explore

Explore My Planet! Fun Fact: Pluto

Objective: Students will learn why scientists no longer consider Pluto a planet in our solar system.

Digital Resources: *Explore My Planet!* Digital Activity, *Explore My Planet! Activity Card* (1 per student)

- Display the *solar system* Flash Card. Divide the class into pairs. Ask students to write a sentence about the solar system. Have volunteers read their sentences to the class.
- Show the *Explore My Planet!*
- On the board, write the sentence *My very energetic mother just served us nine pizzas.* Ask students what this sentence helps people do. If necessary, underline the first letter of each word.
- *What happened in 2006? The number of planets in the solar system was changed from nine to eight. Why is the sentence on the board not useful anymore? Because Pluto is no longer considered a planet.*
- Explain to students that they can use a simple sentence like the one on the board to help them remember the eight planets.
- Have students complete the *Activity Card*. Provide support as needed.
- Ask volunteers to read their sentences to the class.

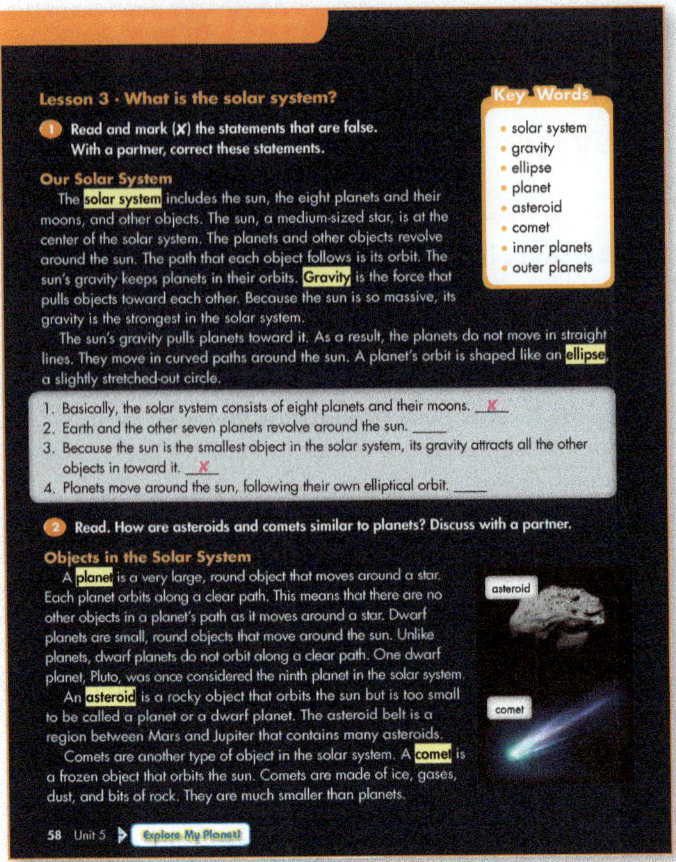

Lesson 3 · What is the solar system?

1 **Read and mark (✗) the statements that are false. With a partner, correct these statements.**

Key Words
- solar system
- gravity
- ellipse
- planet
- asteroid
- comet
- inner planets
- outer planets

Our Solar System

The **solar system** includes the sun, the eight planets and their moons, and other objects. The sun, a medium-sized star, is at the center of the solar system. The planets and other objects revolve around the sun. The path that each object follows is its orbit. The sun's gravity keeps planets in their orbits. **Gravity** is the force that pulls objects toward each other. Because the sun is so massive, its gravity is the strongest in the solar system.

The sun's gravity pulls planets toward it. As a result, the planets do not move in straight lines. They move in curved paths around the sun. A planet's orbit is shaped like an **ellipse**, a slightly stretched-out circle.

1. Basically, the solar system consists of eight planets and their moons. _✗_
2. Earth and the other seven planets revolve around the sun. ____
3. Because the sun is the smallest object in the solar system, its gravity attracts all the other objects in toward it. _✗_
4. Planets move around the sun, following their own elliptical orbit. ____

2 **Read. How are asteroids and comets similar to planets? Discuss with a partner.**

Objects in the Solar System

A **planet** is a very large, round object that moves around a star. Each planet orbits along a clear path. This means that there are no other objects in a planet's path as it moves around a star. Dwarf planets are small, round objects that move around the sun. Unlike planets, dwarf planets do not orbit along a clear path. One dwarf planet, Pluto, was once considered the ninth planet in the solar system.

An **asteroid** is a rocky object that orbits the sun but is too small to be called a planet or a dwarf planet. The asteroid belt is a region between Mars and Jupiter that contains many asteroids.

Comets are another type of object in the solar system. A **comet** is a frozen object that orbits the sun. Comets are made of ice, gases, dust, and bits of rock. They are much smaller than planets.

asteroid

comet

58 Unit 5 Explore My Planet!

Explain

1 **Read and mark (✗) the statements that are false. With a partner, correct these statements.**

Invite students to read the text and mark (✗) the statements that are false. Then have pairs compare their answers and correct the false statements.

2 **Read. How are asteroids and comets similar to planets? Discuss with a partner.**

Ask students to look at the pictures of the asteroid and the comet and describe them. Ask *Do asteroids and comets look like planets?* No. *How are asteroids and comets similar to planets? Read the text to find out.* (Answer: *Asteroids and comets orbit the sun like planets.*)

Elaborate

Science Notebook: Our Solar System

Write the following words on the board: *planet, moon, star, orbit, gravity, dwarf planet, asteroid, comet.* Invite students to illustrate and write a definition for each word in their Science Notebooks.

Ask *What effect does the sun's gravity have on the planets in the solar system?* (Possible answer: *The sun's gravity is so strong that it pulls all the planets toward the sun. This pull causes the planets to move in curved paths around the sun.*)

Lesson 3

What is the solar system?

Objective: Learn why planets shine at night.

Vocabulary: *planets, solar system, stars, moons, satellites, Mercury, Venus*

Digital Resources: Flash Card (*solar system*), *I Will Know...* Digital Activity

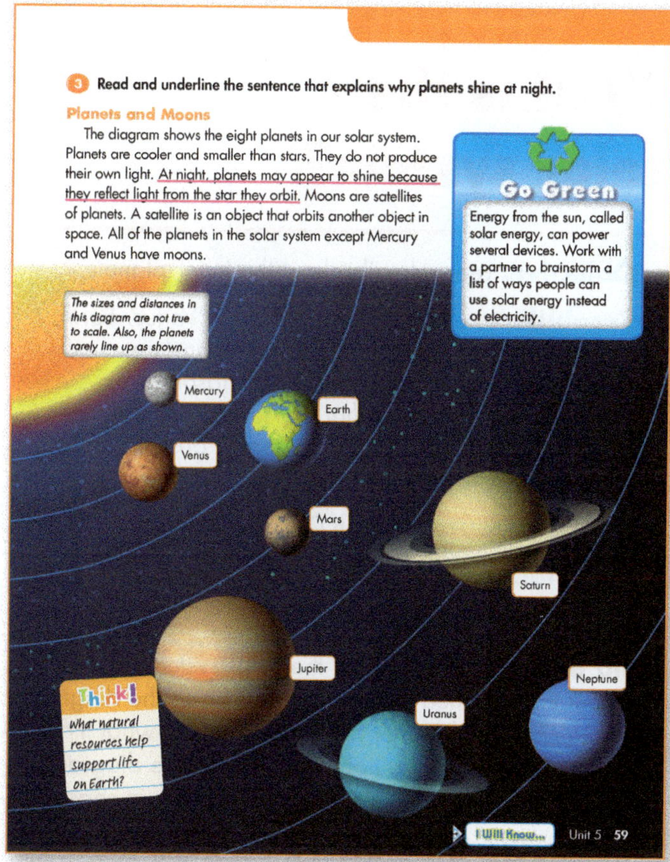

3 Read and underline the sentence that explains why planets shine at night.

Planets and Moons

The diagram shows the eight planets in our solar system. Planets are cooler and smaller than stars. They do not produce their own light. At night, planets may appear to shine because they reflect light from the star they orbit. Moons are satellites of planets. A satellite is an object that orbits another object in space. All of the planets in the solar system except Mercury and Venus have moons.

Go Green

Energy from the sun, called solar energy, can power several devices. Work with a partner to brainstorm a list of ways people can use solar energy instead of electricity.

The sizes and distances in this diagram are not true to scale. Also, the planets rarely line up as shown.

Unit 5 59

Unlock the Big Question

Build Background Write the words *revolve, orbit, planet, moon,* and *ellipse* on the board. Ask one volunteer to play the sun and another student to play a planet. Encourage the student playing the sun to stand still while the student playing the planet revolves around the sun, following an elliptical orbit. Have the student playing the moon revolve around Earth, while that student revolves around the sun. Have other volunteers describe the actions using the words written on the board.

Explain

3 Read and underline the sentence that explains why planets shine at night.

Display the *solar system* Flash Card. Have students read and underline the sentence that explains why planets shine at night.

ELL Content Support

Point to the *solar system* Flash Card. Explain to students that the diagram makes it look like the planets orbit the sun in alignment with each other. However, this alignment is rarely the case. Tell students that, in January and February 2016, Mercury, Venus, Mars, Saturn, and Jupiter were simultaneously visible to the naked eye for the first time in more than a decade. As a follow-up, have students research when an event like this will happen again.

Elaborate

The Sun, Moon, and Stars in Greek and Roman Mythology

Explain to students that the names of most planets originate from ancient Greek and Roman mythology. Have students choose a planet and use the library or the Internet to find the story behind their names. Ask students to share their findings with the class.

Solar Energy

Explain to students that the sun provides energy that helps plants grow and heats Earth. Have pairs brainstorm a list of ways people can use solar energy instead of electricity. Students' lists might contain uses such as heating homes, cooking food, or purifying water.

Think!

Ask *What natural resources help support life on Earth?* Elicit answers from the class, asking students to explain in which ways such resources support life. (Possible answers: *water, oxygen, the sun's energy*)

I Will Know...

Have students do the *I Will Know...* Digital Activity.

Lesson 3

What is the solar system?

Objective: Learn details about the inner planets.

Vocabulary: *inner planets, Mercury, Venus, Earth, Mars, diameter, average surface temperature, moons, craters, dents, closeness, atmosphere, swirling clouds, burning hot, poisonous, Red Planet, minerals, oxygen, support life, winds, dust storms, volcanoes, canyons, flow* (v)

Digital Resources: Flash Card (*solar system*)

Materials: construction paper

Build Background Display the *solar system* Flash Card. Point out the speckled band that appears on the diagram between Mars and Jupiter for students to say what it is. Ask students what the light blue lines in the diagram represent and whether we can see these lines in space.

ELL Reading Strategy Support

Have students discuss how a diagram helps us better understand what we read. Have students describe what else they can learn from diagrams. Invite students to locate other diagrams in this same unit or in previous units and compare and contrast these diagrams with the diagram of the solar system on page 59. Have students share their comparisons.

Explain

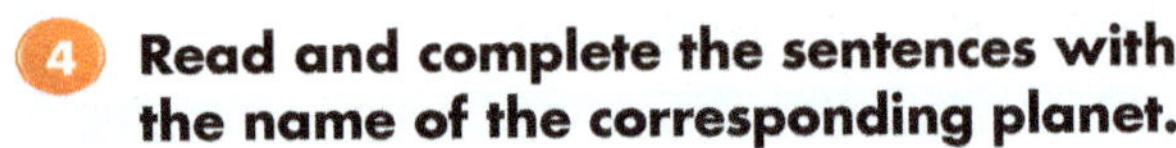

4 Read and complete the sentences with the name of the corresponding planet.

Before reading, have pairs describe the pictures of the four planets and predict what they think their surfaces are like. Have students read and complete the sentences with the names of the corresponding planets. Check answers as a class.

Think!

Ask How are the inner planets alike and different? Elicit answers from the class. (Possible answers: *The inner planets are small and closest to the sun, and they all have rocky surfaces. They differ because some are hotter or colder than the others. They are different sizes, and some have no moons.*)

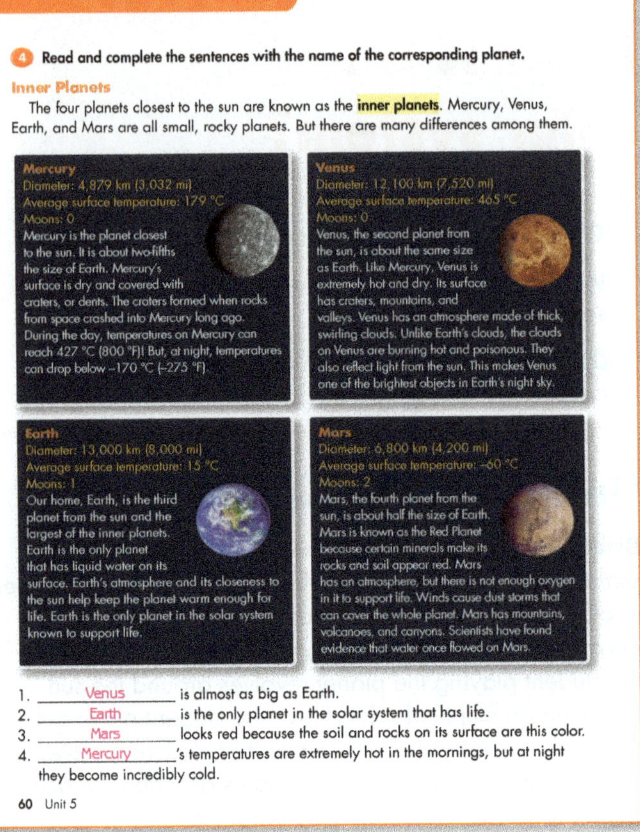

4 Read and complete the sentences with the name of the corresponding planet.

Inner Planets

The four planets closest to the sun are known as the **inner planets**. Mercury, Venus, Earth, and Mars are all small, rocky planets. But there are many differences among them.

Mercury
Diameter: 4,879 km (3,032 mi)
Average surface temperature: 179 °C
Moons: 0
Mercury is the planet closest to the sun. It is about two-fifths the size of Earth. Mercury's surface is dry and covered with craters, or dents. The craters formed when rocks from space crashed into Mercury long ago. During the day, temperatures on Mercury can reach 427 °C (800 °F)! But, at night, temperatures can drop below −170 °C (−275 °F).

Venus
Diameter: 12,100 km (7,520 mi)
Average surface temperature: 465 °C
Moons: 0
Venus, the second planet from the sun, is about the same size as Earth. Like Mercury, Venus is extremely hot and dry. Its surface has craters, mountains, and valleys. Venus has an atmosphere made of thick, swirling clouds. Unlike Earth's clouds, the clouds on Venus are burning hot and poisonous. They also reflect light from the sun. This makes Venus one of the brightest objects in Earth's night sky.

Earth
Diameter: 13,000 km (8,000 mi)
Average surface temperature: 15 °C
Moons: 1
Our home, Earth, is the third planet from the sun and the largest of the inner planets. Earth is the only planet that has liquid water on its surface. Earth's atmosphere and its closeness to the sun help keep the planet warm enough for life. Earth is the only planet in the solar system known to support life.

Mars
Diameter: 6,800 km (4,200 mi)
Average surface temperature: −60 °C
Moons: 2
Mars, the fourth planet from the sun, is about half the size of Earth. Mars is known as the Red Planet because certain minerals make its rocks and soil appear red. Mars has an atmosphere, but there is not enough oxygen in it to support life. Winds cause dust storms that can cover the whole planet. Mars has mountains, volcanoes, and canyons. Scientists have found evidence that water once flowed on Mars.

1. ___Venus___ is almost as big as Earth.
2. ___Earth___ is the only planet in the solar system that has life.
3. ___Mars___ looks red because the soil and rocks on its surface are this color.
4. ___Mercury___'s temperatures are extremely hot in the mornings, but at night they become incredibly cold.

60 Unit 5

ELL Vocabulary Support

Ask students to list all the adjectives that are used to describe the four planets and their main features on page 60. Write the adjectives on the board: *small, rocky, close, hot, dry, thick, swirling, poisonous,* and *red*. Then have volunteers choose an adjective and make a sentence about any of the planets using that adjective.

Elaborate

Science Notebook: Inner Planets

Have students write the four sentences in their Science Notebooks. Ask them to find information in the texts that supports the ideas expressed in each sentence and add them. Example: *Venus is almost as big as Earth. Venus has a diameter of 12,100 km, and Earth has a diameter of 13,000 km.*

Inner Planet Chart

Use a chart to help students organize and categorize the information they have learned about each planet so far. Make an eight-column chart on construction paper and write the name of a planet at the top of each column. Then have students work in pairs to write what they know about the four inner planets in their specific columns. Ask partners to then review the lesson to recall information that they may have missed and add it to the chart. Have partners take turns asking each other questions that can be answered using information in the chart.

Lesson 3

What is the solar system?

Objective: Learn details about the outer planets.

Vocabulary: *outer planets, large, gases, liquids, solid rock, dense clouds, surface, moons, rings, dust, ice, cloudy atmosphere, hydrogen, helium, storm, blue-green, thick, telescope, windy, stormy*

Digital Resources: Flash Card (*solar system*), *Lesson 3 Check* (print out 1 per student), *Got it? 60-Second Video*

Build Background Display the *solar system* Flash Card for students to name the eight planets. Write a number next to each planet's name to clarify the planets' order. Then ask volunteers to identify the planet closest to the sun, farthest from the sun, fifth from the sun, and so on, by calling out a planet's number and saying its name.

Explain

5 Read and underline three characteristics the four outer planets have in common.

Have students look at the pictures and predict what they think the four planets have in common. Ask students to read the introductory paragraph and underline three characteristics the outer planets have in common.

ELL Content Support

New ninth planet?

Tell students that, in January 2016, astronomers in the United States announced that they may have discovered another planet on the outskirts of our solar system. Astronomers hypothesize that it is made mostly of gases and estimate that it is ten times bigger than Earth.

6 Read again and answer the questions with a partner.

Have pairs describe in detail the pictures of the four planets and predict what they think their surfaces are like. Ask pairs to read and answer the questions.

Elaborate

Outer Planet Chart

Display the Planet Chart. This time have students work in pairs to write what they know about the four outer planets in their specific columns. Check the whole chart with the class.

5 Read and underline three characteristics the four outer planets have in common.

Outer Planets

Jupiter, Saturn, Uranus, and Neptune are known as the **outer planets**. They are all very large planets made mostly of gases and liquids. Instead of solid rock, dense clouds cover their surfaces. They all have many moons and rings made of dust, ice, and rocks.

Jupiter
Diameter: 142,984 km (88,846 mi)
Moons: At least 63
Jupiter is the fifth planet from the sun. It is so large that all of the other planets in the solar system could fit inside it! Jupiter's cloudy atmosphere is mainly hydrogen and helium. A huge storm has been raging for hundreds of years in its atmosphere. The storm is called the Great Red Spot.

Saturn
Diameter: 120,540 km (74,900 mi)
Moons: At least 62
Saturn is the sixth planet from the sun. It is the second biggest planet in the solar system, about ten times the size of Earth. Like Jupiter, Saturn has a thick, cloudy atmosphere instead of a solid surface. Saturn's rings can be seen only through a telescope. Saturn has the clearest rings of the outer planets.

Uranus
Diameter: 51,118 km (31,763 mi)
Moons: At least 27
Uranus, the seventh planet from the sun, is a very cold planet. It looks blue-green because of a gas in its cloudy atmosphere. Uranus rotates on its side, like a ball rolling through space.

Neptune
Diameter: 49,528 km (30,775 mi)
Moons: At least 13
Neptune is the farthest planet from the sun. It takes about 165 Earth years to orbit the sun. A gas in its cloudy atmosphere gives Neptune a blue color. Neptune is a windy, stormy, and cold planet.

6 Read again and answer the questions with a partner.

1. What planet rotates on its side like a ball rolling? Uranus
2. What planet is the farthest from the sun? Neptune
3. What is the largest planet in the solar system? Jupiter
4. What planet has rings that can be easily seen through a telescope? Saturn

Lesson 3 Check | Got it? 60-Second Video | Unit 5 61

ELL Language Support

Write *closest* and *brightest* the board. Point out to students that superlatives can be expressed using the suffix *-est*. Write on the board superlative adjectives (*closest, brightest, reddest, hottest, coldest, smallest, largest*) and have students use them to write sentences about the eight planets in their notebooks. You can use the following sentence frame for students to complete: ________ *is the* ________ *planet in the solar system.*

Evaluate

Lesson 3 Check **Assessment for Learning**

Distribute the *Lesson 3 Check* and guide students as they complete it. Check answers as a class. Then ask students to grade their progress on the topic of the solar system from 1 to 3: 3 = *I understand what the solar system is;* 2 = I need to study more; 1 = *I need help!* Encourage students giving themselves a 1 or 2 to describe what they found difficult and what they need to study more.

Review Key Words for Lesson 3 (see Student's Book page 58). Play the *Got it? 60-Second Video* to review the lesson material.

Let's Investigate!

In this unit, students learned about constellations, moon phases, eclipses, and objects in the solar system. In this lab, students will observe the difference between the shape of a circle and the shape of an ellipse.

Let's Investigate! Lab **What is the shape of a planet's path?**

Objective: Observe that ellipses are elongated compared to circles.

Materials: white paper (25 cm x 40 cm), heavy cardboard (30 cm x 45 cm), 2 straight pins, string (30 cm), metric ruler, safety goggles, masking tape

Digital Resources: *Let's Investigate!* Digital Lab, *Let's Investigate! Activity Card* (1 per group)

Advance Preparation: For each group, cut a 30 cm piece of string and a 30 x 45 cm piece of corrugated cardboard.

- Divide students into groups of four and distribute materials.
- Have students tape the paper onto the cardboard and stick a pin in the center. Show them how to tie a knot to make a loop of string.
- Ask students to put the loop over the pin and use a pencil and the string to draw a circle, holding the pencil upright against the stretched string as they draw a circle.
- Then have students follow the instructions to draw the two ellipses.
- Have students measure the largest and smallest diameters, describe the shapes, and record their observations in the chart.
- Students will conclude that the first ellipse is slightly elongated compared to the circle and that the second ellipse is more elongated compared to both the circle and the first ellipse.
- At the end of the activity, have students share their observations with the class. Guide them to conclude that the shapes of the two ellipses are similar to the shape of the planets' orbits around the sun.

Teacher Time-Saving Option: Show the *Let's Investigate!* Digital Lab as an alternative to the hands-on lab activity.

Materials

paper

heavy cardboard

metric ruler

tape

2 pins

piece of string

safety goggles

Let's Investigate!

What is the shape of a planet's path?

1. Tape the paper onto the cardboard. Stick a pin in the center. Tie a knot to make a loop of string.
2. Put the loop over the pin. Use a pencil and the string to draw a circle. Hold the pencil upright against the stretched string as you draw your circle.
3. Measure the largest and smallest diameters. Describe the shape. Record in the chart.
4. Put a second pin about 5 mm away from the first pin. Put the loop of string over both pins. Repeat Steps 2 and 3 with the loop over both pins.
5. Set the second pin about 10 mm away from the first pin. Put the loop of string over both pins. Repeat Steps 2 and 3 with the loop over both pins.

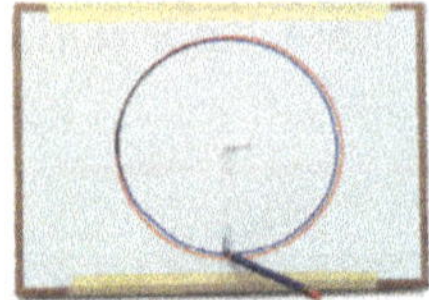

Sample data

Orbit Measurement Chart

Distance Between Pins (millimeters)	Largest Diameter	Smallest Diameter	Shape (circle or ellipse)
0	25 cm	25 cm	circle
5	25 cm	22 cm	ellipse
10	25 cm	19 cm	ellipse

62 Unit 5 Let's Investigate! Lab

Class Project: Make a Sundial

Materials: poster board, drawing compass, modeling clay, unsharpened pencil, ruler, clock, directional compass, marker

Have students cut a circle, 23 cm (9 in) in diameter, from poster board. Ask them to draw a line through the center of the circle and write *noon* at one end. Then have students stand a pencil up in the center of the circle using a ball of clay.

Show students how to place their sundials in the sunlight with the word *noon* pointing north. Have students check the pencil's shadow as it moves around the circle over time. Ask students to mark as many hours as they can.

The sun appears to move from east to west in the sky because Earth is turning on its axis from west to east. As a result, the shadow cast by the pencil appears to move clockwise around the sundial.

Note: It might be helpful to have students demonstrate the apparent movement of the sun with a flashlight.

Unlock the Big Question

Have students refer to the Big Question on the Unit Opener page. In pairs, have them recall what they have learned about the solar system. Invite student pairs to share their answers to questions 6 and 7 on the *Let's Investigate! Activity Card*.

Unit 5 Review

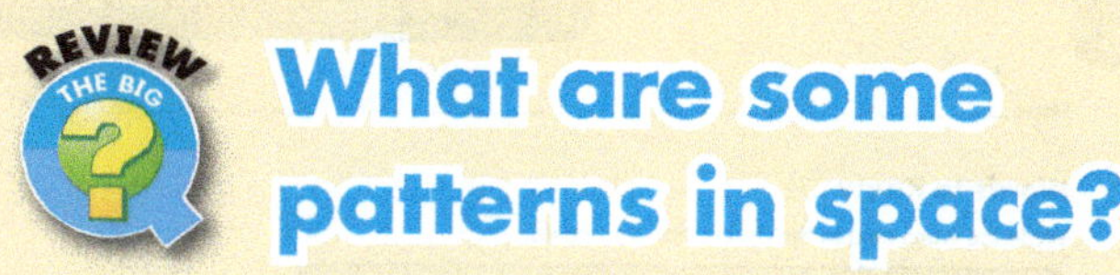

What are some patterns in space?

Digital Resources: Print out 1 of each per student: *Got it? Self Assessment*, *Got it? Quiz*

Evaluate

Strategies for Targeted Review

The following are strategies for providing targeted review for students if they encounter challenges with the content.

Lesson 1 How do star patterns change?

Question 1

If… students are having difficulty describing a star, a constellation, and Polaris, then… direct students to pages 53 and 54. Encourage students to look back at the pictures, describe them, and then find the definitions.

Lesson 2 What are the phases of the moon?

Question 2

If… students are having difficulty deciding whether the statements are true or false, then… direct students to pages 55, 56, and 57 and have them find the answers to the questions.

Lesson 3 What is the solar system?

Question 3

If… students are having difficulty remembering the names of the planets, then… direct students' attention to the *solar system* Flash Card and elicit the names of the eight planets. Then have them name the inner and outer planets.

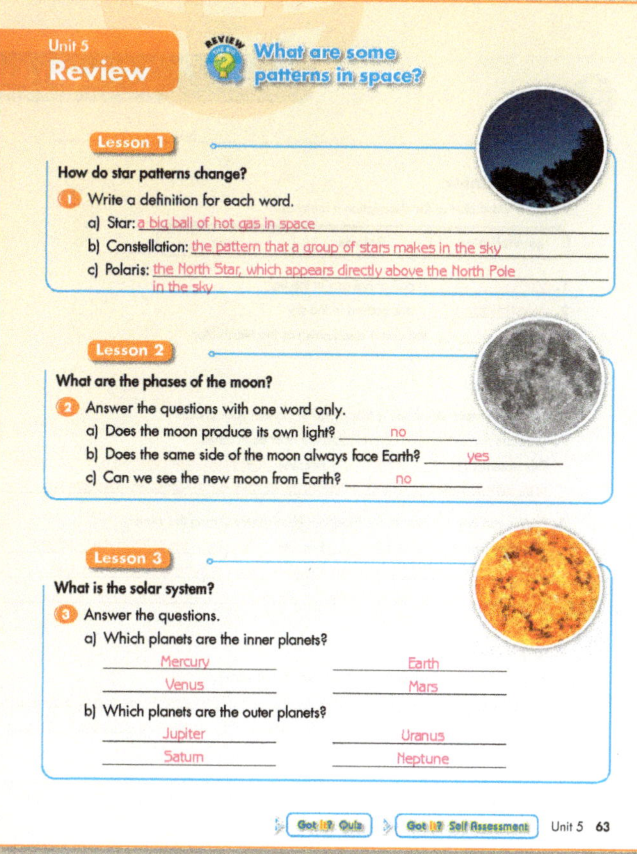

Unit 5 Review

What are some patterns in space?

Lesson 1

How do star patterns change?

1 Write a definition for each word.

a) Star: a big ball of hot gas in space

b) Constellation: the pattern that a group of stars makes in the sky

c) Polaris: the North Star, which appears directly above the North Pole in the sky

Lesson 2

What are the phases of the moon?

2 Answer the questions with one word only.

a) Does the moon produce its own light? no

b) Does the same side of the moon always face Earth? yes

c) Can we see the new moon from Earth? no

Lesson 3

What is the solar system?

3 Answer the questions.

a) Which planets are the inner planets?

Mercury Earth

Venus Mars

b) Which planets are the outer planets?

Jupiter Uranus

Saturn Neptune

Got it? Quiz Got it? Self Assessment Unit 5 63

ELL Language Support

Before students start working on the Review activities, have them read each question aloud along with you.

Got it? Self Assessment

Immediately after students have completed the Review activities, distribute a *Got it? Self Assessment* to each student. Have students complete the *Stop! Wait!* and *Go!* statements for each lesson, allowing them to look back through the lesson material if necessary.

Got it? Quiz

Distribute a Unit 5 *Got it? Quiz* to each student. Quizzes may be used for assessing students' understanding of unit concepts as well as for grading purposes.

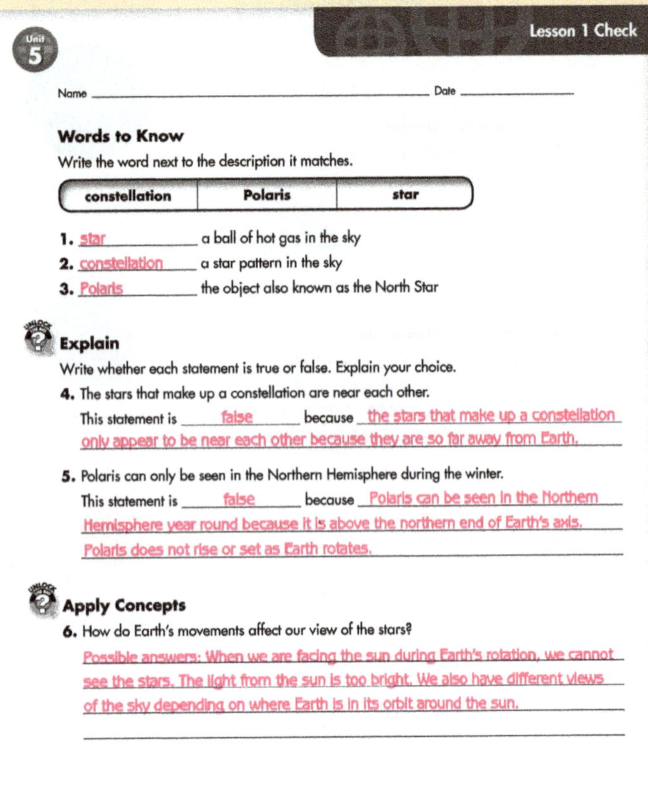

Unit 5

Lesson 1 Check

Name ____________ Date ____________

Words to Know

Write the word next to the description it matches.

constellation	Polaris	star

1. star ____ a ball of hot gas in the sky
2. constellation ____ a star pattern in the sky
3. Polaris ____ the object also known as the North Star

Explain

Write whether each statement is true or false. Explain your choice.

4. The stars that make up a constellation are near each other.

This statement is false because the stars that make up a constellation only appear to be near each other because they are so far away from Earth.

5. Polaris can only be seen in the Northern Hemisphere during the winter.

This statement is false because Polaris can be seen in the Northern Hemisphere year round because it is above the northern end of Earth's axis. Polaris does not rise or set as Earth rotates.

Apply Concepts

6. How do Earth's movements affect our view of the stars?

Possible answers: When we are facing the sun during Earth's rotation, we cannot see the stars. The light from the sun is too bright. We also have different views of the sky depending on where Earth is in its orbit around the sun.

Unit 5, *Lesson 1 Check* • How do star patterns change?

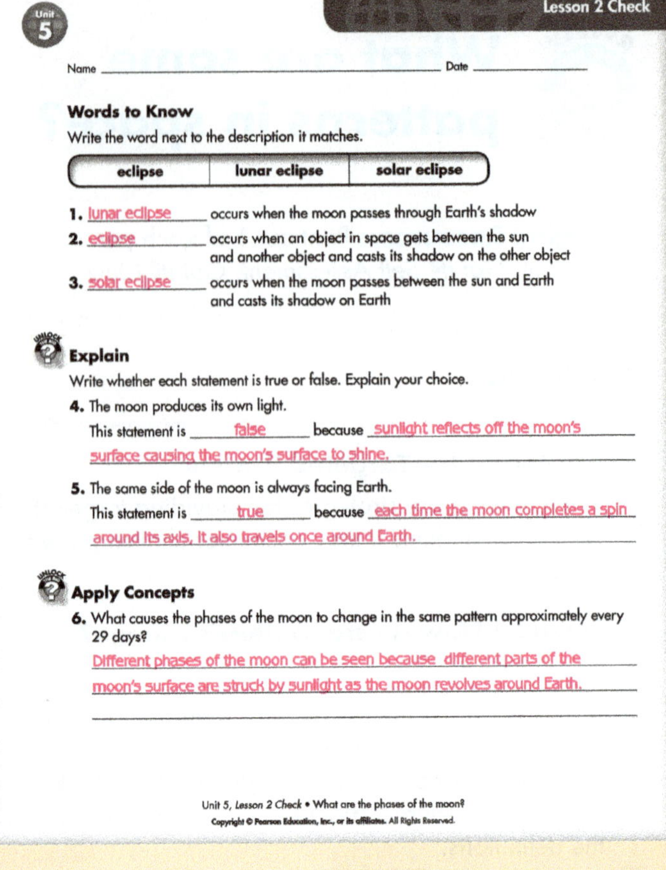

Unit 5

Lesson 2 Check

Name ____________ Date ____________

Words to Know

Write the word next to the description it matches.

eclipse	lunar eclipse	solar eclipse

1. lunar eclipse ____ occurs when the moon passes through Earth's shadow
2. eclipse ____ occurs when an object in space gets between the sun and another object and casts its shadow on the other object
3. solar eclipse ____ occurs when the moon passes between the sun and Earth and casts its shadow on Earth

Explain

Write whether each statement is true or false. Explain your choice.

4. The moon produces its own light.

This statement is false because sunlight reflects off the moon's surface causing the moon's surface to shine.

5. The same side of the moon is always facing Earth.

This statement is true because each time the moon completes a spin around its axis, it also travels once around Earth.

Apply Concepts

6. What causes the phases of the moon to change in the same pattern approximately every 29 days?

Different phases of the moon can be seen because different parts of the moon's surface are struck by sunlight as the moon revolves around Earth.

Unit 5, *Lesson 2 Check* • What are the phases of the moon?

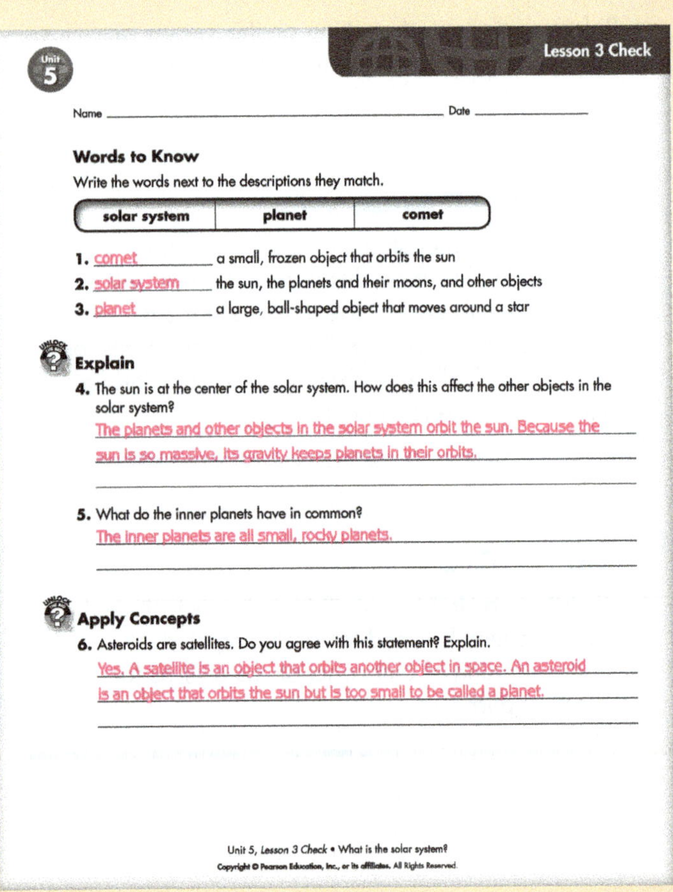

Unit 5

Lesson 3 Check

Name ____________ Date ____________

Words to Know

Write the words next to the descriptions they match.

solar system	planet	comet

1. comet ____ a small, frozen object that orbits the sun
2. solar system ____ the sun, the planets and their moons, and other objects
3. planet ____ a large, ball-shaped object that moves around a star

Explain

4. The sun is at the center of the solar system. How does this affect the other objects in the solar system?

The planets and other objects in the solar system orbit the sun. Because the sun is so massive, its gravity keeps planets in their orbits.

5. What do the inner planets have in common?

The inner planets are all small, rocky planets.

Apply Concepts

6. Asteroids are satellites. Do you agree with this statement? Explain.

Yes. A satellite is an object that orbits another object in space. An asteroid is an object that orbits the sun but is too small to be called a planet.

Unit 5, *Lesson 3 Check* • What is the solar system?

Unit 5

Lesson 1 Let's Explore! Activity Card

Name ____________ Date ____________

Materials

- Star Finder Pattern
- Star Wheel
- scissors
- stapler
- folder
- glue

What star patterns can you see?

1. Use the Star Finder Pattern to make a Star Finder.
2. Set the dial for 7 P.M. on November 1. Record the star groups you could observe at that time. Now set the dial for 11 P.M.

Record the star groups. Possible answer:

7 P.M. Pegasus, Cassiopeia, Big Dipper, Little Dipper, Pleiades

11 P.M. Pegasus, Cassiopeia, Big Dipper, Little Dipper, Pleiades, Orion

Explain Your Results

3. Which star group could you see at 11 P.M. but not at 7 P.M.? Explain why.

Orion. The visible part of the sky changed because Earth rotates on its axis.

Why do some stars look brighter than other stars?

Possible answer: They are closer to Earth.

Unit 5, Lesson 1 *Let's Explore!* Lab • How do star patterns change?

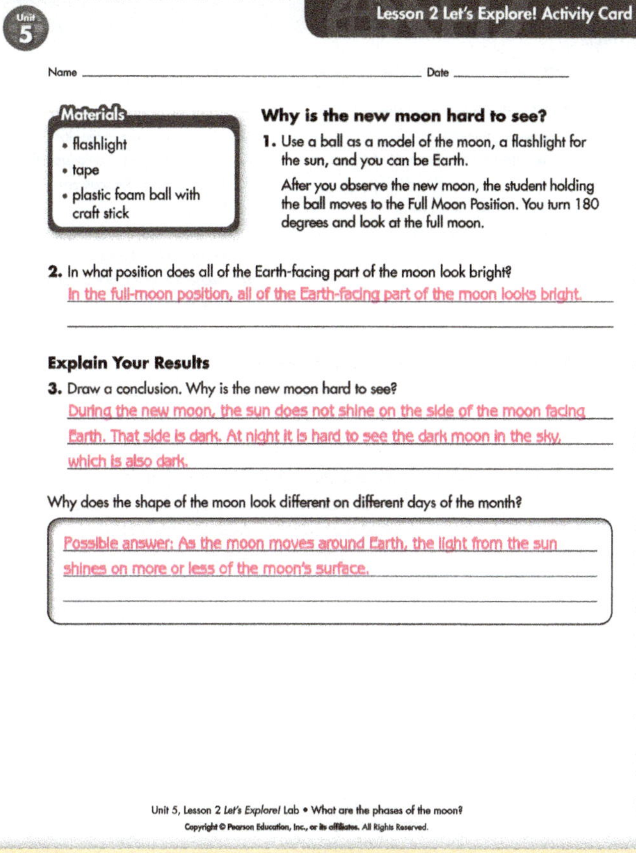

Unit 5

Lesson 2 Let's Explore! Activity Card

Name ______________________ Date ______________

Materials
- flashlight
- tape
- plastic foam ball with craft stick

Why is the new moon hard to see?

1. Use a ball as a model of the moon, a flashlight for the sun, and you can be Earth.

   After you observe the new moon, the student holding the ball moves to the Full Moon Position. You turn 180 degrees and look at the full moon.

2. In what position does all of the Earth-facing part of the moon look bright?

   In the full-moon position, all of the Earth-facing part of the moon looks bright.

Explain Your Results

3. Draw a conclusion. Why is the new moon hard to see?

   During the new moon, the sun does not shine on the side of the moon facing Earth. That side is dark. At night it is hard to see the dark moon in the sky, which is also dark.

Why does the shape of the moon look different on different days of the month?

Possible answer: As the moon moves around Earth, the light from the sun shines on more or less of the moon's surface.

Unit 5, Lesson 2 *Let's Explore!* Lab • What are the phases of the moon?

Unit 5

Lesson 3 Explore My Planet! Activity Card

Name ______________________ Date ______________

Fun Fact: Pluto

For a long time, people used simple sentences to remember the order of the planets in the solar system. "My very energetic mother just served us nine pizzas" is one example. Each word starts with the same letter as the name of a planet: Mercury, Venus, Earth, Mars, Jupiter, Saturn, Uranus, Neptune, and Pluto.

In 2006, however, the number of planets in the solar system changed from nine to eight. Scientists had observed that Pluto is much smaller than the other planets. Scientists decided that Pluto is actually a dwarf planet.

A dwarf planet is a small, round body that orbits a star. Pluto is no longer considered a planet like Earth or Saturn.

1. Write a sentence that will help you remember the order of the remaining eight planets.

   Answers will vary.

What is one way that Pluto is similar to Earth?

Possible answer: It has a moon.

Unit 5, Lesson 3 *Explore My Planet!* Activity • What is the solar system?

Unit 5

Let's Investigate! Activity Card

Name ______________________ Date ______________

Analyze and Conclude

6. How did the shape of the orbits change in your model?

   The farther apart the pins were, the more stretched out the shape.

7. Relate the pattern you observed in this activity to the observed orbits of the planets.

   Possible answer: The orbit I created was closer to the center of the paper in one area and farther from the center in other areas. This is the shape of an ellipse. The planets orbit the sun in an ellipse, too.

Unit 5, *Let's Investigate!* Lab • Earth and Space

Unit 5

Lessons 1–3 Got it? Self Assessment

Name ______________________ Date ______________

Got it? Self Assessment

Complete the statements for each lesson.

Lesson 1 How do star patterns change?

Stop! I need help with ______________________

Wait! I have a question about ______________________

Go! Now I know ______________________

Lesson 2 What are the phases of the moon?

Stop! I need help with ______________________

Wait! I have a question about ______________________

Go! Now I know ______________________

Lesson 3 What is the solar system?

Stop! I need help with ______________________

Wait! I have a question about ______________________

Go! Now I know ______________________

Unit 5, *Got it? Self Assessment* • Earth and Space

Unit 5

Name ______________________ Date ____________

Got it? Quiz

Circle the choice you think is correct for each multiple choice question.

1. What is a constellation?
   - (A) a pattern of stars
   - B a group of planets
   - C the remains of stars
   - D a star that does not rise or set

2. Which is the best description of the sun?
   - A a giant-sized star near Earth
   - (B) a medium-sized star near Earth
   - C a giant-sized star far from Earth
   - D a medium-sized star far from Earth

3. What phase of the moon is shown?
   - A new moon
   - B full moon
   - (C) first-quarter moon
   - D crescent moon

4. What event will occur when the sun, moon, and Earth are in the positions shown in the illustration?
   - A full moon
   - B new moon
   - (C) solar eclipse
   - D lunar eclipse

5. What keeps Earth revolving around the sun?
   - (A) the force of gravity
   - B the pull of the moon
   - C the tilt of Earth's axis
   - D the rotation of Earth

6. Which planet is NOT an outer planet?
   - A Saturn
   - (B) Mars
   - C Neptune
   - D Jupiter

Unit 5

Got it? Quiz

Name ______________________ Date ____________

7. What do the four inner planets have in common?
   - A They are mostly made of gases.
   - (B) They are small and rocky.
   - C They all have moons.
   - D They all have rings.

8. Why does the North Star seem to stay in the same place in the sky?
   - A because it is part of a constellation
   - B because it is in the southern hemisphere
   - (C) because it appears directly above the North Pole
   - D because it rotates at the same rate as Earth does

9. Why do stars appear to move across the sky? Explain.

Possible answers: Earth's rotation on its axis makes stars appear to move across the sky.

10. Explain the difference between a lunar eclipse and a solar eclipse.

Possible answers: During a lunar eclipse, Earth passes between the sun and the moon and casts its shadow on the moon. During a solar eclipse, the moon passes between the sun and Earth and casts its shadow on Earth.

Unit 5, *Got it? Quiz* • Earth and Space

Copyright © Pearson Education, Inc., or its affiliates. All Rights Reserved.

Teacher's Notes

Unit 5 Study Guide

What are some patterns in space?

Lesson 1
How do star patterns change?

- The apparent patterns of stars are called constellations.
- Polaris, the North Star, does not seem to move in the sky, and stars appear to revolve around it.

Lesson 2
What are the phases of the moon?

- The moon's phases depend on where the sun, moon, and Earth are.
- Each time the moon rotates on its axis, it revolves once around Earth.
- During an eclipse, an object in space casts its shadow on another.

Lesson 3
What is the solar system?

- The solar system includes the sun, planets, moons, and other objects.
- Inner planets are small and rocky.
- Outer planets are large and made of gas and liquids.

Review the Big Question

What are some patterns in space?

Have students use what they have learned from the unit to answer the question in their own words.

How has your answer to the Big Question changed since the beginning of the unit? What are some things you learned that caused your answer to change?

Make a Concept Map

Have students make a concept map like the one shown on this page to help them organize key concepts.

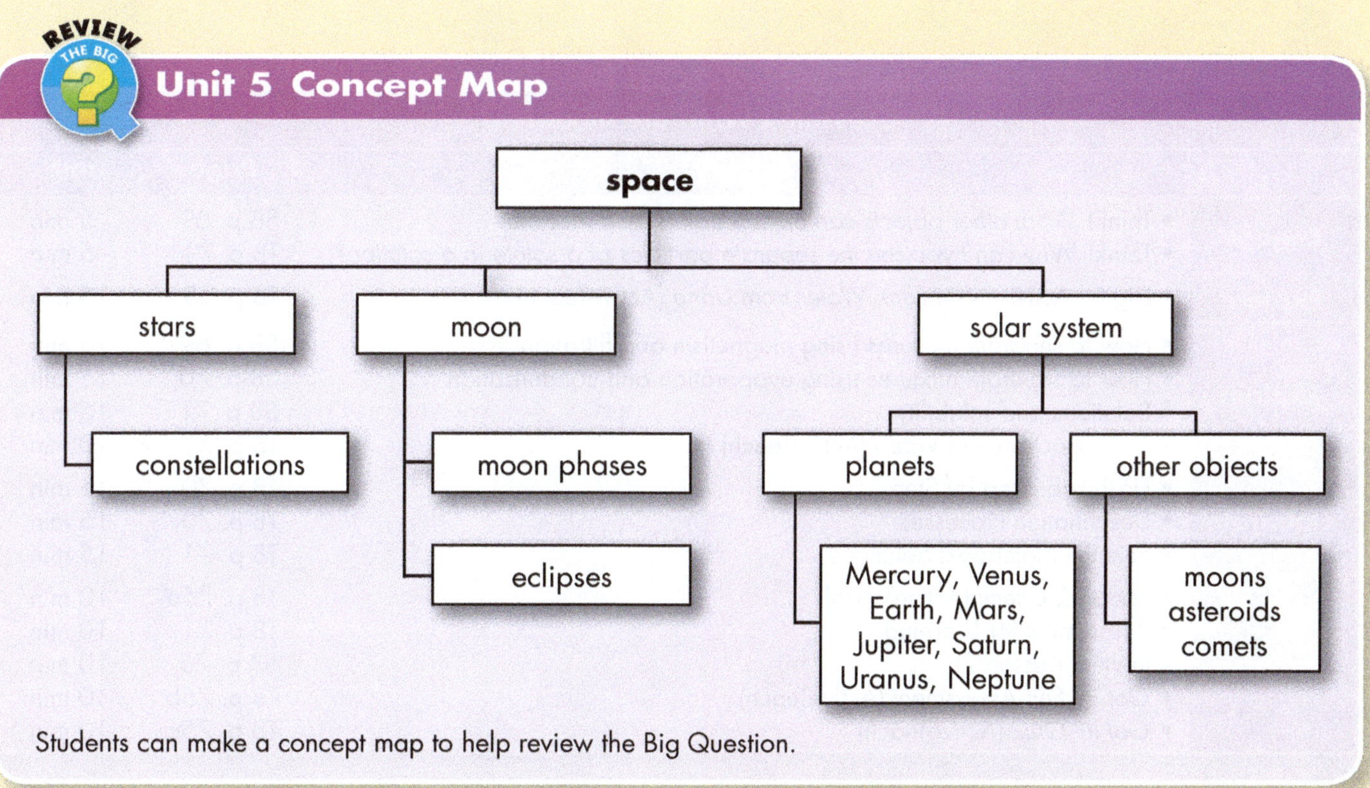

Students can make a concept map to help review the Big Question.

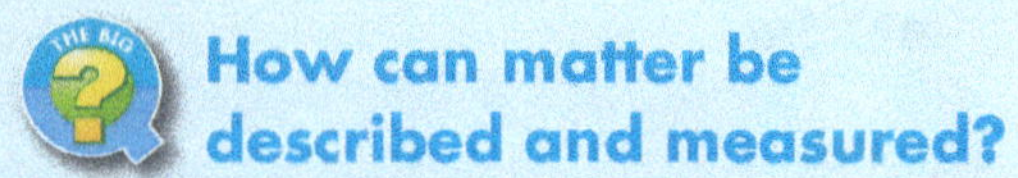

Lesson Plan

Unit Opener & Lesson 1 How is matter measured?

	Activity	Pages	Time
Engage	• Unit Opener: Think! *Can you unpop popped popcorn? Why or why not?*	SB p. 64	5 min
	• Unit Opener: Identify how water changes states.	SB p. 64	10 min
	• Unit Opener: Identify ways of measuring solids and liquids.	SB p. 64	10 min
	• Think! *A magazine is put through a shredder. Does it have more mass after it is shredded? Why?*	SB p. 65	5 min
	• Think! *What is the difference between volume and mass?*	SB p. 67	5 min
Explore	• Digital Lab: *How does dividing clay affect its mass?* (ActiveTeach)	TB p. 65	15 min
Explain	• Law of conservation of mass	SB p. 65	15 min
	• How to measure mass using metric units	SB p. 66	15 min
	• How to measure the volume of rectangular objects	SB p. 67	25 min
	• How to measure the volume of liquids and solids	SB p. 68	25 min
	• *Got it? 60-Second Video* (ActiveTeach)	TB p. 68	10 min
Elaborate	• Law of Conservation of Mass	TB p. 65	15 min
	• Science Notebook: Metric System Units	TB p. 66	15 min
	• At-Home Lab: Measure Up Matter	SB p. 66	10 min
	• Science Notebook: Measuring Volume	TB p. 67	20 min
	• Science Notebook: Cubic Inches	TB p. 67	15 min
	• Flash Lab: Splash!	SB p. 68	20 min
	• Story of Archimedes and the Golden Crown	TB p. 68	10 min
Evaluate	• *Lesson 1 Check* (ActiveTeach)	TB p. 75a	10 min
	• Assessment for Learning	TB p. 68	10 min
	• Review (Lesson 1)	SB p. 75	10 min
	• *Got it? Self Assessment* (ActiveTeach)	TB p. 75b	10 min
	• *Got it? Quiz* (ActiveTeach)	TB p. 75c	10 min

Lesson 2 What are mixtures?

	Activity	Pages	Time
Engage	• Think! *What other objects can be separated by a magnet?*	SB p. 69	5 min
	• Think! *Why can't you see the separate particles of a solute in a solution?*	TB p. 71	5 min
Explore	• Digital Activity: *Fun Fact: Water from Urine* (ActiveTeach)	TB p. 69	15 min
Explain	• How to separate mixtures using magnetism and filtration	SB p. 69	15 min
	• How to separate mixtures using evaporation and condensation	SB p. 70	15 min
	• Solutions and solubility	SB p. 71	15 min
	• *Got it? 60-Second Video* (ActiveTeach)	TB p. 71	10 min
Elaborate	• Flash Lab: Step by Step	TB p. 70	15 min
	• Desalination Processes	TB p. 70	15 min
	• Science Notebook: Solutions	TB p. 71	15 min
Evaluate	• *Lesson 2 Check* (ActiveTeach)	TB p. 75a	10 min
	• Assessment for Learning	TB p. 71	10 min
	• Review (Lesson 2)	SB p. 75	10 min
	• *Got it? Self Assessment* (ActiveTeach)	TB p. 75b	10 min
	• *Got it? Quiz* (ActiveTeach)	TB p. 75c	10 min

Lesson 3 How does matter change?			
	Activity	**Pages**	**Time**
Engage	• Think! *Does photosynthesis cause a physical or chemical change?*	TB p. 73	5 min
Explore	• Digital Lab: *How can you tell if a change has occurred?* (ActiveTeach)	TB p. 72	15 min
Explain	• Physical changes • Chemical changes • *Got it? 60-Second Video* (ActiveTeach)	SB p. 72 SB p. 73 TB p. 73	15 min 15 min 10 min
Elaborate	• Science Notebook: Chemical Changes • At-Home Lab: Shiny Pennies	TB p. 73 SB p. 73	15 min 15 min
Evaluate	• *Lesson 3 Check* (ActiveTeach) • Assessment for Learning • Review (Lesson 3) • *Got it? Self Assessment* (ActiveTeach) • *Got it? Quiz* (ActiveTeach)	TB p. 75a TB p. 73 SB p. 75 TB p. 75b TB p. 75c	10 min 10 min 10 min 10 min 10 min
Lab	• *Let's Investigate! Does steel wool rust faster in water or in vinegar?* (ActiveTeach)	SB p. 74	30 min

Flash Cards

mass

volume

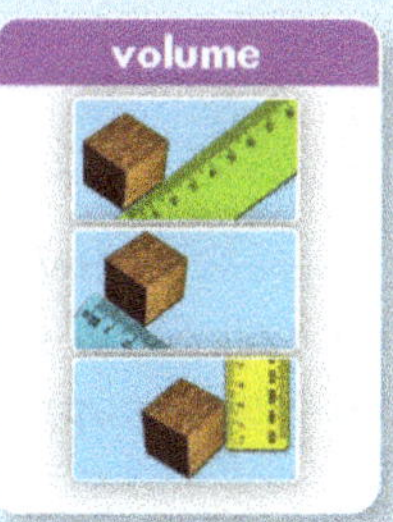

mixture

magnetism

filtration

evaporation

solution

physical change

chemical change

Lesson 1

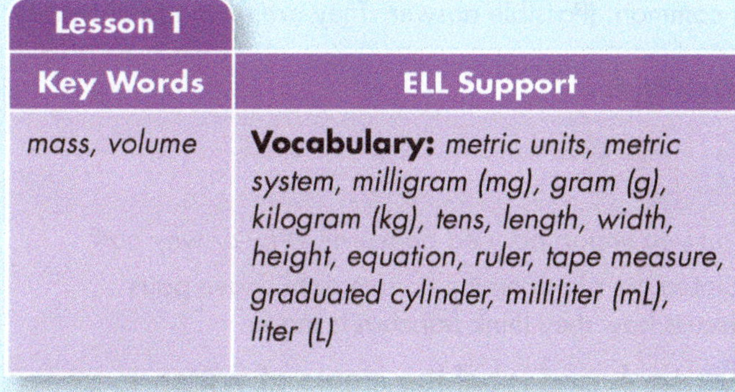

Key Words	ELL Support
mass, volume	**Vocabulary:** *metric units, metric system, milligram (mg), gram (g), kilogram (kg), tens, length, width, height, equation, ruler, tape measure, graduated cylinder, milliliter (mL), liter (L)*

Lesson 2

Key Words	ELL Support
mixture, filtration, evaporation, condensation, solution, solute, solvent, solubility	**Vocabulary:** *substances, properties, magnetism, nails, filter, solids, liquids, gas, vapor, stir, break down, dissolve, raise the temperature, crush (v), drop (v)*

Lesson 3

Key Words	ELL Support
physical change, chemical change	**Vocabulary:** *matter, size, state, substance, block of wood, properties, coals, ashes, origami sculpture, shape, particles, arrangement, iron nail, hardness, particles, trash, evidence, spoil, decay, rusting*

Unit 6 Matter

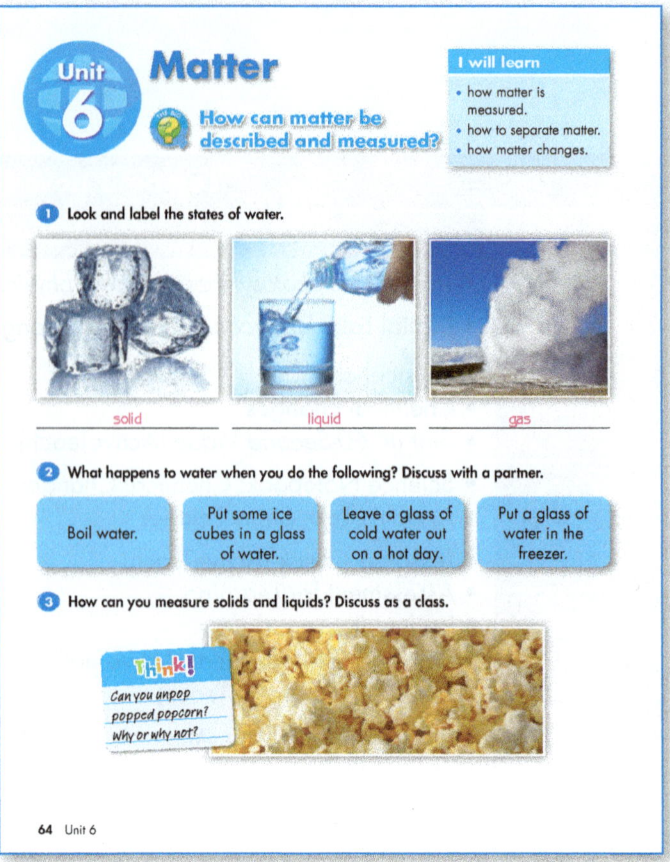
Unit 6 Matter

How can matter be described and measured?

I will learn
- how matter is measured.
- how to separate matter.
- how matter changes.

1 Look and label the states of water.

solid | liquid | gas

2 What happens to water when you do the following? Discuss with a partner.

Boil water. | Put some ice cubes in a glass of water. | Leave a glass of cold water out on a hot day. | Put a glass of water in the freezer.

3 How can you measure solids and liquids? Discuss as a class.

Think! Can you unpop popped popcorn? Why or why not?

64 Unit 6

Unit Objectives

Lesson 1: Students will demonstrate an understanding of how matter is measured.

Lesson 2: Students will explain how mixtures can be separated.

Lesson 3: Students will learn how matter changes into materials with different characteristics.

Vocabulary: *states of water, solid, liquid, gas, boil* (v), *ice cubes, freezer, unpop* (v), *pop* (v), *popcorn*

Materials: pictures of a popsicle, a river, and a cloud, some kernels of popcorn, some popped popcorn

Introduce the Big Question

How can matter be described and measured?

Build Background Display pictures of a popsicle, a river, and a cloud. Have students discuss what they have in common. (Possible answer: *They are all made of water.*)

Engage

Think!

Can you unpop popped popcorn? Why or why not?

Point to the photo on the bottom right. Have pairs discuss how they think popcorn is made.

1 Look and label the states of water.

Use the photos to elicit vocabulary. Guide students to say that they represent the three states of water. Have students label the three states.

2 What happens to water when you do the following? Discuss with a partner.

Have pairs read and discuss what happens to water in each situation. (Possible answers: *Water evaporates when it boils. Ice cubes melt when they are put in a glass of water. Cold water warms up when you leave it out on a hot day. Water freezes when you put it in the freezer.*)

3 How can you measure solids and liquids? Discuss as a class.

On the board, make a two-column chart with the headings *Solids* and *Liquids*. *Which objects in our classroom are solids, and which are liquids?* Write students' ideas in the corresponding columns. Discuss with the class how solids and liquids can be measured. (Possible answers: *using a balance, using a graduated cylinder*)

Think! Again!

How can you make popcorn? Kernels of popcorn pop when you put them in hot oil. Show students some kernels of popcorn and some pieces of popped popcorn. *So, if you put some kernels in hot oil, they pop and become popcorn?* Yes! *Can you unpop popped popcorn? Why or why not?* (Possible answer: *No, the pieces of popcorn cannot be kernels again because they have already popped.*)

ELL Content Support

Inside every popcorn kernel is a starchy substance and a small amount of water. The outer shell of the kernel is called the hull. The hull is nonporous. This helps to keep the water sealed inside of the kernel. When a popcorn kernel is heated, the water inside builds up pressure as it turns into steam. The steam gelatinizes the starch, changing its chemical composition. When too much pressure is trapped inside the kernel, the hull explodes and the gelatinized starch inside the kernel breaks out. The starch cools, forming the fluffy white substance known as popcorn.

Lesson 1

How is matter measured?

Objective: Learn about the relationship between matter and mass and about the law of conservation of mass.

Vocabulary: *matter, mass, space, make up, property, measure* (v), *measurement, location, remove, crushed, solid objects, liquids, air, breathe, pan balance, level* (adj), *equal*

Digital Resources: Flash Card (*mass*), *Let's Explore!* Digital Lab

Materials: picture of a pan balance, two bars of clay, pages of a magazine (one per student), picture of a paper shredder

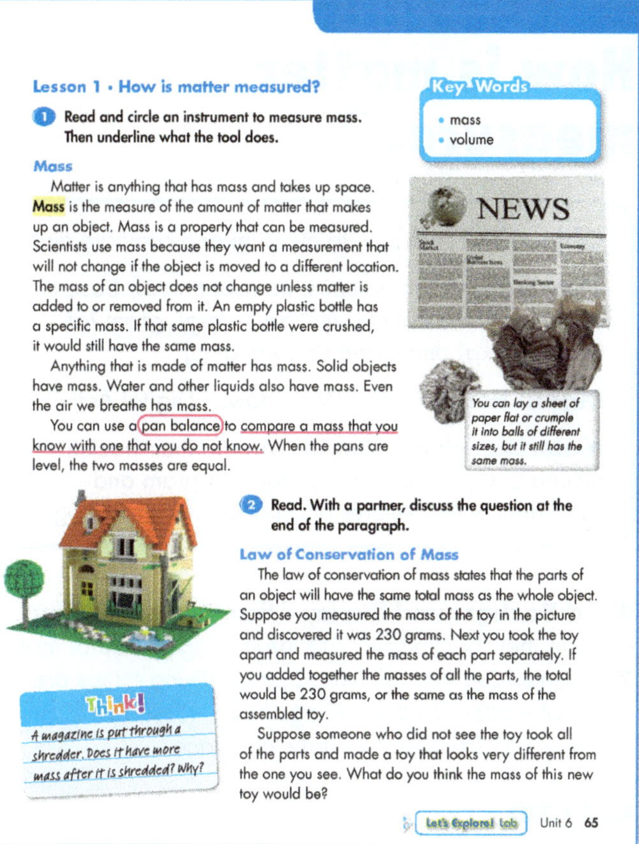

Lesson 1 • How is matter measured?

1 Read and circle an instrument to measure mass. Then underline what the tool does.

Key Words
- mass
- volume

Mass

Matter is anything that has mass and takes up space. **Mass** is the measure of the amount of matter that makes up an object. Mass is a property that can be measured. Scientists use mass because they want a measurement that will not change if the object is moved to a different location. The mass of an object does not change unless matter is added to or removed from it. An empty plastic bottle has a specific mass. If that same plastic bottle were crushed, it would still have the same mass.

Anything that is made of matter has mass. Solid objects have mass. Water and other liquids also have mass. Even the air we breathe has mass.

You can use a pan balance to compare a mass that you know with one that you do not know. When the pans are level, the two masses are equal.

You can lay a sheet of paper flat or crumple it into balls of different sizes, but it still has the same mass.

2 Read. With a partner, discuss the question at the end of the paragraph.

Law of Conservation of Mass

The law of conservation of mass states that the parts of an object will have the same total mass as the whole object. Suppose you measured the mass of the toy in the picture and discovered it was 230 grams. Next you took the toy apart and measured the mass of each part separately. If you added together the masses of all the parts, the total would be 230 grams, or the same as the mass of the assembled toy.

Suppose someone who did not see the toy took all of the parts and made a toy that looks very different from the one you see. What do you think the mass of this new toy would be?

Think!
A magazine is put through a shredder. Does it have more mass after it is shredded? Why?

Let's Explore! Lab Unit 6 65

Unlock the Big Question

Write the following on the board:
I will learn how matter can be measured.

Build Background Display a picture of a pan balance. Show students two bars of clay. *Do these bars of clay have the same weight? We don't know! How can you tell? You can use a pan balance. If you put a bar of clay on each pan, how can you know if their masses are equal?* Have students use the words *level* and *equal* in their answers.

Explore

Let's Explore! Lab How does dividing clay affect its mass?

Objective: Students will measure the mass of clay pieces to explore how the mass of an object compares to the mass of its pieces.

Digital Resources: *Let's Explore!* Digital Lab, *Let's Explore! Activity Card* (1 per student) (*Optional*: Do the lab in class; refer to the *Activity Card* for materials and steps.)

- Show the Digital Lab and have students complete the *Activity Card*. Have students check their answers in small groups or pairs. Provide support as needed.
- Guide students in a discussion about the conservation of mass.

Explain

1 Read and circle an instrument to measure mass. Then underline what the tool does.

Write the words *mass* and *matter* on the board. Read the first paragraph with students and elicit the definitions of the two words. Ask students to read and circle an instrument that is used to measure mass and then underline how it does so.

2 Read. With a partner, discuss the question at the end of the paragraph.

Ask students to look at the picture of the toy house and describe it. *How can you make a house like this? By putting together small plastic bricks.* Write on the board *Law of Conservation of Mass*. Have students read and underline what the law states. Discuss with the class what the law is about. Then have pairs read the text again and discuss the question at the end of the paragraph.

Elaborate

Law of Conservation of Mass

Direct students' attention to the *mass* Flash Card. *Is the mass of the three forms of paper the same or different? The same!* Give each student a page from a magazine. Ask them to give the paper the shape they want without tearing it apart. Display all the paper shapes on a table. *Has the mass of each sheet of paper stayed the same or has it changed? It's the same!*

Display the picture of a paper shredder and explain what it is for. *A magazine is put through a shredder. Does it have more mass after it is shredded? Why?* (Possible answer: *The magazine has the same mass before and after it is shredded. No matter has been removed, so it has the same mass.*)

Lesson 1

How is matter measured?

Objective: Learn how to measure and compare mass.

Vocabulary: *metric units, measure mass, base unit of mass, metric system, milligram (mg), gram (g), kilogram (kg), tens, prefixes, equal* (v)

Digital Resources: *I Will Know...* Digital Activity

Materials: objects students carry in their backpacks, picture of a pan balance, pictures of gram and kilogram cubes, pictures of a horse and a ladybug

Build Background Display different objects students carry in their backpacks. *Are these things made of matter? Yes! Remember that all living and nonliving things are made of matter. Do these things have mass? Yes! How can you tell? Because they take up space.*

Explain

3 Read and underline the metric units that are used to measure mass.

What is the mass of the toy house on page 65? 230 grams. What units can we use to measure the mass of objects? Ask students to read and underline the metric units that are used to measure mass.

ELL Content Support

Elicit the metric units and write them on the board. *What are these metric units based on? Tens!* Write the following equivalences for students to fill in:

1 gram = ________ milligrams.

1 kilogram = ________ grams.

4 Look at the pictures and answer the questions with a partner.

Display the picture of a pan balance. Show students two school objects. *How can you estimate which of these two objects has more mass? By putting the objects on opposite sides of a balance. Now imagine one side of the balance is lower than the other side. What can you conclude about the masses of the two objects? The object on the lower side of the balance has a greater mass than the other object. What can you do to measure the mass of one object?* Direct students' attention to the picture in the center of the page. *Each cube on this balance has a mass of 1 g. Let's count the cubes to measure the mass of the eraser. What is its mass? 22 grams!* Ask pairs to study the two pictures and answer the questions.

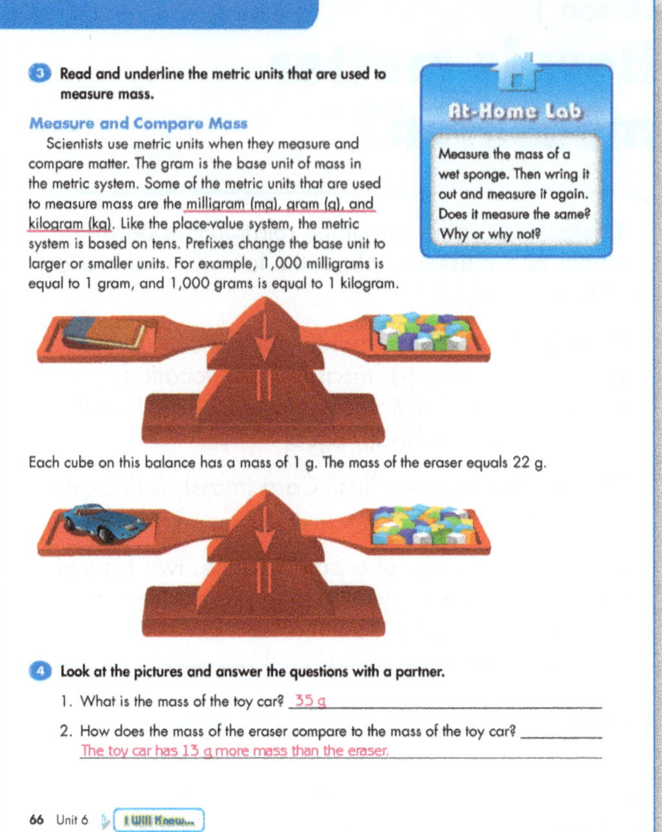

3 Read and underline the metric units that are used to measure mass.

Measure and Compare Mass

Scientists use metric units when they measure and compare matter. The gram is the base unit of mass in the metric system. Some of the metric units that are used to measure mass are the milligram (mg), gram (g), and kilogram (kg). Like the place-value system, the metric system is based on tens. Prefixes change the base unit to larger or smaller units. For example, 1,000 milligrams is equal to 1 gram, and 1,000 grams is equal to 1 kilogram.

At-Home Lab

Measure the mass of a wet sponge. Then wring it out and measure it again. Does it measure the same? Why or why not?

Each cube on this balance has a mass of 1 g. The mass of the eraser equals 22 g.

4 Look at the pictures and answer the questions with a partner.

1. What is the mass of the toy car? 35 g
2. How does the mass of the eraser compare to the mass of the toy car? The toy car has 13 g more mass than the eraser.

66 Unit 6 I Will Know...

Elaborate

Science Notebook: Metric System Units

Show students a set of gram and kilogram cubes. Have them draw and label both cubes in their Science Notebooks. Then have students draw and label pictures of three classroom objects that can be measured with a balance. Have them estimate the masses of the three objects and represent those masses with drawings of gram cubes.

At-Home Lab

Measure Up Matter

Materials: a sponge, bowl with water, a scale

Have students measure the mass of a wet sponge. Then tell them to wring the sponge out and measure it again. Guide students to conclude that the mass is different because the mass of the wet sponge included water.

I Will Know...

Have students do the *I Will Know...* Digital Activity.

Lesson 1

How is matter measured?

Objective: Learn to calculate the volume of rectangular solid objects.

Vocabulary: *mass, volume, property of matter, measure, space, take up, solid, rectangular, ruler, length, width, height, multiply, measurements, cube, equation*

Digital Resources: Flash Card (*mass*)

Materials: tools for measuring length (ruler, tape measure, meterstick), assorted square and rectangular boxes, a cereal box

Build Background Display the *mass* Flash Card and have students discuss what mass is. Remind students that mass is a property of matter that can be measured. *What tool can you use to measure the mass of an object? A balance!*

Explain

5 With a partner, calculate the volume of the objects in the table.

Display and have students name tools used for measuring length (*ruler, tape measure, meterstick*). *Look at these tools. What can you measure with them?* Show students a cereal box. Ask a volunteer to come to the front and measure the box's length, width, and height with a ruler. Write the measurements on the board. Write *volume* on the board and have students read and underline its definition. Then elicit the equation used to calculate volume and write it on the board. Draw students' attention to the three pictures of the cube and the equations. Finally, have students calculate the volume of the three objects in the chart. Allow students to compare their answers with a partner. Check answers as a class.

6 Will the volume of a wooden cube change if you cut the cube in half? Why or why not? Discuss as a class.

Read the question for the class to discuss the answer. (Possible answer: *No. The two pieces would still take up the same amount of space.*)

Elaborate

Science Notebook: Measuring Volume

Divide the class into small groups. Provide students with assorted square and rectangular boxes and rulers. Ask them to measure the boxes to find the volume of each. Have them draw the boxes and record the measurements and volumes in their Science Notebooks.

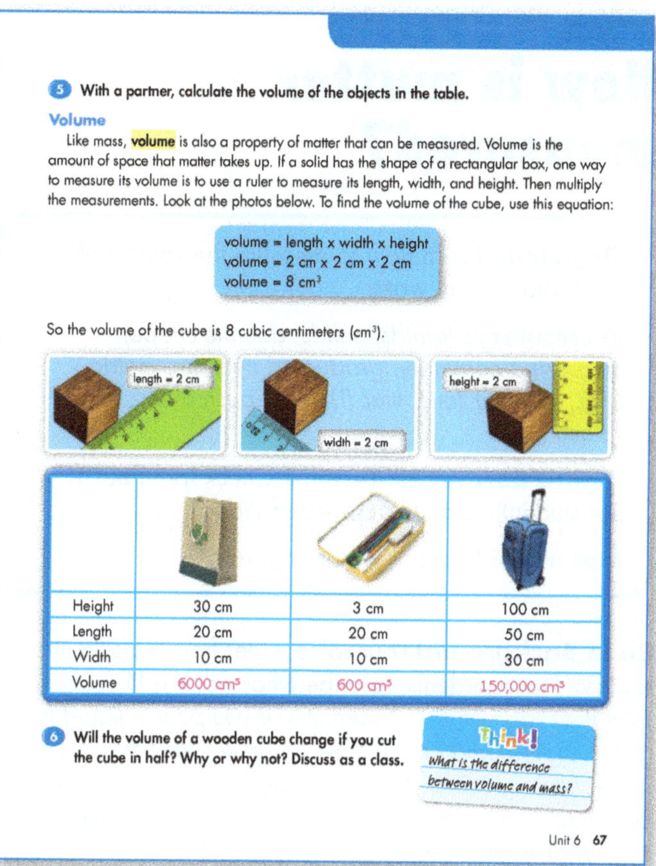

5 With a partner, calculate the volume of the objects in the table.

Volume

Like mass, **volume** is also a property of matter that can be measured. Volume is the amount of space that matter takes up. If a solid has the shape of a rectangular box, one way to measure its volume is to use a ruler to measure its length, width, and height. Then multiply the measurements. Look at the photos below. To find the volume of the cube, use this equation:

volume = length x width x height
volume = 2 cm x 2 cm x 2 cm
volume = 8 cm^3

So the volume of the cube is 8 cubic centimeters (cm^3).

Height	30 cm	3 cm	100 cm
Length	20 cm	20 cm	50 cm
Width	10 cm	10 cm	30 cm
Volume	6000 cm^3	600 cm^3	150,000 cm^3

6 Will the volume of a wooden cube change if you cut the cube in half? Why or why not? Discuss as a class.

Think!
What is the difference between volume and mass?

Unit 6 67

Science Notebook: Cubic Inches

Explain to students that the United States uses units including cubic inches (in^3) and cubic feet (ft^3) to measure the volume of a solid. *One way to find the volume of a solid is to use a ruler with inches to measure the length, width, and height of the solid.* Give pairs a box and ask them to measure it using a ruler with inches. Have them draw the box and record the box's measurements and volume using inches in their Science Notebooks.

ELL Content Support

U.S. System of Measurement

Scientists and federal agencies all over the world use the metric system. However, the United States is the only industrialized country where the metric system is not the official system of measurement. Ask *Should the United States convert to the metric system? Why or why not?* Invite students to discuss as a class.

Ask *What is the difference between volume and mass?* (*Volume measures the amount of space matter takes up. Mass measures the amount of matter in an object.*) Encourage students to discuss the question in small groups. Circulate among the groups to monitor discussion and to provide support as needed. Invite groups to share their ideas with the class.

Lesson 1

How is matter measured?

Objective: Learn how to measure the volume of solids that sink in water.

Vocabulary: *liquids, solids, definite shape, measuring container, graduated cylinder, metric units, milliliter (mL), liter (L), sink, push away, risen, splash* (n)

Digital Resources: *Lesson 1 Check* (print out 1 per student), *Got it? 60-Second Video*

Materials: bottle of water, a glass

Build Background Pour some water from a bottle into a glass. Have students describe what water is like. *Liquids take the shape of their containers. If you pour a liquid from one container into another, the liquid will take the shape of the new container. You can easily measure the volume of liquids. How can you measure the volume of liquids?* Accept all logical answers.

Explain

7 Read and underline the two metric system units used to measure volume.

Have students look at the pictures and say what the graduated cylinders are for. Students read and underline the two metric system units to measure volume.

8 Look at the pictures. Calculate the volume of the ball with a partner.

Read the captions of the two pictures out loud. *What are the two steps needed to measure the volume of a solid object using a graduated cylinder?* Have pairs calculate the volume of the ball.

Elaborate

Splash!

Materials: small objects of different weights, container with water

Divide the class into small groups. Distribute materials. Ask students to drop a few objects in the water to see which one makes the biggest splash and record their observations. Guide students to conclude that the heaviest object is the one that makes the biggest splash.

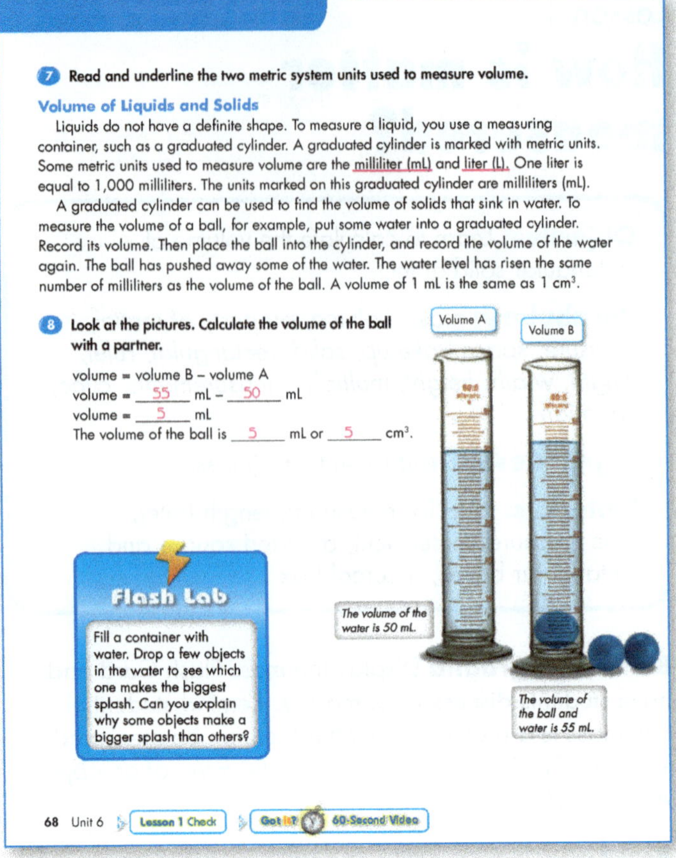

7 Read and underline the two metric system units used to measure volume.

Volume of Liquids and Solids

Liquids do not have a definite shape. To measure a liquid, you use a measuring container, such as a graduated cylinder. A graduated cylinder is marked with metric units. Some metric units used to measure volume are the milliliter (mL) and liter (L). One liter is equal to 1,000 milliliters. The units marked on this graduated cylinder are milliliters (mL).

A graduated cylinder can be used to find the volume of solids that sink in water. To measure the volume of a ball, for example, put some water into a graduated cylinder. Record its volume. Then place the ball into the cylinder, and record the volume of the water again. The ball has pushed away some of the water. The water level has risen the same number of milliliters as the volume of the ball. A volume of 1 mL is the same as 1 cm³.

8 Look at the pictures. Calculate the volume of the ball with a partner.

volume = volume B – volume A
volume = 55 mL – 50 mL
volume = 5 mL
The volume of the ball is 5 mL or 5 cm³.

Flash Lab
Fill a container with water. Drop a few objects in the water to see which one makes the biggest splash. Can you explain why some objects make a bigger splash than others?

68 Unit 6 Lesson 1 Check Got it? 60-Second Video

Story of Archimedes and the Golden Crown

Have students research in books or on the Internet what the story of Archimedes and the Golden Crown is about. Discuss in class the relationship between the story and the topic of this class.

Evaluate

Lesson 1 Check Assessment for Learning

Distribute the *Lesson 1 Check* and guide students as they complete it. Check answers as a class. Then ask students to grade their progress on the topic of measuring matter from 1 to 3: 3 = *I understand how matter is measured*; 2 = *I need to study more*; 1 = *I need help!* Encourage students giving themselves a 1 or 2 to describe what they found difficult and what they need to study more.

Got it?  **60-Second Video**

Review Key Words for Lesson 1 (see Student's Book page 65). Play the *Got it? 60-Second Video* to review the lesson material.

Lesson 2

What are mixtures?

Objective: Learn what mixtures are and how some mixtures can be separated using magnetism and filtration.

Vocabulary: *mixture, substances, chemically, sorted, piles, joined, properties, magnetism, magnet, attracted, nails, filtration, filter, solids, liquids, sand*

Digital Resources: Flash Card (*mixture*), *Explore My Planet!* Digital Activity

Materials: glass of water, pictures of an astronaut and a spacecraft, mixtures of desk supplies

Unlock the Big Question

Write the following on the board: *I will learn how to separate mixtures.*

Build Background Show students a glass of water and have them discuss what process water goes through to become drinkable.

Explore

Explore My Planet! Fun Fact: Water from Urine

Objective: Students will learn why astronauts need a system to get drinkable water.

Digital Resources: *Explore My Planet!* Digital Activity, *Explore My Planet! Activity Card* (1 per student)

- Display pictures of an astronaut and a spacecraft. Have students discuss what astronauts drink and where they get water.
- Show the *Explore My Planet!*
- Ask *Where does the water astronauts drink come from? From the astronauts themselves.*
- Have pairs complete the *Activity Card*. Provide support as needed. Ask volunteers to read their answers to the class.

Explain

1 Read and circle *T* (true) or *F* (false). With a partner, correct the false statements.

Display the *mixture* Flash Card. *A salad is a mixture of different ingredients. A mixture is made up of substances that are combined but can be separated.* Ask students to read and decide whether the sentences are true or false. Before checking the activity with the class, have pairs compare their answers.

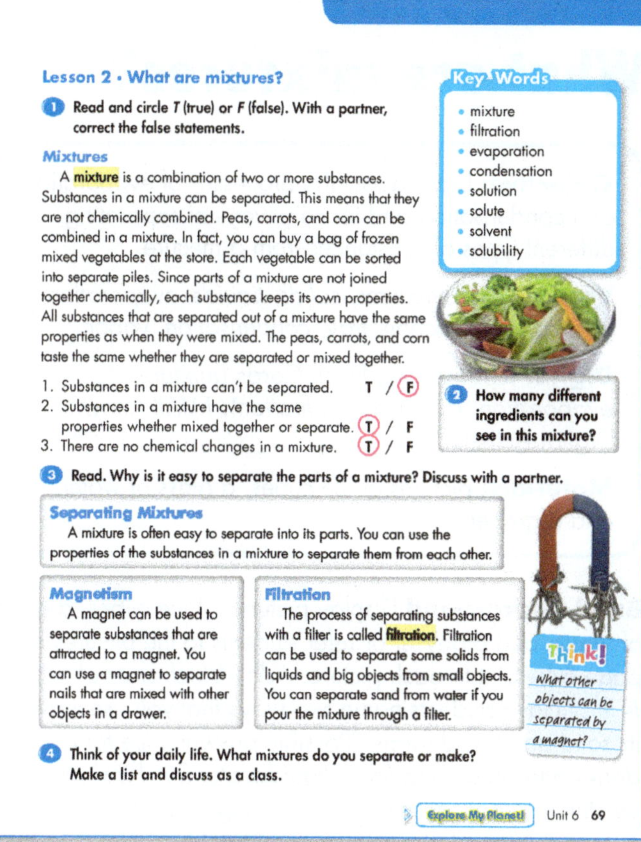

Lesson 2 • What are mixtures?

1 Read and circle *T* (true) or *F* (false). With a partner, correct the false statements.

Mixtures

A **mixture** is a combination of two or more substances. Substances in a mixture can be separated. This means that they are not chemically combined. Peas, carrots, and corn can be combined in a mixture. In fact, you can buy a bag of frozen mixed vegetables at the store. Each vegetable can be sorted into separate piles. Since parts of a mixture are not joined together chemically, each substance keeps its own properties. All substances that are separated out of a mixture have the same properties as when they were mixed. The peas, carrots, and corn taste the same whether they are separated or mixed together.

1. Substances in a mixture can't be separated. T / F
2. Substances in a mixture have the same properties whether mixed together or separate. T / F
3. There are no chemical changes in a mixture. T / F

Key Words
- mixture
- filtration
- evaporation
- condensation
- solution
- solute
- solvent
- solubility

2 How many different ingredients can you see in this mixture?

3 Read. Why is it easy to separate the parts of a mixture? Discuss with a partner.

Separating Mixtures

A mixture is often easy to separate into its parts. You can use the properties of the substances in a mixture to separate them from each other.

Magnetism

A magnet can be used to separate substances that are attracted to a magnet. You can use a magnet to separate nails that are mixed with other objects in a drawer.

Filtration

The process of separating substances with a filter is called **filtration**. Filtration can be used to separate some solids from liquids and big objects from small objects. You can separate sand from water if you pour the mixture through a filter.

Think! What other objects can be separated by a magnet?

4 Think of your daily life. What mixtures do you separate or make? Make a list and discuss as a class.

Explore My Planet! Unit 6 69

2 How many different ingredients can you see in this mixture?

Have pairs look at the picture of the salad and list the different ingredients it has. (Possible answers: *Six: lettuce, radish, peppers, tomato, broccoli, and cucumber.*) Ask students what they do when they do not like any of the ingredients.

3 Read. Why is it easy to separate the parts of a mixture? Discuss with a partner.

Have students read the three texts to find the answer to the question. Then have pairs discuss why it is easy to separate the parts of a mixture.

4 Think of your daily life. What mixtures do you separate or make? Make a list and discuss as a class.

Point to the *mixture* Flash Card. Have students list the mixtures they separate or make on a daily basis. (Possible answers: *Most kinds of food are mixtures that you make, like pizza, sandwiches, milk with cereal, and lemonade. You can separate laundry, recycling, and your favorite candy from mixed candy.*)

Think!

Ask *What other objects can be separated by a magnet?* (Possible answers: *Objects made of steel or iron: staples, paper clips, screws, pins, safety pins, needles, keys, spoons, forks, knives*) Guide students to conclude that magnets do not attract all metals.

Lesson 2

What are mixtures?

Objective: Learn about the processes of evaporation and condensation and identify ways of separating different types of substances from a mixture.

Vocabulary: *mixtures, solids, liquids, evaporation, gas, salt, salty ocean, tray, condensation, vapor*

Digital Resources: Flash Cards (*mixture, evaporation, magnetism, filtration*), *I Will Know...* Digital Activity

Materials: pictures of ice cubes, a glass of water, and a geyser

Build Background Display pictures of ice cubes, a glass of water, and a geyser or refer students to the pictures on page 64. Point to each picture. Ask: *Which state of matter is this?* Have students name two ways that you use water in its solid and liquid forms. (Possible answers: *ice for cooling drinks and preserving food; liquid water for drinking and washing*) *What causes water to change from one state to another?* Allow students to respond freely.

Explain

5 Read. Use your own words to express the processes of evaporation and condensation. Work in small groups.

Write on the board the words *evaporation* and *condensation*. Ask students to read the two texts to find out what they mean. Check answers as a class. Divide the class into small groups. Have students explain to one another what the two processes are. Invite students to use drawings if necessary.

6 Look and complete the statements.

Display the *mixture* Flash Card. Ask *How easy is it to separate the ingredients of this mixture?* Very easy! *What would you do to "unmix" this salad?* (*You can pick out the different ingredients by hand.*) Elicit other properties or methods that can be used to separate mixtures and write them on the board. (Possible answers: *magnetism, filtration, evaporation,* and *condensation*)

Direct students' attention to the picture of the mixture in the center of the page. Have pairs read and complete the sentences using the words written on the board. As a round up, display the *evaporation, magnetism,* and *filtration* Flash Cards to elicit situations when each technique might be useful.

I Will Know...

Have students do the *I Will Know...* Digital Activity.

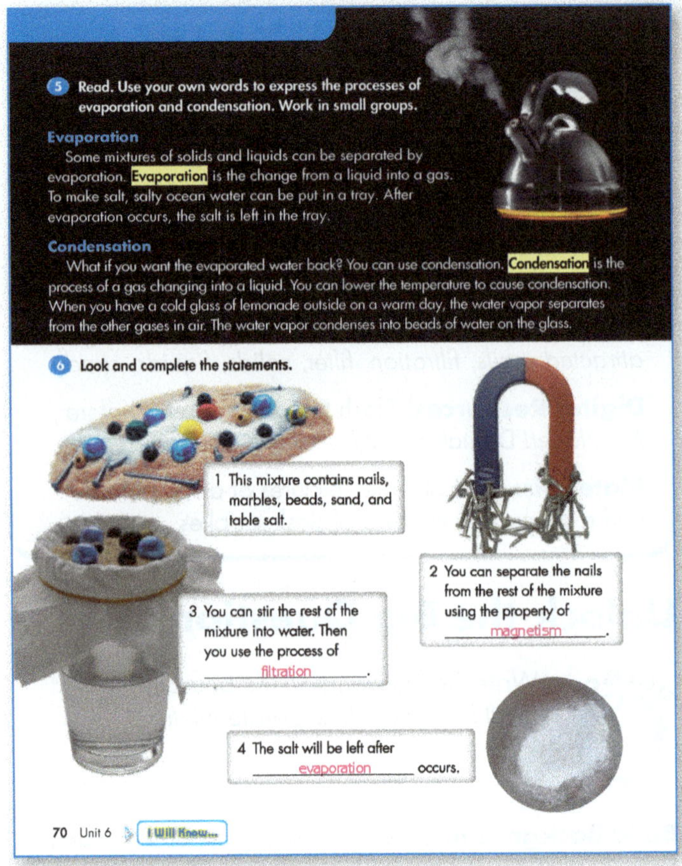

5 Read. Use your own words to express the processes of evaporation and condensation. Work in small groups.

Evaporation

Some mixtures of solids and liquids can be separated by evaporation. **Evaporation** is the change from a liquid into a gas. To make salt, salty ocean water can be put in a tray. After evaporation occurs, the salt is left in the tray.

Condensation

What if you want the evaporated water back? You can use condensation. **Condensation** is the process of a gas changing into a liquid. You can lower the temperature to cause condensation. When you have a cold glass of lemonade outside on a warm day, the water vapor separates from the other gases in air. The water vapor condenses into beads of water on the glass.

6 Look and complete the statements.

70 Unit 6 I Will Know...

Elaborate

Flash Lab

Step by Step

Materials: paper and pencil

Encourage students to identify the method for separating a mixture of paper clips, wood chips, gravel, and sugar into its components before they write the instructions.

Help students organize their instructions in numbered steps. For example, the first step might describe how to use magnetism to separate paper clips from the mixture, and the second step might describe how to use filtration to separate the wood chips. Students should show an understanding of the properties of the substances as well as the methods used in separating mixtures.

Desalination Processes

Ask *Why do you think scientists would want to develop new techniques for removing salt from seawater?* (Possible answer: *There is not enough clean water for drinking, irrigation, and other uses.*) *The process of removing salt from seawater is called desalination.* Have students work with a partner to research desalination processes. Have partners choose one technique that they think might be the most efficient. Have them present their findings to the class.

Lesson 2

What are mixtures?

Objective: Learn about solutions and solubility.

Vocabulary: *mixture, broken down, dissolve, solution, solvent, solute, solubility, crush* (v), *crystals*

Digital Resources: Flash Card (*solution*), *Lesson 2 Check* (print out 1 per student), *Got it? 60-Second Video*

Build Background Display the *solution* Flash Card. Elicit the ingredients that students think were used to make that solution. Have students use their own words to explain what a solution is.

Explain

7 Read. Which substance is the solvent and which one is the solute? Label the pictures.

Ask students to read both paragraphs to find out what a solution is. Have them identify the things labeled in the picture at the top of the page. (Possible answers: *salt/sugar/baking soda* and *water*) Then have pairs read the second paragraph again to find out which substance is the solvent and which one is the solute.

8 Read and underline two ways to make the solute dissolve more quickly.

Read the first paragraph with the class. Have students discuss why the solubility of sand in water is very low. Continue reading the remaining paragraphs for students to underline two ways to make a solute dissolve more quickly.

9 How can you make chocolate powder dissolve more quickly?

Direct students' attention to the picture on the bottom left and have them say what the boy is doing. Discuss with the class how the boy can make the chocolate powder dissolve more quickly.

Elaborate

Science Notebook: Solutions

Have students make a chart titled *Solutions* with two columns labeled *Solute* and *Solvent* in their Science Notebooks. Have them identify four solutions they have used and their solutes and solvents.

7 Read. Which substance is the solvent and which one is the solute? Label the pictures.

Solutions

If you stir salt and water together, you make a mixture. You cannot see the salt in this mixture because it has broken down into very small particles. The salt has dissolved in the water. The salt and the water form a special kind of mixture called a **solution**.

solute

solvent

In a solution, one or more substances are dissolved in another substance. The most common kind of solution is a solid dissolved in a liquid, such as salt in water. In a solution, the substance that is dissolved is the **solute**. A **solvent** is the substance that takes in, or dissolves, the other substance. Usually there is more solvent than solute. In the ocean, salt and other minerals are dissolved in water. Ocean water is a solution. But a solution does not have to be a liquid. The air you breathe, for example, is a solution made up of gases.

8 Read and underline two ways to make the solute dissolve more quickly.

9 How can you make chocolate powder dissolve more quickly?

Solubility

The ability of one substance to dissolve in another is called its solubility. **Solubility** is a measure of the amount of a substance that will dissolve in another substance. The solubility of materials can be high or low. The solubility of sand in water is very low.

Sometimes raising the temperature of the solvent can speed up the process of dissolving the solute. This is true for most solids. For example, you can dissolve sugar more quickly in warm water than you can in cold water.

Another way to make a solute dissolve more quickly is to crush it. If you drop a sugar cube into a cup of water, it may take a while to dissolve. If you crush the sugar cube into tiny crystals, the crystals will dissolve very quickly. More of the sugar particles are touching the water when the sugar is in tiny crystals than when it is in a sugar cube.

Unit 6 71

Think!

Direct students' attention to the *solution* Flash Card. Ask *Why can't you see the separate particles of a solute in a solution?* Invite them to discuss the question as a class. (*Because the solute has broken into very small particles.*)

Evaluate

Lesson 2 Check **Assessment for Learning**

Distribute the *Lesson 2 Check* and guide students as they complete it. Check answers as a class. Then ask students to grade their progress on the topic of mixtures and solutions from 1 to 3: 3 = *I can describe what mixtures and solutions are*; 2 = *I need to study more*; 1 = *I need help!* Encourage students giving themselves a 1 or 2 to describe what they found difficult and what they need to study more.

60-Second Video

Review Key Words for Lesson 2 (see Student's Book page 69). Play the *Got it? 60-Second Video* to review the lesson material.

Lesson 3

How does matter change?

Objective: Learn characteristics of physical changes.

Vocabulary: *matter, size, state, substance, carve, block of wood, properties, burn, coals, ashes, physical change, origami sculpture, shape, particles, arrangement*

Digital Resources: Flash Card (*solution*), *Let's Explore!* Digital Lab, *I Will Know...* Digital Activity

Materials: baking soda, white vinegar, picture of a block of wood, various classroom materials

Unlock the Big Question

Write the following on the board: *I will learn how some materials can change and become materials with different characteristics.*

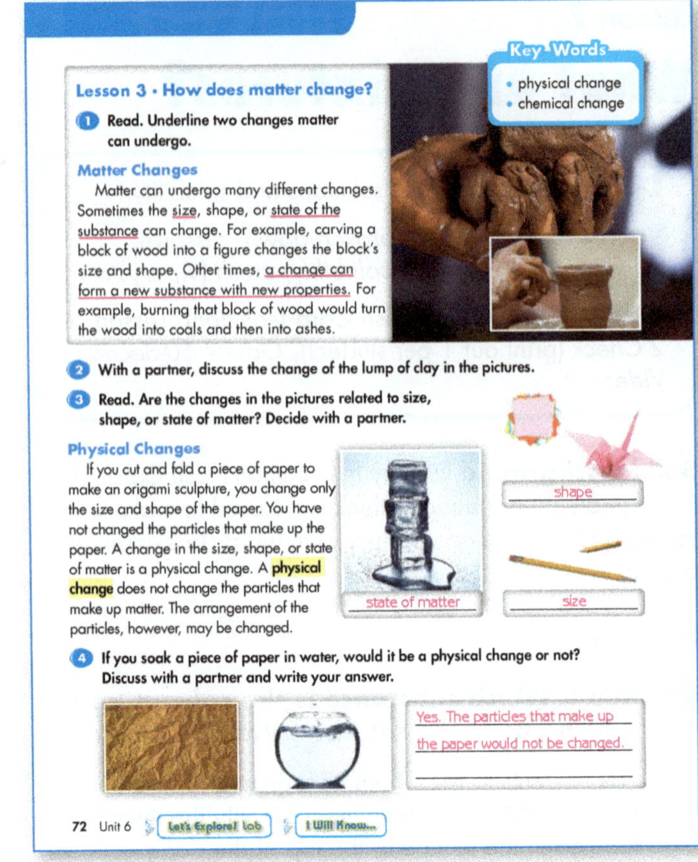

Lesson 3 • How does matter change?

1 Read. Underline two changes matter can undergo.

Key Words
- physical change
- chemical change

Matter Changes

Matter can undergo many different changes. Sometimes the size, shape, or state of the substance can change. For example, carving a block of wood into a figure changes the block's size and shape. Other times, a change can form a new substance with new properties. For example, burning that block of wood would turn the wood into coals and then into ashes.

2 With a partner, discuss the change of the lump of clay in the pictures.

3 Read. Are the changes in the pictures related to size, shape, or state of matter? Decide with a partner.

Physical Changes

If you cut and fold a piece of paper to make an origami sculpture, you change only the size and shape of the paper. You have not changed the particles that make up the paper. A change in the size, shape, or state of matter is a physical change. A **physical change** does not change the particles that make up matter. The arrangement of the particles, however, may be changed.

4 If you soak a piece of paper in water, would it be a physical change or not? Discuss with a partner and write your answer.

Yes. The particles that make up the paper would not be changed.

72 Unit 6 Let's Explore! Lab I Will Know...

Build Background Display the *solution* Flash Card. Discuss with the class what the solvent and the solute were like before becoming a solution and the characteristics of the new solution.

Explore

Let's Explore! Lab How can you tell if a change has occurred?

Objective: Students will observe and describe the changes that occur when baking soda and vinegar are combined.

Digital Resources: *Let's Explore!* Digital Lab, *Let's Explore! Activity Card* (1 per student) (*Optional*: Do the lab in class; refer to the *Activity Card* for materials and steps.)

- Display some baking soda and a glass with vinegar. Allow students to observe both substances and describe them.
- Show the Digital Lab. Guide students in a discussion about what type of change occurs when baking soda and vinegar are combined.
- Have students form pairs and complete the *Activity Card*. Provide support as needed. Review answers as a class.

Explain

1 Read. Underline two changes matter can undergo.

Display a picture of a block of wood. Discuss with the class different things they can do with the block of wood. Read the first paragraph with the class. Ask students to underline two changes matter can undergo.

2 With a partner, discuss the change of the lump of clay in the pictures.

Have pairs describe the two pictures at the top of the page. Then have them discuss how a lump of clay may become a vase.

3 Read. Are the changes in the pictures related to size, shape, or state of matter? Decide with a partner.

Ask three volunteers to describe how the matter changed in each picture. Then have pairs read and decide what kinds of change the objects in the pictures underwent.

4 If you soak a piece of paper in water, would it be a physical change or not? Discuss with a partner and write your answer.

Have students describe in their own words what *physical change* means. *What properties of matter are affected by physical changes?* (*Size, shape, or state of matter are affected by physical changes.*) Ask pairs to discuss the question. Check answers as a class.

I Will Know...

Have students do the *I Will Know...* Digital Activity.

Lesson 3

How does matter change?

Objective: Learn characteristics of chemical changes.

Vocabulary: *iron nail, damp, rust, properties, hardness, substance, chemical change, matter, particles, trash, evidence, spoil, decay, rusting, burning, give off energy*

Digital Resources: Flash Cards (*solution, physical change, chemical change*), *Lesson 3 Check* (print out 1 per student), *Got it? 60-Second Video*

Materials: picture of a glass of water with ice cubes, a penny

Build Background Display the *solution* Flash Card and a picture of a glass of water with an ice cube. Discuss with the class the differences between the types of changes occurring in the pictures. Accept all logical answers.

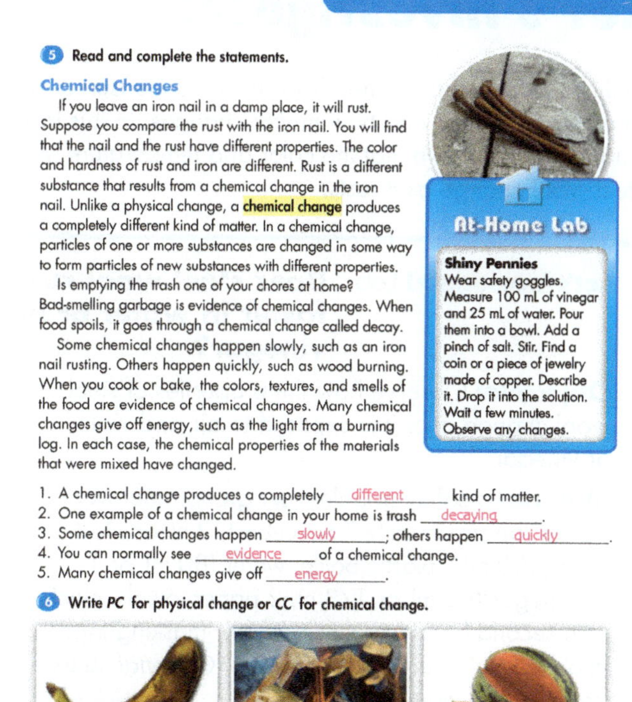

5 Read and complete the statements.

Chemical Changes

If you leave an iron nail in a damp place, it will rust. Suppose you compare the rust with the iron nail. You will find that the nail and the rust have different properties. The color and hardness of rust and iron are different. Rust is a different substance that results from a chemical change in the iron nail. Unlike a physical change, a **chemical change** produces a completely different kind of matter. In a chemical change, particles of one or more substances are changed in some way to form particles of new substances with different properties.

Is emptying the trash one of your chores at home? Bad-smelling garbage is evidence of chemical changes. When food spoils, it goes through a chemical change called decay.

Some chemical changes happen slowly, such as an iron nail rusting. Others happen quickly, such as wood burning. When you cook or bake, the colors, textures, and smells of the food are evidence of chemical changes. Many chemical changes give off energy, such as the light from a burning log. In each case, the chemical properties of the materials that were mixed have changed.

At-Home Lab

Shiny Pennies
Wear safety goggles. Measure 100 mL of vinegar and 25 mL of water. Pour them into a bowl. Add a pinch of salt. Stir. Find a coin or a piece of jewelry made of copper. Describe it. Drop it into the solution. Wait a few minutes. Observe any changes.

1. A chemical change produces a completely different kind of matter.
2. One example of a chemical change in your home is trash decaying.
3. Some chemical changes happen slowly; others happen quickly.
4. You can normally see evidence of a chemical change.
5. Many chemical changes give off energy.

6 Write PC for physical change or CC for chemical change.

CC CC PC

Lesson 3 Check Got it? 60-Second Video Unit 6 73

Explain

5 Read and complete the statements.

Explain to students that, in addition to physical changes, matter can undergo chemical changes. Read the first paragraph with the class and ask students to underline what a chemical change is. Have pairs read the whole text and complete the sentences. Check answers as a class.

6 Write *PC* for physical change or *CC* for chemical change.

Elicit the difference between physical and chemical changes. On the board, write *A chemical change produces a completely different kind of matter with different chemical properties. A physical change does not change the particles that make up matter.* Elicit examples of both types of changes.

Have pairs decide whether the objects in the pictures represent physical or chemical changes.

Elaborate

Science Notebook: Chemical Changes

Have students list all the chemical changes mentioned in the text and write them on the board. (Answers: *iron nails rusting, trash decaying, wood burning, cooking, baking*) Ask students to choose three chemical changes and illustrate what they think the matter looked like before and after the chemical changes.

At-Home Lab

Shiny Pennies

Materials: safety goggles, vinegar, salt, graduated cylinder, oxidized pennies or piece of copper jewelry, bowl or other container

Students should note that the penny or the piece of jewelry starts out brown but then turns copper colored and shiny.

Evaluate

Lesson 3 Check Assessment for Learning

Distribute the *Lesson 3 Check* and guide students as they complete it. Check answers as a class. Then ask students to grade their progress on the topic of how matter changes from 1 to 3: 3 = *I understand more about physical and chemical changes;* 2 = *I need to study more;* 1 = *I need help!* Encourage students giving themselves a 1 or 2 to describe what they found difficult and what they need to study more.

Got it? 60-Second Video

Review Key Words for Lesson 3 (see Student's Book page 72). Play the *Got it? 60-Second Video* to review the lesson material.

Let's Investigate!

In this unit, students learned how matter is measured, how to separate matter, and how matter changes. In this lab, students will observe that the color and temperature of steel wool change as it rusts.

Let's Investigate! Lab **Does steel wool rust faster in water or in vinegar?**

Objective: Observe how the properties of steel wool change after it is exposed to air and water or vinegar.

Materials: safety goggles, per group: steel wool (2 pieces, approximately 3 cm cubes), 2 clear plastic cups (300 mL), plastic bottle, water (80 mL or 1/3 c), vinegar (80 mL or 1/3 c), 2 paper towels, clock with second hand (whole class use), masking tape (teacher use), marker (teacher use) (*Optional*: latex-free gloves, 1 pair per student)

Digital Resources: *Let's Investigate!* Digital Lab, *Let's Investigate! Activity Card* (1 per group)

Advance Preparation: For each group, cut 2 pieces of steel wool (approximately 3 cm cubes). The shape does not matter. Do not use scouring pads that contain detergents.

- Divide students into small groups and distribute materials.
- Have students observe the properties of the steel wool and record them.
- Ask students to fill one cup 1/3 full with water and the other cup 1/3 full with vinegar. Then have students put a piece of steel wool in each cup and wait 1 minute.
- Show students how to take the steel wool out of each cup and squeeze out any extra liquid. Have students place each piece of wool on a paper towel in front of its cup and wait 5 minutes.
- Students will observe that the water-soaked steel wool does not rust noticeably in 5 minutes. The vinegar soaked steel wool starts changing to yellow-orange and becomes warm as it rusts.
- At the end of the activity, have students share their observations with the class.

Teacher Time-Saving Option: Show the *Let's Investigate!* Digital Lab as an alternative to the hands-on lab activity.

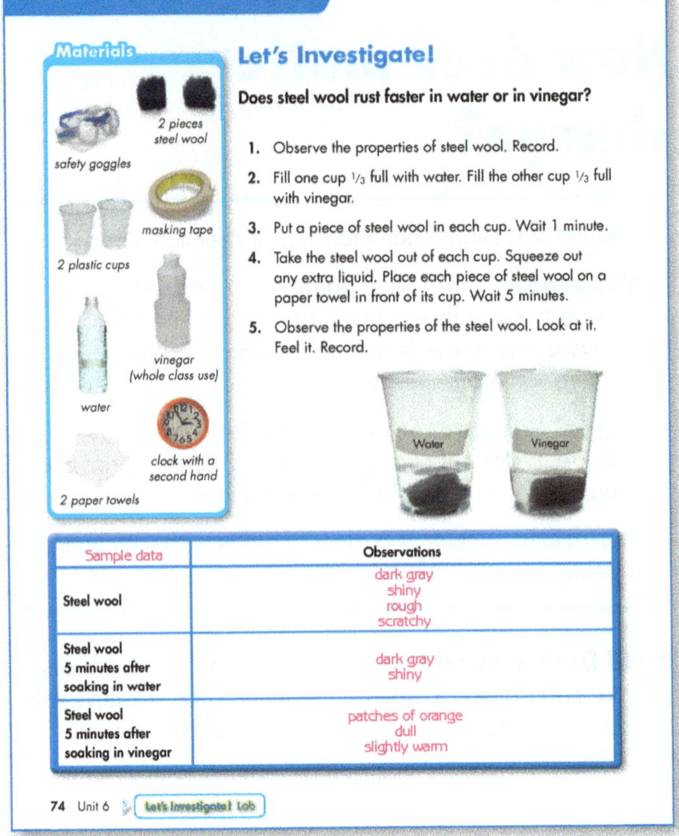

Materials

2 pieces steel wool

safety goggles

masking tape

2 plastic cups

vinegar (whole class use)

water

clock with a second hand

2 paper towels

Let's Investigate!

Does steel wool rust faster in water or in vinegar?

1. Observe the properties of steel wool. Record.
2. Fill one cup 1/3 full with water. Fill the other cup 1/3 full with vinegar.
3. Put a piece of steel wool in each cup. Wait 1 minute.
4. Take the steel wool out of each cup. Squeeze out any extra liquid. Place each piece of steel wool on a paper towel in front of its cup. Wait 5 minutes.
5. Observe the properties of the steel wool. Look at it. Feel it. Record.

Sample data	Observations
Steel wool	dark gray shiny rough scratchy
Steel wool 5 minutes after soaking in water	dark gray shiny
Steel wool 5 minutes after soaking in vinegar	patches of orange dull slightly warm

74 Unit 6 Let's Investigate! Lab

Unlock the Big Question

Have students refer to the Big Question on the Unit Opener page. In pairs, have them recall what they have learned about how matter can be described and measured and how it changes. Invite student pairs to share their answers to questions 6, 7, and 8 on the *Let's Investigate! Activity Card.*

Class Project: Harmful Chemicals

Materials: large sheet of construction paper (1 per group), art supplies

Explain to students that there are lots of chemical products that harm the environment, such as detergents, pesticides, oil, and fertilizers. Divide the class into small groups. Have students choose one chemical and research on the Internet how it can harm the environment and ways to solve the problem. Have students make a poster, illustrating how the chemicals harm the environment. Encourage groups to present their posters to the class and give solutions to the problems.

Unit 6 Review

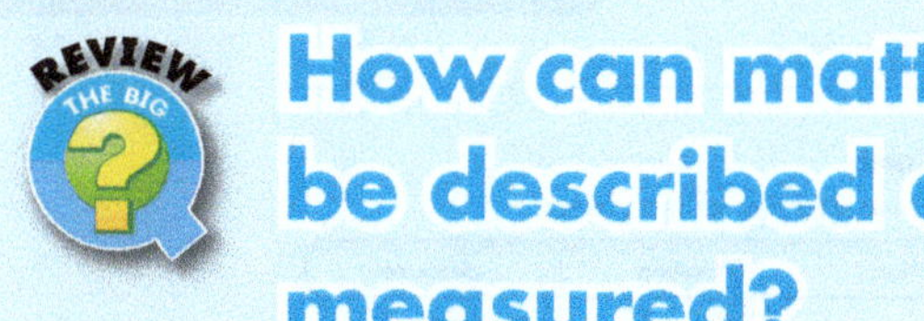

How can matter be described and measured?

Digital Resources: Print out 1 of each per student: *Got it? Self Assessment, Got it? Quiz*

Evaluate

Strategies for Targeted Review

The following are strategies for providing targeted review for students if they encounter challenges with the content.

Lesson 1 How is matter measured?

Question 1

If... students are having difficulty matching concepts with their meanings, then... direct students to pages 65, 66, and 67. Encourage students to circle the words and their definitions.

Lesson 2 What are mixtures?

Question 2

If... students are having difficulty matching the sentence halves, then... direct students to Lesson 2 and have them find the information about the concepts.

Lesson 3 How does matter change?

Question 3

If... students are having difficulty identifying the types of changes, then... direct students' attention to pages 72 and 73 to reread what physical and chemical changes are.

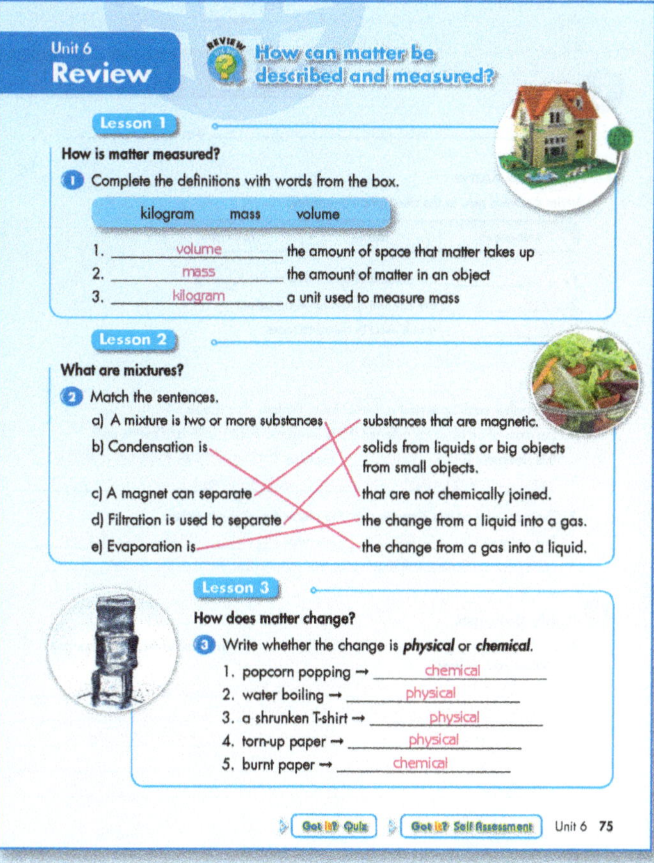
Unit 6 Review

How can matter be described and measured?

Lesson 1

How is matter measured?

1 Complete the definitions with words from the box.

kilogram mass volume

1. volume the amount of space that matter takes up
2. mass the amount of matter in an object
3. kilogram a unit used to measure mass

Lesson 2

What are mixtures?

2 Match the sentences.

a) A mixture is two or more substances — that are not chemically joined.
b) Condensation is — the change from a gas into a liquid.
c) A magnet can separate — substances that are magnetic.
d) Filtration is used to separate — solids from liquids or big objects from small objects.
e) Evaporation is — the change from a liquid into a gas.

Lesson 3

How does matter change?

3 Write whether the change is ***physical*** or ***chemical***.

1. popcorn popping → chemical
2. water boiling → physical
3. a shrunken T-shirt → physical
4. torn-up paper → physical
5. burnt paper → chemical

Got it? Quiz Got it? Self Assessment Unit 6 75

ELL Language Support

Before students start working on the Review activities, have them read each question aloud along with you.

Got it? Self Assessment

Immediately after students have completed the Review activities, distribute a *Got it? Self Assessment* to each student. Have students complete the *Stop! Wait!* and *Go!* statements for each lesson, allowing them to look back through the lesson material if necessary.

Got it? Quiz

Distribute a Unit 6 *Got it? Quiz* to each student. Quizzes may be used for assessing students' understanding of unit concepts as well as for grading purposes.

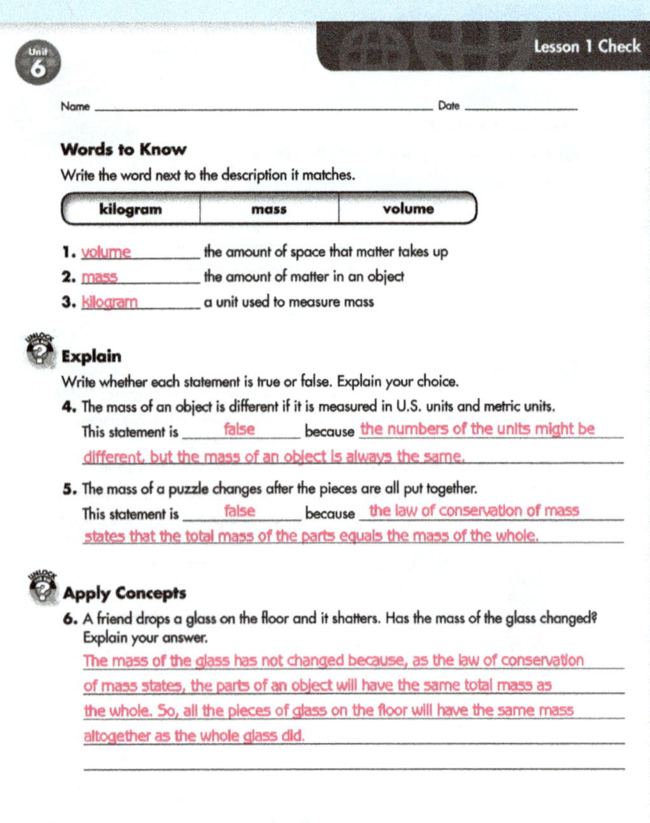

Unit 6

Lesson 1 Check

Name ______________________ Date ____________

Words to Know

Write the word next to the description it matches.

kilogram	mass	volume

1. volume ________ the amount of space that matter takes up
2. mass ________ the amount of matter in an object
3. kilogram ________ a unit used to measure mass

Explain

Write whether each statement is true or false. Explain your choice.

4. The mass of an object is different if it is measured in U.S. units and metric units.
   This statement is false because the numbers of the units might be different, but the mass of an object is always the same.
5. The mass of a puzzle changes after the pieces are all put together.
   This statement is false because the law of conservation of mass states that the total mass of the parts equals the mass of the whole.

Apply Concepts

6. A friend drops a glass on the floor and it shatters. Has the mass of the glass changed? Explain your answer.
   The mass of the glass has not changed because, as the law of conservation of mass states, the parts of an object will have the same total mass as the whole. So, all the pieces of glass on the floor will have the same mass altogether as the whole glass did.

Unit 6, *Lesson 1 Check* • How is matter measured?

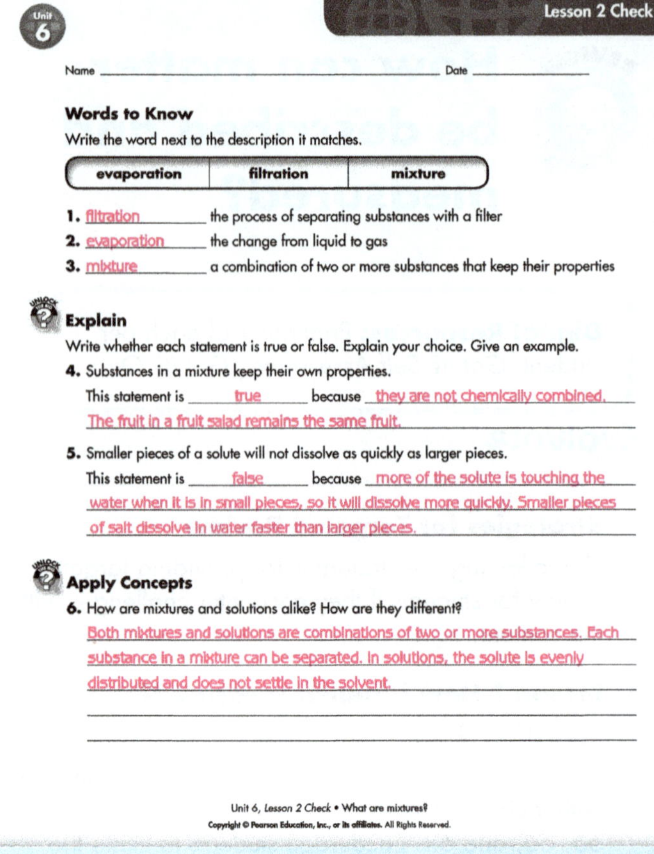

Unit 6

Lesson 2 Check

Name ______________________ Date ____________

Words to Know

Write the word next to the description it matches.

evaporation	filtration	mixture

1. filtration ________ the process of separating substances with a filter
2. evaporation ________ the change from liquid to gas
3. mixture ________ a combination of two or more substances that keep their properties

Explain

Write whether each statement is true or false. Explain your choice. Give an example.

4. Substances in a mixture keep their own properties.
   This statement is true because they are not chemically combined. The fruit in a fruit salad remains the same fruit.
5. Smaller pieces of a solute will not dissolve as quickly as larger pieces.
   This statement is false because more of the solute is touching the water when it is in small pieces, so it will dissolve more quickly. Smaller pieces of salt dissolve in water faster than larger pieces.

Apply Concepts

6. How are mixtures and solutions alike? How are they different?
   Both mixtures and solutions are combinations of two or more substances. Each substance in a mixture can be separated. In solutions, the solute is evenly distributed and does not settle in the solvent.

Unit 6, *Lesson 2 Check* • What are mixtures?

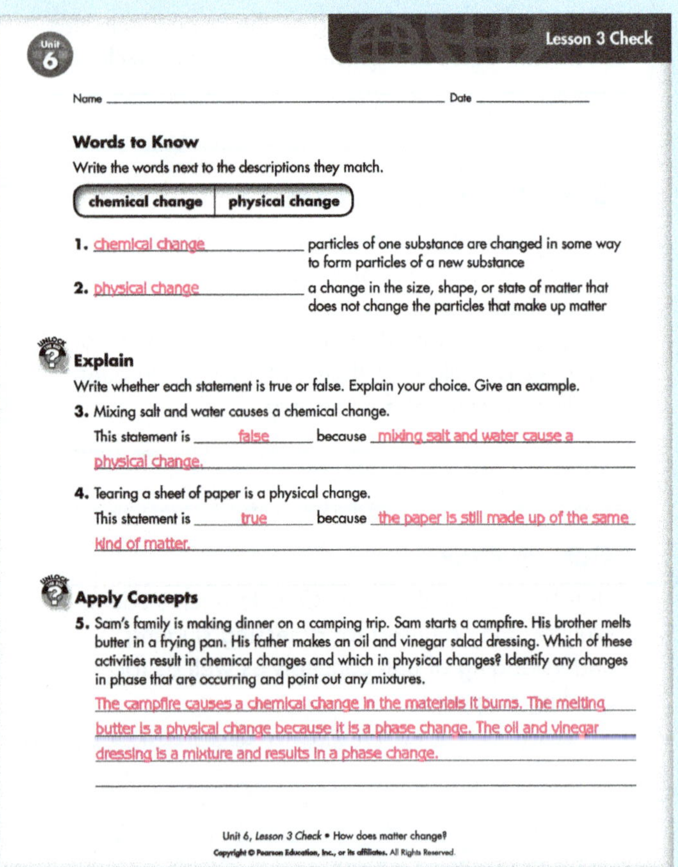

Unit 6

Lesson 3 Check

Name ______________________ Date ____________

Words to Know

Write the words next to the descriptions they match.

chemical change	physical change

1. chemical change ________ particles of one substance are changed in some way to form particles of a new substance
2. physical change ________ a change in the size, shape, or state of matter that does not change the particles that make up matter

Explain

Write whether each statement is true or false. Explain your choice. Give an example.

3. Mixing salt and water causes a chemical change.
   This statement is false because mixing salt and water cause a physical change.
4. Tearing a sheet of paper is a physical change.
   This statement is true because the paper is still made up of the same kind of matter.

Apply Concepts

5. Sam's family is making dinner on a camping trip. Sam starts a campfire. His brother melts butter in a frying pan. His father makes an oil and vinegar salad dressing. Which of these activities result in chemical changes and which in physical changes? Identify any changes in phase that are occurring and point out any mixtures.
   The campfire causes a chemical change in the materials it burns. The melting butter is a physical change because it is a phase change. The oil and vinegar dressing is a mixture and results in a phase change.

Unit 6, *Lesson 3 Check* • How does matter change?

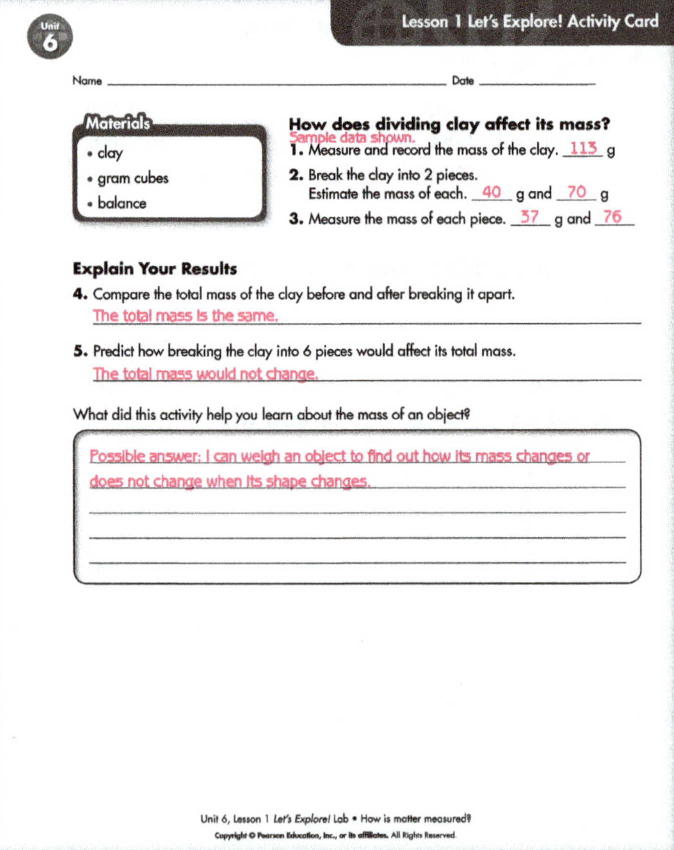

Unit 6

Lesson 1 Let's Explore! Activity Card

Name ______________________ Date ____________

Materials
- clay
- gram cubes
- balance

How does dividing clay affect its mass?

Sample data shown.

1. Measure and record the mass of the clay. 113 g
2. Break the clay into 2 pieces. Estimate the mass of each. 40 g and 70 g
3. Measure the mass of each piece. 37 g and 76

Explain Your Results

4. Compare the total mass of the clay before and after breaking it apart.
   The total mass is the same.
5. Predict how breaking the clay into 6 pieces would affect its total mass.
   The total mass would not change.

What did this activity help you learn about the mass of an object?

Possible answer: I can weigh an object to find out how its mass changes or does not change when its shape changes.

Unit 6, Lesson 1 *Let's Explore!* Lab • How is matter measured?

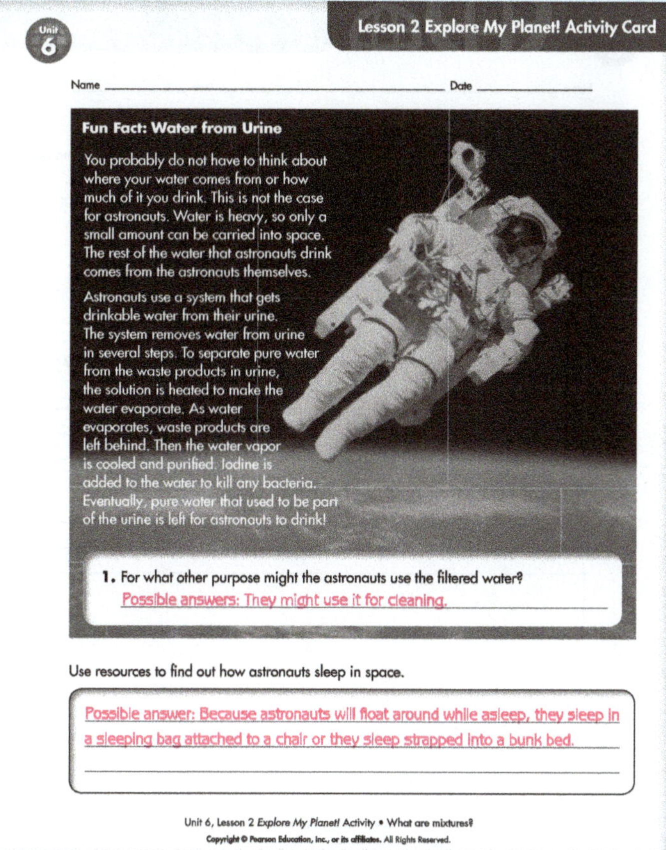

Unit 6

Lesson 2 Explore My Planet! Activity Card

Name ______________________ Date ____________

Fun Fact: Water from Urine

You probably do not have to think about where your water comes from or how much of it you drink. This is not the case for astronauts. Water is heavy, so only a small amount can be carried into space. The rest of the water that astronauts drink comes from the astronauts themselves.

Astronauts use a system that gets drinkable water from their urine. The system removes water from urine in several steps. To separate pure water from the waste products in urine, the solution is heated to make the water evaporate. As water evaporates, waste products are left behind. Then the water vapor is cooled and purified. Iodine is added to the water to kill any bacteria. Eventually, pure water that used to be part of the urine is left for astronauts to drink!

1. For what other purpose might the astronauts use the filtered water?
Possible answers: They might use it for cleaning.

Use resources to find out how astronauts sleep in space.

Possible answer: Because astronauts will float around while asleep, they sleep in a sleeping bag attached to a chair or they sleep strapped into a bunk bed.

Unit 6, Lesson 2 *Explore My Planet!* Activity • What are mixtures?

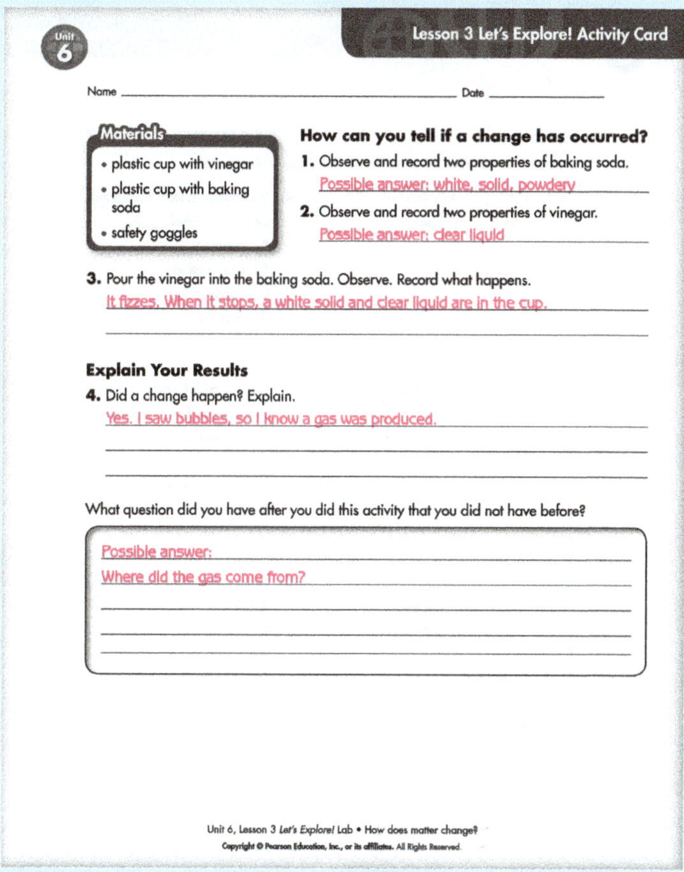

Unit 6

Lesson 3 Let's Explore! Activity Card

Name ______________________ Date ____________

Materials

- plastic cup with vinegar
- plastic cup with baking soda
- safety goggles

How can you tell if a change has occurred?

1. Observe and record two properties of baking soda.
Possible answer: white, solid, powdery
2. Observe and record two properties of vinegar.
Possible answer: clear liquid
3. Pour the vinegar into the baking soda. Observe. Record what happens.
It fizzes. When it stops, a white solid and clear liquid are in the cup.

Explain Your Results

4. Did a change happen? Explain.
Yes. I saw bubbles, so I know a gas was produced.

What question did you have after you did this activity that you did not have before?

Possible answer:
Where did the gas come from?

Unit 6, Lesson 3 *Let's Explore!* Lab • How does matter change?

Unit 6

Let's Investigate! Activity Card

Name ______________________ Date ____________

Analyze and Conclude

6. Which piece of steel wool rusted faster?
The piece that was in the vinegar rusted faster.
7. Based on your observations, is rusting a physical change or chemical change? Explain using what you observed.
Possible answer: The steel wool in the vinegar rusted more and felt slightly warm. Heat is a sign of a chemical reaction. The color changed and the texture changed. These can also be signs of a chemical reaction.
8. Why might it be important for scientists to describe how the properties of matter change during an investigation?
Possible answer:
It may help scientists understand the processes that cause the changes.

Unit 6, *Let's Investigate!* Lab • Matter

Unit 6

Lessons 1–3 Got it? Self Assessment

Name ______________________ Date ____________

Got it? Self Assessment

Complete the statements for each lesson.

Lesson 1 How is matter measured?

Stop! I need help with ____________

Wait! I have a question about ____________

Go! Now I know ____________

Lesson 2 What are mixtures?

Stop! I need help with ____________

Wait! I have a question about ____________

Go! Now I know ____________

Lesson 3 How does matter change?

Stop! I need help with ____________

Wait! I have a question about ____________

Go! Now I know ____________

Unit 6, *Got it? Self Assessment* • Matter

Unit 6

Got it? Quiz

Name ____________ Date ____________

Got it? Quiz

Circle the choice you think is correct for each multiple choice question.

1. Measuring ______ tells you how much matter is in an object.
   - **(A)** mass
   - B weight
   - C volume
   - D height

2. Peas, carrots, and lettuce can be combined into ________.
   - A a solution
   - **(B)** a mixture
   - C a solvent
   - D a solute

3. What happens when you take ice cream out of the freezer?
   - A It freezes.
   - B It evaporates.
   - **(C)** It melts.
   - D It condenses.

4. The change in state from liquid into gas is called ______.
   - A filtration
   - **(B)** evaporation
   - C condensation
   - D magnetism

5. Which method would you use to separate soil that is dissolved in water?
   - A magnetism
   - B solution
   - C condensation
   - **(D)** filtration

6. What is the volume of the box?
   - A 9 cm^3
   - B 11 cm^3
   - **(C)** 6 cm^3
   - D 3 cm^3

1 cm^3

Unit 6, *Got it? Quiz* • Matter

Unit 6

Got it? Quiz

Name ____________ Date ____________

7. What is the volume of a box that is 11 cm long, 3 cm wide, and 7 cm high?
   - A 231 cm
   - **(B)** 231 cm^3
   - C 23.1 cm^3
   - D 231 m

8. Suppose you have 50 mL of water in a graduated cylinder. After you place a marble in the cylinder, the water rises to 78 mL. What is the volume of the marble?
   - A 78 mL
   - B 50 mL
   - C 128 mL
   - **(D)** 28 mL

9. A friend cuts an apple into small pieces. Is any of the apple's mass lost? Why or why not? Explain.

   Possible answer: No, because the law of the conservation of matter states that the parts of an object will have the same total mass as the whole object.

10. Which type of change occurs when a nail rusts? Explain.

    Possible answer: When a nail rusts, the iron in the nail reacts chemically with the oxygen in the air. This is a chemical change called oxidation.

Unit 6, *Got it? Quiz* • Matter

Teacher's Notes

Unit 6 Study Guide

How can matter be described and measured?

Lesson 1
How is matter measured?

- Mass is the measure of the amount of matter in an object.
- Volume is the amount of space taken up by an object.

Lesson 2
What are mixtures?

- A mixture is two or more substances that are not chemically joined.
- Methods to separate a mixture into its parts include filtration, magnetism, evaporation, and condensation.

Lesson 3
How does matter change?

- Matter can undergo physical and chemical changes.
- A physical change does not change the particles that make up matter.
- Chemical changes produce new substances with different chemical properties.

Review the Big Question

How can matter be described and measured?

Have students use what they have learned from the unit to answer the question in their own words. *How has your answer to the Big Question changed since the beginning of the unit? What are some things you learned that caused your answer to change?*

Make a Concept Map

Have students make a concept map like the one shown on this page to help them organize key concepts.

Unit 6 Concept Map

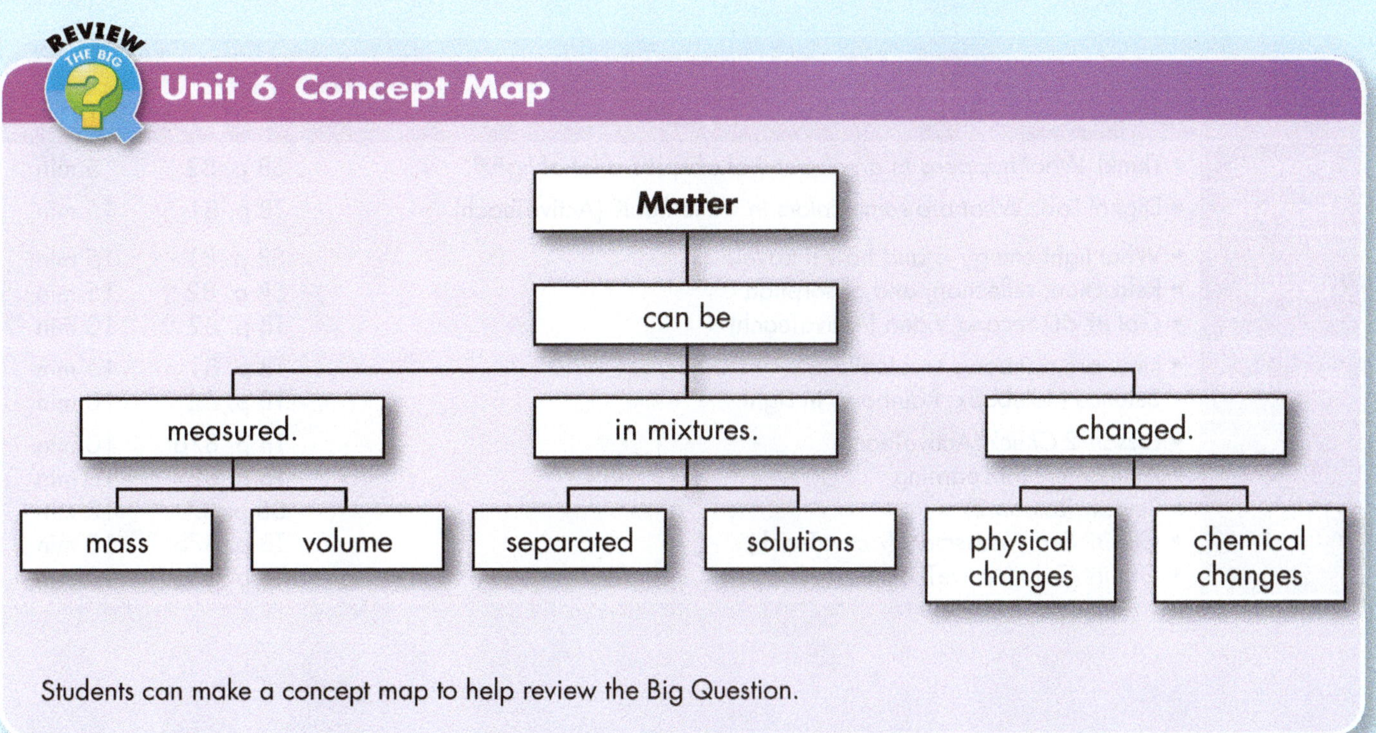

Students can make a concept map to help review the Big Question.

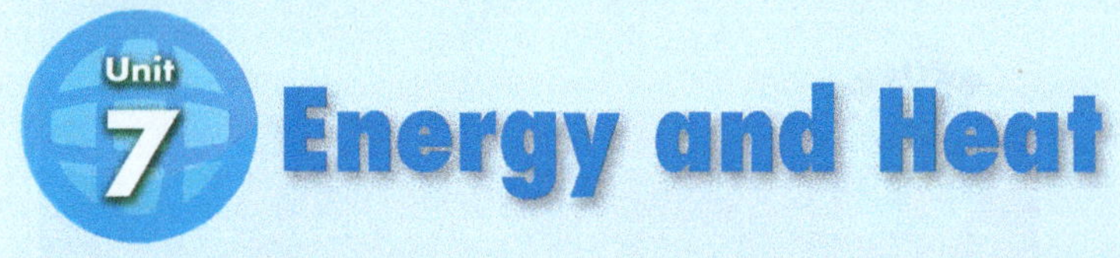

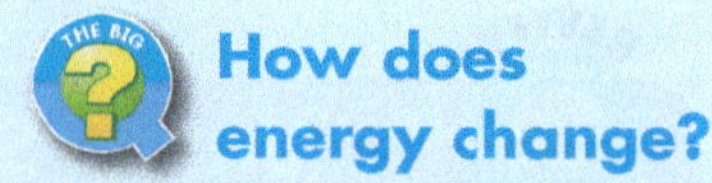

Lesson Plan

Unit Opener & Lesson 1 What is sound energy?			
	Activity	**Pages**	**Time**
Engage	• Unit Opener: Think! *What puts the boom in the fireworks?* • Unit Opener: List examples of sound, thermal, and light energy. • Unit Opener: Identify forms of energy produced by a computer. • Think! *When will the strings of the guitar stop making sounds?* • Think! *You stand outside a room with doors and windows closed. Why can you still hear sounds made in the room?*	SB p. 76 SB p. 76 SB p. 76 SB p. 77 SB p. 79	5 min 10 min 5 min 5 min 5 min
Explore	• Digital Activity: *Fun Fact: Guitar* (ActiveTeach)	TB p. 77	15 min
Explain	• What sound energy is and how it travels • Frequency and wavelength • Pitch and volume • Sound and temperature • *Got it? 60-Second Video* (ActiveTeach)	SB p. 77 SB p. 78 SB p. 79 SB p. 80 TB p. 80	15 min 15 min 15 min 15 min 10 min
Elaborate	• Science Notebook: Objects That Produce Sound • Wave Patterns • Science Notebook: Sounds I Hear • Noise Pollution • At-Home Lab: Water Music	TB p. 77 TB p. 78 TB p. 79 TB p. 79 SB p. 80	10 min 15 min 10 min 20 min 15 min
Evaluate	• *Lesson 1 Check* (ActiveTeach) • Assessment for Learning • Review (Lesson 1) • *Got it? Self Assessment* (ActiveTeach) • *Got it? Quiz* (ActiveTeach)	TB p. 87a TB p. 80 SB p. 87 TB p. 87b TB p. 87c	10 min 10 min 10 min 10 min 10 min

Lesson 2 What is light energy?			
	Activity	**Pages**	**Time**
Engage	• Think! *What happens to an object that absorbs a lot of light?*	SB p. 82	5 min
Explore	• Digital Lab: *What are some colors in white light?* (ActiveTeach)	TB p. 81	15 min
Explain	• What light energy is and how it travels • Refraction, reflection, and absorption • *Got it? 60-Second Video* (ActiveTeach)	SB p. 81 SB p. 82 TB p. 82	15 min 15 min 10 min
Elaborate	• How are rainbows created? • Science Notebook: Rainbows in Light	TB p. 81 TB p. 82	15 min 10 min
Evaluate	• *Lesson 2 Check* (ActiveTeach) • Assessment for Learning • Review (Lesson 2) • *Got it? Self Assessment* (ActiveTeach) • *Got it? Quiz* (ActiveTeach)	TB p. 87a TB p. 82 SB p. 87 TB p. 87b TB p. 87c	10 min 10 min 10 min 10 min 10 min

Lesson 3 What is heat?			
	Activity	**Pages**	**Time**
Engage	• Think! *What could be other important uses of solar panels?*	SB p. 85	5 min
Explore	• Digital Lab: *How does heat move?* (ActiveTeach)	TB p. 83	15 min
Explain	• What heat is and how it is transferred through conduction • Convection and radiation • Other forms of energy that give off heat • *Got it? 60-Second Video* (ActiveTeach)	SB p. 83 SB p. 84 SB p. 85 TB p. 85	15 min 15 min 15 min 10 min
Elaborate	• At-Home Lab: Heat on the Move • Conduction, Convection, and Radiation • Fossil Fuels	SB p. 83 TB p. 84 TB p. 85	10 min 15 min 10 min
Evaluate	• *Lesson 3 Check* (ActiveTeach) • Assessment for Learning • Review (Lesson 3) • *Got it? Self Assessment* (ActiveTeach) • *Got it? Quiz* (ActiveTeach)	TB p. 87a TB p. 85 SB p. 87 TB p. 87b TB p. 87c	10 min 10 min 10 min 10 min 10 min
Lab	• *Let's Investigate! Which material is the better heat conductor?* (ActiveTeach)	SB p. 86	30 min

Flash Cards

sound energy

wavelength

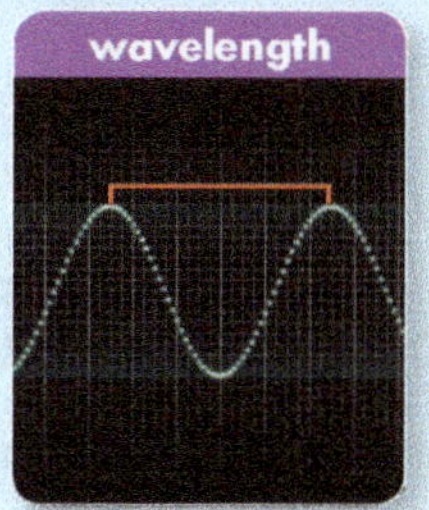

light energy

refraction

reflection

absorption

conduction

convection

radiation

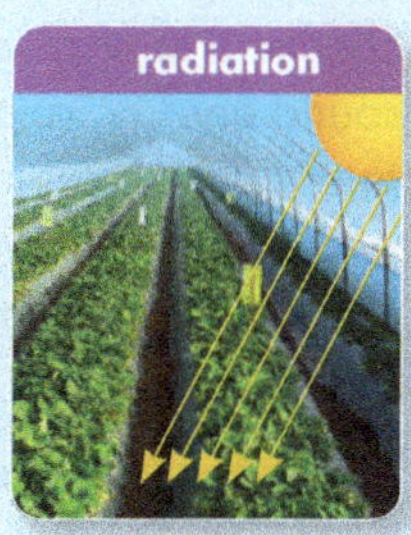

Lesson 1

Key Words	ELL Support
sound, vibration, sound wave, frequency, wavelength, pitch, volume, amplitude	**Vocabulary:** *blare (n), beep (n), car horn, quack (n), rumble (n), thunder (n), sound energy, pass through, matter, empty space, guitar string, disturbance, eardrum, particles, compressions, cycles, oscilloscope, device, display (v), screen, crests, troughs, tuba, brass tubing, whistle*

Lesson 2

Key Words	ELL Support
refraction, reflection, absorption	**Vocabulary:** *source, light energy, chemical energy, give off light, bioluminescence, chemical reaction, campfire, blend of colors, light spectrum, prism, strike (v), reflect off, speed (n), bend (v), bounce off, smooth*

Lesson 3

Key Words	ELL Support
heat, conduction, convection, radiation	**Vocabulary:** *thermal energy, matter, transfer (v), heat source, oven, warm up, radiant energy, send out, fossil fuels, coal, natural gas, oil, laser light, beam of light, solar panel, solar heat system, flow (v), pump (v), rub (v), friction* **Words That Can Be Verbs and Nouns:** *flow, heat, transfer, warm, cut, drill, bond*

Unit Objectives

Lesson 1: Students will describe sound energy and explain how it is produced.

Lesson 2: Students will learn about visible light and the refraction, reflection, and absorption of light.

Lesson 3: Students will recognize that heat flows from hot objects to cold ones and give examples of good and bad conductors of heat.

Vocabulary: *light energy, waves, empty space, sun, sunflowers, thermal energy, randomly, matter, flow* (n), *heat* (n/v), *cook* (v), *sound energy, vibrations, alarm clock, play a musical instrument, fireworks*

Materials: pictures of things that make sounds (car, bird, dog, guitar, cat, alarm clock, watch, cell phone, bell, jet, fireworks, etc.)

How does energy change?

Build Background Display the pictures of things that make sounds. Pairs discuss which sounds they like and which they dislike. Have volunteers say why they like or dislike those sounds.

Engage

Point to the photo on the bottom right. *When fireworks go off, they can be louder than a jet engine. Even if you watch from a safe distance, a fireworks display can be loud enough to make you shake! What forms of energy are at work in a fireworks display?* Write students' ideas on the board. Have them explain their reasoning. (Possible answers: *sound energy, heat, light energy*)

1 Look and read. Match the pictures to the texts. Then write the type of energy illustrated in the pictures.

Draw a three-column chart on the board with the headings *Light Energy, Thermal Energy,* and *Sound Energy.* Elicit examples of each type of energy and write them in the corresponding columns. Have students match the descriptions to the texts and write the types of energy illustrated in the pictures.

2 How have you used the types of energy above? With a partner, write examples of these forms of energy.

Refer to the chart on the board. Have pairs discuss how

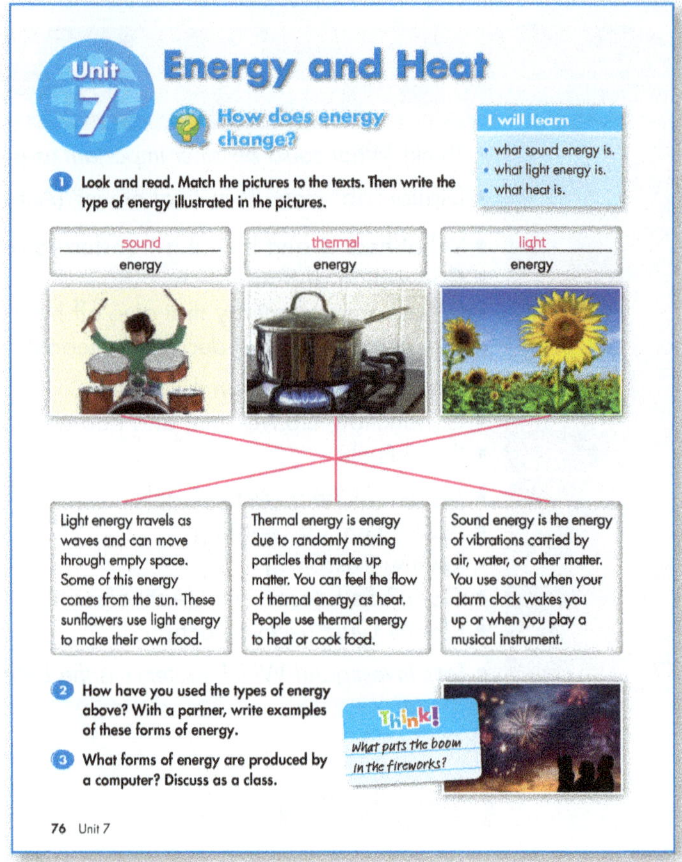

Unit 7 Energy and Heat

How does energy change?

I will learn
- what sound energy is.
- what light energy is.
- what heat is.

1 Look and read. Match the pictures to the texts. Then write the type of energy illustrated in the pictures.

sound energy	thermal energy	light energy
Light energy travels as waves and can move through empty space. Some of this energy comes from the sun. These sunflowers use light energy to make their own food.	Thermal energy is energy due to randomly moving particles that make up matter. You can feel the flow of thermal energy as heat. People use thermal energy to heat or cook food.	Sound energy is the energy of vibrations carried by air, water, or other matter. You use sound when your alarm clock wakes you up or when you play a musical instrument.

2 How have you used the types of energy above? With a partner, write examples of these forms of energy.

3 What forms of energy are produced by a computer? Discuss as a class.

Think! What puts the boom in the fireworks?

76 Unit 7

they have used the types of energy mentioned in the previous activity.

3 What forms of energy are produced by a computer? Discuss as a class.

Draw a computer on the board and have students guess what it is. Discuss with the class what forms of energy are produced by a computer. Have students justify their answers.

Think! Again!

What do you think puts the boom in the fireworks? Ask volunteers to share their ideas with the class along with their reasoning. Explain to students that, when fireworks explode, chemical energy is transformed quickly into light energy, sound energy, and heat. *The quick release of energy into the air around the explosion makes the surrounding air expand faster than the speed of sound. This produces a shock wave of sound energy you hear as the boom.*

ELL Content Support

Fireworks and Speed

Fireworks are seen before they are heard because the speed of light is much faster than the speed of sound. While the speed of light is 300,000,000 meters per second, the speed of sound in dry air is only about 343 meters per second. So, if fireworks explode 1,000 meters away, it will only take three-millionths of a second for the light to reach the viewer. However, the loud boom will take about three seconds.

Lesson 1

What is sound energy?

Objective: Learn what sound energy is and how sound travels.

Vocabulary: *blare* (n), *alarm clock, beep* (n), *car horn, quack* (n), *rumble* (n), *thunder* (n), *storm, vibrations, pass through, matter, vibrate, solids, liquids, gases, empty space, guitar string, sound waves, disturbance, move through, eardrum, particles, pattern, bunched together, compressions*

Digital Resources: Flash Card (*sound energy*), *Explore My Planet!* Digital Activity

Materials: pictures of different musical instruments, paper plate, rice, sound source (table radio, stereo speaker, etc.), guitar (if possible)

Unlock the Big Question

Write the following on the board:
I will learn what sound energy is and how it is produced.

Build Background Display the *sound energy* Flash Card. Have students describe what the boy is doing. Use the opportunity to review or pre-teach key words.

Explore

Explore My Planet! Fun Fact: Guitar

Objective: Students will learn about different parts of a guitar and how they produce sound.

Digital Resources: *Explore My Planet!* Digital Activity, *Explore My Planet! Activity Card* (1 per student)

- Show the *Explore My Planet!* and have students complete the *Activity Card*.
- Ask students to check their answers in small groups or pairs. Provide support as needed.
- Then invite volunteers to share their responses with the class.
- If there is access to an acoustic guitar, use it to show students how it produces sound.

Explain

1 Read and underline the three statements that describe what sound is. Then circle five objects that produce sound.

Write *sound* on the board. Students read and underline the three statements that describe what sound is. Then have students circle five objects that produce sound. Ask students to imitate the sounds the five objects make.

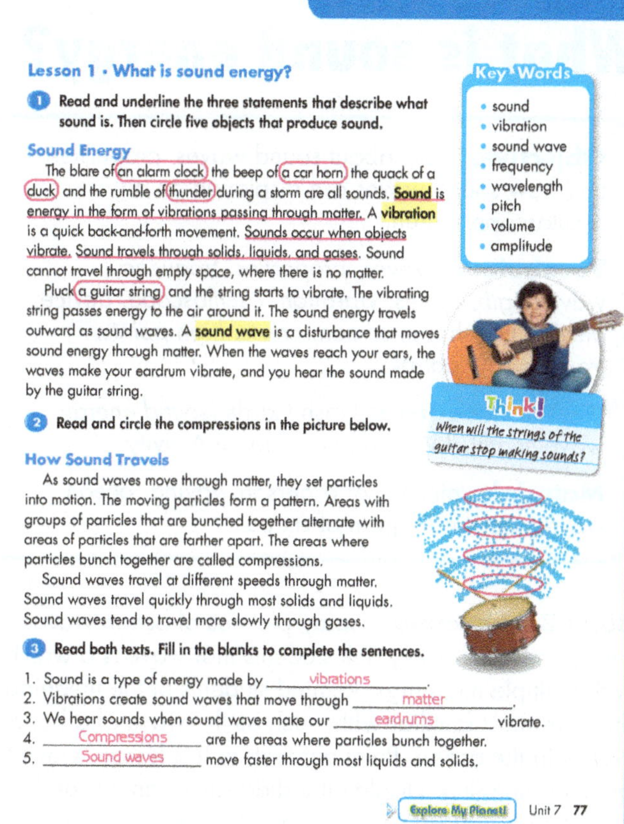

Lesson 1 • What is sound energy?

1 Read and underline the three statements that describe what sound is. Then circle five objects that produce sound.

Key Words
- sound
- vibration
- sound wave
- frequency
- wavelength
- pitch
- volume
- amplitude

Sound Energy

The blare of an alarm clock, the beep of a car horn, the quack of a duck, and the rumble of thunder during a storm are all sounds. **Sound** is energy in the form of vibrations passing through matter. A **vibration** is a quick back-and-forth movement. Sounds occur when objects vibrate. Sound travels through solids, liquids, and gases. Sound cannot travel through empty space, where there is no matter.

Pluck a guitar string and the string starts to vibrate. The vibrating string passes energy to the air around it. The sound energy travels outward as sound waves. A **sound wave** is a disturbance that moves sound energy through matter. When the waves reach your ears, the waves make your eardrum vibrate, and you hear the sound made by the guitar string.

Think! When will the strings of the guitar stop making sounds?

2 Read and circle the compressions in the picture below.

How Sound Travels

As sound waves move through matter, they set particles into motion. The moving particles form a pattern. Areas with groups of particles that are bunched together alternate with areas of particles that are farther apart. The areas where particles bunch together are called compressions.

Sound waves travel at different speeds through matter. Sound waves travel quickly through most solids and liquids. Sound waves tend to travel more slowly through gases.

3 Read both texts. Fill in the blanks to complete the sentences.

1. Sound is a type of energy made by __vibrations__.
2. Vibrations create sound waves that move through __matter__.
3. We hear sounds when sound waves make our __eardrums__ vibrate.
4. __Compressions__ are the areas where particles bunch together.
5. __Sound waves__ move faster through most liquids and solids.

Explore My Planet! Unit 7 77

Think!

When will the strings of the guitar stop making sounds?

Refer students to the statements about sound written on the board. Then ask the question and have students discuss. (Possible answer: *When they stop vibrating.*)

2 Read and circle the compressions in the picture below.

Ask students to read and circle the compressions in the picture of the sound waves coming from the drum.

3 Read both texts. Fill in the blanks to complete the sentences.

Have students read both texts and complete the sentences. Check answers as a class.

Elaborate

Science Notebook: Objects That Produce Sound

Write *alarm clock, car horn, duck,* and *thunder* on the board. Ask students to illustrate the objects in their Science Notebooks. Have students label their illustrations using the full phrases with their collocations (*blare of an alarm clock, beep of a car horn, quack of a duck, rumble of thunder*).

Lesson 1

What is sound energy?

Objective: Learn about sound waves' properties and how sound waves are displayed on an oscilloscope screen.

Vocabulary: *waves, properties, frequency, wavelength, cycles, vibration, oscilloscope, device, display* (v), *shape, screen, crests, high points, troughs, low points*

Digital Resources: Flash Cards (*sound energy, wavelength*), *I Will Know...* Digital Activity

Materials: pictures of a person waving, ocean waves, and a diagram of an ear

Build Background Display pictures of ocean waves and someone waving. Tell students that *wave* is a word with multiple meanings. Remind students of the meanings they are most familiar with, such as the wave of a hand or waves in the ocean. Point out that, in this lesson, *wave* has another meaning. Display the diagram of an ear and the *sound energy* Flash Card. *A sound wave is a disturbance that moves sound energy through matter.* Point to the diagram of the ear first and then to the guitar. *When the waves reach your ear, the waves make your eardrum vibrate and you hear the sound made by the guitar string.*

Explain

4 Read and match the words to their definitions.

Read both paragraphs with students. Write *frequency, wavelength, oscilloscope,* and *crest* on the board. Have volunteers use their own words to explain the meanings of those words. Finally, have students read and match the words to their definitions.

ELL Content Support

How are wavelength and frequency related?

Students might mistakenly think that wavelength and frequency are directly proportional. However, wavelength and frequency are actually inversely proportional. In other words, shorter wavelengths have higher frequencies, and longer wavelengths have lower frequencies.

On the board, draw a wave that has crests and troughs very close together. *This is a wave with short wavelengths and a high frequency.* Then draw a wave that has crests and troughs far apart from one another. *This is a wave with long wavelengths and a low frequency.*

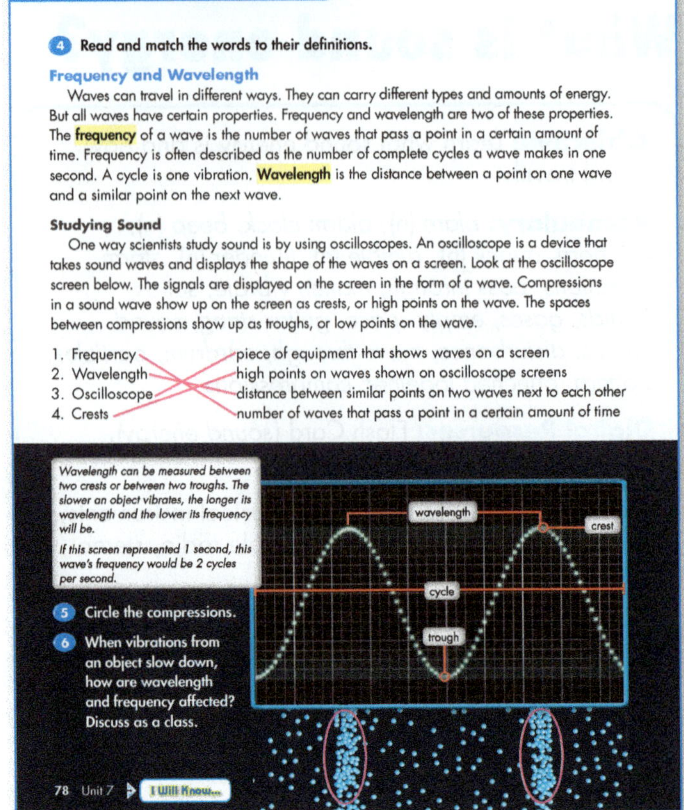

4 Read and match the words to their definitions.

Frequency and Wavelength

Waves can travel in different ways. They can carry different types and amounts of energy. But all waves have certain properties. Frequency and wavelength are two of these properties. The **frequency** of a wave is the number of waves that pass a point in a certain amount of time. Frequency is often described as the number of complete cycles a wave makes in one second. A cycle is one vibration. **Wavelength** is the distance between a point on one wave and a similar point on the next wave.

Studying Sound

One way scientists study sound is by using oscilloscopes. An oscilloscope is a device that takes sound waves and displays the shape of the waves on a screen. Look at the oscilloscope screen below. The signals are displayed on the screen in the form of a wave. Compressions in a sound wave show up on the screen as crests, or high points on the wave. The spaces between compressions show up as troughs, or low points on the wave.

1. Frequency
2. Wavelength
3. Oscilloscope
4. Crests

piece of equipment that shows waves on a screen
high points on waves shown on oscilloscope screens
distance between similar points on two waves next to each other
number of waves that pass a point in a certain amount of time

Wavelength can be measured between two crests or between two troughs. The slower an object vibrates, the longer its wavelength and the lower its frequency will be.

If this screen represented 1 second, this wave's frequency would be 2 cycles per second.

5 Circle the compressions.

6 When vibrations from an object slow down, how are wavelength and frequency affected? Discuss as a class.

78 Unit 7 I Will Know...

5 Circle the compressions.

Direct students' attention to the sound wave diagram. Have them describe the meanings of each word. Then ask students to circle the compressions.

6 When vibrations from an object slow down, how are wavelength and frequency affected? Discuss as a class.

Display the *wavelength* Flash Card. Ask *When vibrations from an object slow down, how are wavelength and frequency affected?* Help students understand that, when vibrations slow down, wavelength becomes longer and frequency becomes lower.

Elaborate

Wave Patterns

Divide the class into small groups. Provide groups with three different wave patterns on graph paper. Be sure that the patterns vary in wavelength but not wave height. For each pattern, have students label the crest and trough and measure the wavelength. Have students discuss which wave has the highest frequency and justify their answer.

I Will Know...

Have students do the *I Will Know...* Digital Activity.

Lesson 1

What is sound energy?

Objective: Learn the difference between pitch and volume.

Vocabulary: *frequency, pitch, vibrate, higher pitch, lower pitch, tuba, brass tubing*

Materials: pictures of different musical instruments, including a tuba and a saxophone, recording of a tuba, pictures of a jet and a car

Build Background Brainstorm different musical instruments and write their names on the board. *Which of these instruments do you think make high sounds? Which of them make low sounds?* Display pictures of a tuba and a saxophone. *How do the sounds of a saxophone and tuba compare?* (*A tuba makes a lower sound than a saxophone.*)

Explain

7 Read and complete the sentences.

Write *pitch* and *volume* on the board. Read both texts with the class. While you read, ask students to underline their definitions. Have volunteers explain the meanings in their own words.

Refer to the pictures of the tuba and saxophone on the board. Have students discuss which of these instruments makes low-pitched sounds. Direct students' attention to the picture on the top right. Have students make high-pitched sounds like dolphins make. Discuss with the class the difference between the sounds a tuba and a dolphin make to ensure they understand the difference between high- and low-pitched sounds.

Display the pictures of a jet and a car. Have students compare the sound each object makes and discuss which makes louder sounds. Have students discuss why they think the boy in the picture is plugging his ears.

Finally have pairs read and complete the sentences with the words written on the board.

ELL Content Support

High and Low Pitch

Have students put their hand on their throats. Ask them to make a low-pitched sound. Then ask them to make a high-pitched sound. Have students explain how their throat adjusts while making these sounds. (Possible answers: *I tuck my chin in and my throat feels wider when making low-pitched sounds. When making high-pitched sounds, I stretch my neck out and raise my head.*)

7 Read and complete the sentences.

Pitch

People experience the frequency of sound as its pitch. **Pitch** is how high or low a sound is. Pitch depends on the frequency of the sound wave. Objects that vibrate more quickly have higher frequencies. Those objects have a higher pitch. Objects that vibrate more slowly have a lower frequency and a lower pitch.

The material of the object making the sound and its size and shape affect the pitch you hear. A tuba, for example, is a musical instrument made of several meters of brass tubing. Tubas make low-pitched sounds. Some sounds are too low-pitched or too high-pitched for humans to hear.

A dolphin communicates by making high-pitched sounds.

Volume

When you describe a sound, probably one of the first things you think about is loudness, or volume. You know that some sounds are louder than others. A jet engine, for example, makes much more noise than a car. What, exactly, is volume? **Volume** is a measure of how strong a sound seems to us. The more energy there is in the sound wave, the louder the sound and the higher the volume.

The volume of a sound is related to its amplitude. **Amplitude** is the height of a wave measured from its midline. The higher the amplitude of a wave, the more energy it has, and the louder it sounds. Suppose you turn down the volume on the TV as soft as it will go. You can barely hear it. Then you turn it up as high as it will go. Now you have to plug your ears so they don't hurt! You have not changed the pitch of the sound. You have changed its amplitude and its volume.

1. If you change the frequency or wavelength of a sound wave, you change its ______pitch______.
2. If you change the amplitude, or height, of a sound wave, you change its ______volume______.

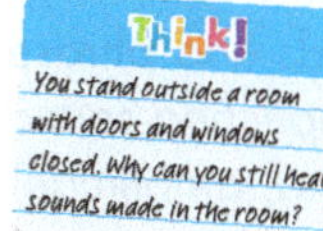

Think!
You stand outside a room with doors and windows closed. Why can you still hear sounds made in the room?

Unit 7 79

Elaborate

Science Notebook: Sounds I Hear

Ask students to sit quietly and listen for a few minutes. Then have them write a paragraph describing the sounds they heard. Ask them to classify the sounds as low-pitched or high-pitched.

Noise Pollution

The sounds from machinery, traffic, construction, and airplanes cause noise pollution. Many cities and towns have rules that limit noise levels. Ask students to identify different rules they would propose to their local government to keep noise from becoming a problem.

Think!

You stand outside a room with doors and windows closed. Why can you still hear sounds made in the room?

Have students discuss how sound travels. Close windows and doors, and ask students to sit quietly. Ask *Why can you still hear sounds made outside the room?* (Possible answer: *Because sounds travel through solids, not just through gases.*)

Lesson 1

What is sound energy?

Objective: Learn how to interpret wave patterns and how temperature affects sound.

Vocabulary: *wavelength, amplitude, midline, properties of sound, whistle, volume, temperature, decrease (v), speed, direction, sound waves, bottles, blow across*

Digital Resources: *Lesson 1 Check* (print out 1 per student), *Got it? 60-Second Video*

Materials: bottle of water, a glass, pictures of daytime and nighttime

Build Background Draw a two-column chart labeled *Volume* and *Pitch*. Around the chart, write the following words and phrases randomly: *frequency, tuba, dolphin, low-pitched sound, high-pitched sound, loudness, strong sound, amplitude, loud sound, plug your ears, turn up or down the volume*. Have volunteers come to the board and write the words or phrases in the corresponding columns. As a review, have students make sentences using the phrases in the chart.

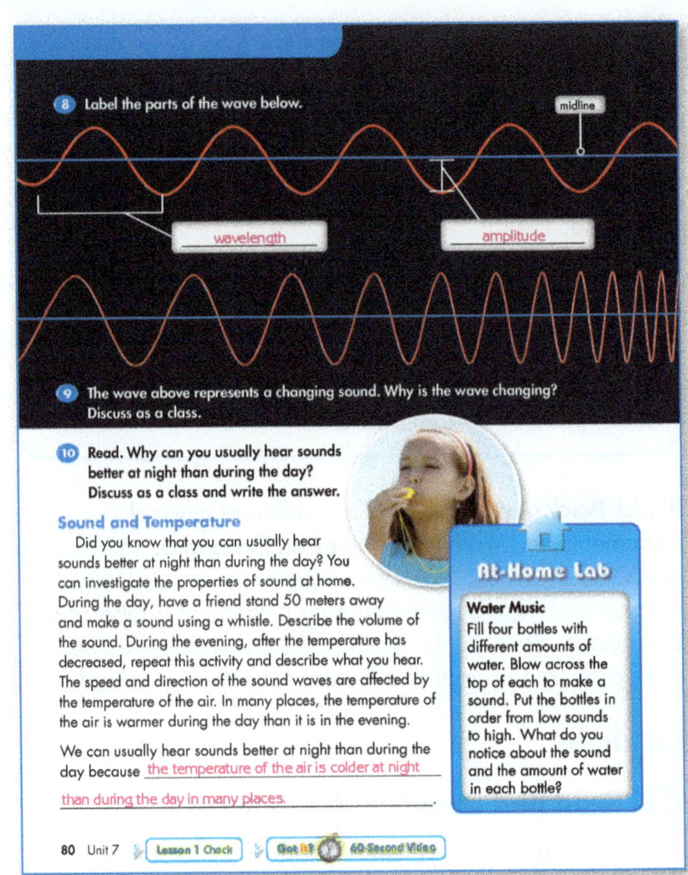

8 Label the parts of the wave below.

9 The wave above represents a changing sound. Why is the wave changing? Discuss as a class.

10 Read. Why can you usually hear sounds better at night than during the day? Discuss as a class and write the answer.

Sound and Temperature

Did you know that you can usually hear sounds better at night than during the day? You can investigate the properties of sound at home. During the day, have a friend stand 50 meters away and make a sound using a whistle. Describe the volume of the sound. During the evening, after the temperature has decreased, repeat this activity and describe what you hear. The speed and direction of the sound waves are affected by the temperature of the air. In many places, the temperature of the air is warmer during the day than it is in the evening.

We can usually hear sounds better at night than during the day because the temperature of the air is colder at night than during the day in many places.

At-Home Lab

Water Music

Fill four bottles with different amounts of water. Blow across the top of each to make a sound. Put the bottles in order from low sounds to high. What do you notice about the sound and the amount of water in each bottle?

80 Unit 7 Lesson 1 Check Got it? 60-Second Video

Explain

8 Label the parts of the wave below.

Write *wavelength* and *amplitude* on the board. Ask students to look up the meanings of these words in the Unit Glossary if they have difficulty remembering what they mean. Have students look at the wave patterns and label the parts of the wave.

9 The wave above represents a changing sound. Why is the wave changing? Discuss as a class.

Have students look at both sound waves and discuss their differences. Write on the board: *If you change the frequency or wavelength of a sound wave, you change its pitch. If you change the amplitude, or height, of a sound wave, you change its volume.* Discuss with the class why the wave above is changing. (Possible answer: *Because the sound is getting higher-pitched.*)

10 Read. Why can you usually hear sounds better at night than during the day? Discuss as a class and write the answer.

Display pictures of daytime and nighttime. Have students discuss when they think they can hear sounds better, at night or during the day. *Did you know that you can usually hear sounds better at night than during the day? Why is that?* Read the text with the class. Then ask the question and have students discuss. Finally, students complete the statement using their own words.

Elaborate

At-Home Lab

Water Music

Materials: 4 identical bottles, jug with water

Students should notice that the more water there is in the bottle, the higher the sound's pitch is, and that the smaller the air space, the faster the air vibrates and the higher the sound's pitch is.

Evaluate

Lesson 1 Check **Assessment for Learning**

Distribute the *Lesson 1 Check* and guide students as they complete it. Check answers as a class. Then ask students to grade their progress on the topic of sound energy from 1 to 3: 3 = *I understand what sound energy is and how it is produced*; 2 = *I need to study more*; 1 = *I need help!* Encourage students giving themselves a 1 or 2 to describe what they found difficult and what they need to study more.

Got it? 60-Second Video

Review Key Words for Lesson 1 (see Student's Book page 77). Play the *Got it? 60-Second Video* to review the lesson material.

Lesson 2

What is light energy?

Objective: Learn what things can be sources of light and how light energy travels.

Vocabulary: *source, light energy, sunlight, chemical energy, give off light, bioluminescence, chemical reaction, campfire, waves, wavelengths, frequencies, white light, blend of colors, light spectrum*

Digital Resources: Flash Card (*light energy*), *Let's Explore!* Digital Lab, *I Will Know...* Digital Activity

Materials: picture of a rainbow

Unlock the Big Question

Write the following on the board:
I will learn about white light and how light can refract, reflect, and be absorbed.

Build Background Draw the sun on the board. Brainstorm things that depend on sunlight and write them on the board.

Explore

Let's Explore! Lab What are some colors in white light?

Objective: Students will learn how to use water and a mirror to refract and reflect light to make a rainbow-like effect.

Digital Resources: *Let's Explore!* Digital Lab, *Let's Explore! Activity Card* (1 per student) (*Optional*: Do the lab in class; refer to the *Activity Card* for materials and steps.)

- Display a picture of a rainbow. Hold a class discussion of what a rainbow is and what causes rainbows. *Do you think we can make a rainbow? How?*
- Show the Digital Lab.
- Have students form pairs and complete the *Activity Card*. Review answers as a class.

Explain

1 Read and underline the main idea in red and two facts that support the main idea in blue.

Elicit examples of sources of light and write them on the board. Discuss with the class how important light is. Have students read and underline the main idea in red and two facts that support the main idea in blue.

2 Read again. How can this firefly be a source of light energy? Discuss as a class.

Display the *light energy* Flash Card. Have students say what it is and what makes fireflies different from

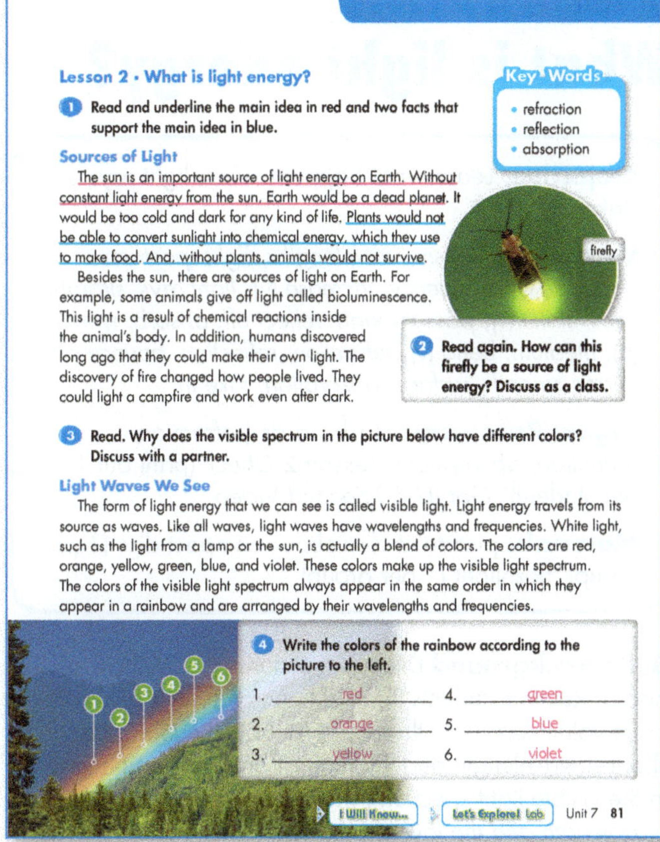
Lesson 2 • What is light energy?

Key Words
- refraction
- reflection
- absorption

1 Read and underline the main idea in red and two facts that support the main idea in blue.

Sources of Light

The sun is an important source of light energy on Earth. Without constant light energy from the sun, Earth would be a dead planet. It would be too cold and dark for any kind of life. Plants would not be able to convert sunlight into chemical energy, which they use to make food. And, without plants, animals would not survive.

Besides the sun, there are sources of light on Earth. For example, some animals give off light called bioluminescence. This light is a result of chemical reactions inside the animal's body. In addition, humans discovered long ago that they could make their own light. The discovery of fire changed how people lived. They could light a campfire and work even after dark.

firefly

2 Read again. How can this firefly be a source of light energy? Discuss as a class.

3 Read. Why does the visible spectrum in the picture below have different colors? Discuss with a partner.

Light Waves We See

The form of light energy that we can see is called visible light. Light energy travels from its source as waves. Like all waves, light waves have wavelengths and frequencies. White light, such as the light from a lamp or the sun, is actually a blend of colors. The colors are red, orange, yellow, green, blue, and violet. These colors make up the visible light spectrum. The colors of the visible light spectrum always appear in the same order in which they appear in a rainbow and are arranged by their wavelengths and frequencies.

4 Write the colors of the rainbow according to the picture to the left.

1.	red	4.	green
2.	orange	5.	blue
3.	yellow	6.	violet

I Will Know... Let's Explore! Lab Unit 7 81

other beetles. Hold a class discussion about how fireflies can be a source of light energy.

3 Read. Why does the visible spectrum in the picture below have different colors? Discuss with a partner.

Have students read and answer the question in pairs. (Possible answer: *It is made up of waves with different wavelengths and frequencies.*)

4 Write the colors of the rainbow according to the picture to the left.

Ask students to write the colors of the rainbow according to the picture. Explain that many scientists today no longer include the color indigo in the visible spectrum because it is difficult to see as a distinct color in between blue and violet.

Elaborate

How are rainbows created?

Divide the class into small groups. Have students investigate and make a poster that depicts how rainbows are created. Invite groups to present their posters to the class.

I Will Know...

Have students do the *I Will Know...* Digital Activity.

Lesson 2

What is light energy?

Objective: Learn what happens when light rays strike an object.

Vocabulary: *prism, wavelengths, light spectrum, decrease (v), frequency, increase (v), light rays, straight lines, strike (v), pass through, reflect off, absorb, speed (n), medium, bend (v), refraction, reflection, bounce off, mirror, smooth, surface, absorption, take in*

Digital Resources: Flash Cards (*refraction, reflection, absorption*), *Lesson 2 Check* (print out 1 per student), *Got it? 60-Second Video*

Materials: picture of a rainbow, compact disc (1 per group), flashlight (1 per group)

Build Background Display a picture of a rainbow on the board. Remind students that light energy travels as waves. Invite volunteers up to the board to label the longest and shortest wavelength and the highest and lowest frequency in the visible light spectrum. (*Red light has the longest wavelength and the lowest frequency, and violet light has the shortest wavelength and highest frequency.*) Review with the class how rainbows are formed.

Explain

5 Read and look at the picture to the right. How are raindrops and prisms similar? Discuss as a class.

Use board drawings to elicit from students how rainbows are created. Have students read and underline what a prism is. Discuss with the class how raindrops and prisms are similar.

6 Read and write the effect that happens when light strikes each of the objects below.

Direct students' attention to the picture of the rainbow displayed on the board. *Remember that rainbows appear when raindrops refract sunlight. This effect happens when sunlight strikes raindrops.* Have students read and write the effect that happens when light strikes the objects mentioned. Display the *refraction*, *reflection*, and *absorption* Flash Cards. Have three volunteers explain each effect using their own words.

Think!

Draw on the board an object that is receiving direct sunlight. Ask *What happens to an object that absorbs a lot of light?* Invite students to discuss the question as a class. (Possible answer: *It becomes hot.*)

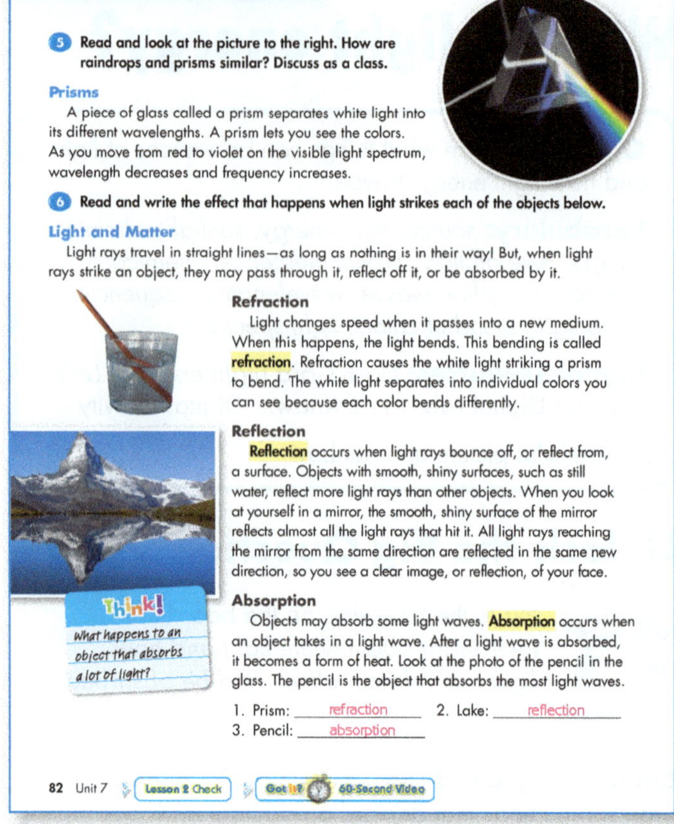

5 Read and look at the picture to the right. How are raindrops and prisms similar? Discuss as a class.

Prisms

A piece of glass called a prism separates white light into its different wavelengths. A prism lets you see the colors. As you move from red to violet on the visible light spectrum, wavelength decreases and frequency increases.

6 Read and write the effect that happens when light strikes each of the objects below.

Light and Matter

Light rays travel in straight lines—as long as nothing is in their way! But, when light rays strike an object, they may pass through it, reflect off it, or be absorbed by it.

Refraction

Light changes speed when it passes into a new medium. When this happens, the light bends. This bending is called **refraction**. Refraction causes the white light striking a prism to bend. The white light separates into individual colors you can see because each color bends differently.

Reflection

Reflection occurs when light rays bounce off, or reflect from, a surface. Objects with smooth, shiny surfaces, such as still water, reflect more light rays than other objects. When you look at yourself in a mirror, the smooth, shiny surface of the mirror reflects almost all the light rays that hit it. All light rays reaching the mirror from the same direction are reflected in the same new direction, so you see a clear image, or reflection, of your face.

Absorption

Objects may absorb some light waves. **Absorption** occurs when an object takes in a light wave. After a light wave is absorbed, it becomes a form of heat. Look at the photo of the pencil in the glass. The pencil is the object that absorbs the most light waves.

Think!
What happens to an object that absorbs a lot of light?

1. Prism: refraction 2. Lake: reflection
3. Pencil: absorption

82 Unit 7 Lesson 2 Check Got it? 60-Second Video

Elaborate

Science Notebook: Rainbows in Light

Divide the class into small groups. Give each group a white sheet of paper, a compact disc, and a flashlight. Tell students to hold the compact disc near a window or a flashlight. Suggest that students move the disc under the light source. Write the question *How does a compact disc act like a prism?* on the board for students to answer in their Science Notebooks. Direct students to write a description and answer the question. Students should understand that, like a prism, the compact disc separates white light into its separate colors.

Evaluate

Lesson 2 Check Assessment for Learning

Distribute the *Lesson 2 Check* and guide students as they complete it. Check answers as a class. Then ask students to grade their progress on the topic of light energy from 1 to 3: 3 = *I can describe the effects that light energy can produce*; 2 = *I need to study more*; 1 = *I need help!* Encourage students giving themselves a 1 or 2 to describe what they found difficult and what they need to study more.

Review Key Words for Lesson 2 (see Student's Book page 81). Play the *Got it? 60-Second Video* to review the lesson material.

Lesson 3

What is heat?

Objective: Learn what heat is and how heat energy can be transferred.

Vocabulary: *thermal energy, flow* (v), *matter, heat* (n), *give off energy, pillow, transfer* (v), *heat source, conduction, warm* (v), *nest, oatmeal, bowl*

Digital Resources: Flash Card (*conduction*), *Let's Explore!* Digital Lab

Unlock the Big Question

Write the following on the board:
I will know that heat flows from hot objects to cold ones. I will know that some materials are good conductors of heat and others are not.

Build Background Have students discuss what they and their family do when they are cold.

Explore

Let's Explore! Lab How does heat move?

Objective: Record temperatures to observe how heat moves from a warm object to a cold one.

Digital Resources: *Let's Explore!* Digital Lab, *Let's Explore! Activity Card* (1 per student) (*Optional*: Do the lab in class; refer to the *Activity Card* for materials and steps.)

- Brainstorm liquids that are usually drunk very hot or very cold. *What do you usually do when these drinks are too hot or too cold for you to drink?*
- Show the Digital Lab. Have students form pairs and complete the *Activity Card*.
- Review answers as a class. Guide students to conclude that heat moves from a warmer object to a cooler one.

Explain

1 Read and underline two examples of conduction.

Have students read and find out what conduction is. Then ask them to read and underline two examples of conduction. Have volunteers explain using board drawings how heat is transferred in each situation.

2 Read. How do you know that heat from the oatmeal has moved? Discuss with a partner.

Have students say what the boy in the picture is eating and whether the oatmeal is hot or cold. Read the text with the class. Have pairs discuss how they know that heat from the oatmeal has moved.

Lesson 3 • What is heat?

1 Read and underline two examples of conduction.

Conduction

Thermal energy flows from something warm to something cool. The transfer of thermal energy between matter of different temperatures is **heat**. A heat source is anything that gives off energy that particles of matter can take in.

When you go to bed at night, does your pillow feel cool on your face? Is the pillow warm when you wake up? That is thermal energy moving! Your body is the heat source.

Thermal energy transfers from your body to your pillow. When solids touch, thermal energy moves by conduction. **Conduction** is the transfer of heat that occurs when one thing touches another.

A bird warming its eggs in a nest is another example of conduction. The bird's body is the heat source. Conduction transfers thermal energy from the bird to the eggs, which are cooler.

2 Read. How do you know that heat from the oatmeal has moved? Discuss with a partner.

A Conduction Example

Have you ever eaten hot oatmeal for breakfast? Suppose that you eat a bowl of hot oatmeal with a metal spoon. Why does the metal spoon begin to feel warmer? The particles of the spoon that touch the oatmeal start to move. As they move more quickly, they crash into other particles in the spoon. Soon, thermal energy from the oatmeal moves throughout the spoon. Heat transfer continues until the oatmeal and the spoon are at the same temperature.

Key Words

- heat
- conduction
- convection
- radiation

At-Home Lab

Heat on the Move

Put a thermometer in various places around your home. Record your findings. Think about why some places are cooler or warmer than others. Where might heat be moving in or out of a home?

Let's Explore! Lab Unit 7 83

ELL Language Support

Verbs and Nouns Write *flow, heat, transfer,* and *warm* on the board. Ask *Are these words verbs or nouns, or are they both verbs and nouns?* (Answer: *They can be both verbs and nouns.*) Have students find these words in the text about conduction and invite them to write the full sentences in their notebooks. Students should identify whether these words are being used as verbs or nouns.

Elaborate

At-Home Lab

Heat on the Move

Brainstorm places around the students' homes where the air temperature might be taken. Possibilities include near doors and windows, near the ceiling, near the floor, in the center of a room, along walls, and so forth. Students put a thermometer in the various places and record their findings. Students should conclude that cooler areas in a warm home are losing heat or that they are farther from a heat source.

Lesson 3

What is heat?

Objective: Learn how heat is transferred through convection and radiation.

Vocabulary: *heat, convection, transfer* (v), *thermal energy, matter, gas, liquid, oven, particles, warm up, radiant energy, radiation, send out, waves*

Digital Resources: Flash Cards (*conduction, convection, radiation*), *I Will Know…* Digital Activity

Materials: a cup of very warm water, food coloring, a small aquarium or a large wide-mouth jar of cold water

Build Background Display the *conduction* Flash Card. Have students identify the heat source and explain how heat is transferred from the bird's body to the eggs. *Remember that conduction occurs when two objects at different temperatures are in contact with each other. Heat flows from the warmer object to the cooler one until they are both at the same temperature. Today we will learn two other ways heat can be transferred.*

Explain

3 Read and complete the statements with the words in the box.

Draw a person baking a cake in an oven on the board. Have students describe what is happening and say what the heat source is. Read the text about convection aloud for students. Refer to the drawing on the board and have students explain what *convection* means.

Next, draw a person sunbathing. Students describe what is happening and discuss what the heat source is. Read the text about radiation with the class. Direct students' attention to the drawing of the person in the sun. *What is radiation or radiant energy? It is a kind of energy that is sent out in waves.*

Display the *convection* and *radiation* Flash Cards. *What do the red and blue arrows show? That heat is moving through the greenhouse. Particles in the warm air are moving upward. What do the yellow arrows show? That the heat from the sun is warming the ground.*

Finally have pairs complete the sentences with the words in the box. Check answers as a class.

I Will Know…

Have students do the *I Will Know…* Digital Activity.

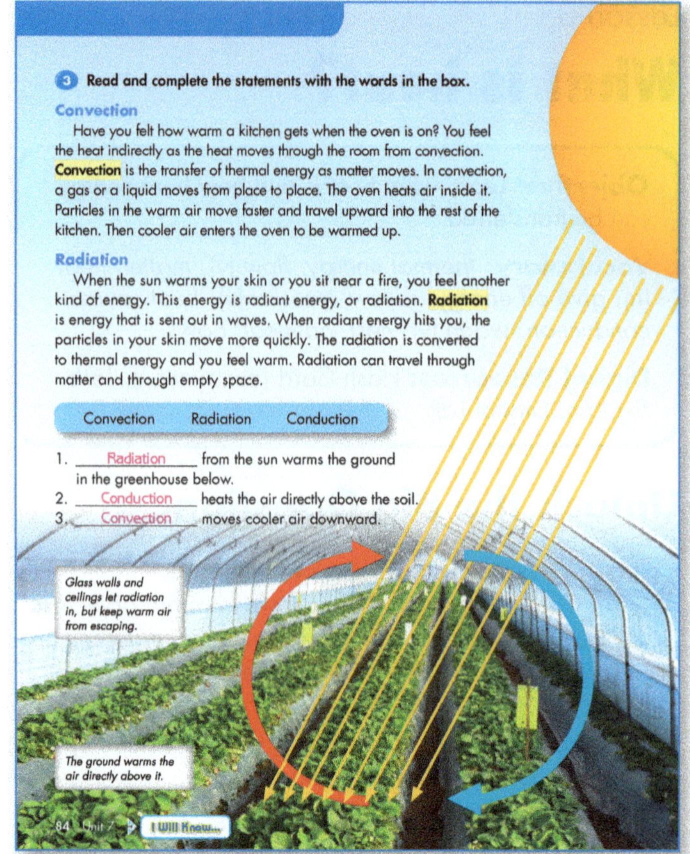

3 **Read and complete the statements with the words in the box.**

Convection

Have you felt how warm a kitchen gets when the oven is on? You feel the heat indirectly as the heat moves through the room from convection. **Convection** is the transfer of thermal energy as matter moves. In convection, a gas or a liquid moves from place to place. The oven heats air inside it. Particles in the warm air move faster and travel upward into the rest of the kitchen. Then cooler air enters the oven to be warmed up.

Radiation

When the sun warms your skin or you sit near a fire, you feel another kind of energy. This energy is radiant energy, or radiation. **Radiation** is energy that is sent out in waves. When radiant energy hits you, the particles in your skin move more quickly. The radiation is converted to thermal energy and you feel warm. Radiation can travel through matter and through empty space.

Convection Radiation Conduction

1. __Radiation__ from the sun warms the ground in the greenhouse below.
2. __Conduction__ heats the air directly above the soil.
3. __Convection__ moves cooler air downward.

ELL Content Support

Write *convection current* on the board. Explain that a convection current suggests a circular movement. To demonstrate, put a few drops of food coloring in a cup of very warm water and pour it slowly into a small aquarium or a large wide-mouth jar of cold water. Have students observe the formation of the convection current and how it curls up and around to form a circle.

Elaborate

Conduction, Convection, and Radiation

Make a three-column chart on construction paper. Label the three columns *Conduction, Convection,* and *Radiation*. Divide the class into three groups: *Conduction, Convection,* and *Radiation*. Have groups investigate two examples of their topic. Ask volunteers to draw two pictures of the type of heat transfer they were assigned on separate sheets of construction paper and then paste them in the correct columns. Invite volunteers to present their drawing to the class.

Lesson 3

What is heat?

Objective: Learn about other forms of energy that can change and give off heat.

Vocabulary: *give off heat, fossil fuels, coal, natural gas, oil, burn, laser light, beam of light, drill* (v), *bond* (v), *treat* (v), *solar panel, solar heat system, flow* (v), *pump* (v), *rub* (v), *friction, skate on ice*

Digital Resources: *Lesson 3 Check* (print out 1 per student), *Got it? 60-Second Video*

Materials: pictures of a gas stove, a car, a charcoal grill

Build Background Display pictures of a gas stove, a car, and a charcoal grill. Have students say what they are used for and how they work.

Explain

4 Read and underline three types of fossil fuels.

Ask *What is heat? The transfer of thermal energy from warmer objects to cooler objects. Where does heat come from? From the sun, our own bodies, animals, and objects that are hot.* Read the first paragraph aloud for students. *Where else can heat come from? From other forms of energy.* Have students read the second paragraph and underline three types of fossil fuels.

5 Read. What evidence would show that the friction of the ice skates sliding on ice causes heat? Discuss with a partner.

Have students rub their hands together. Ask volunteers to describe what happens. (*The hands feel warmer.*) Explain to students that rubbing things together can create heat from friction.

Ask students to read and discuss in pairs what evidence would show that the friction of the ice skates sliding on ice causes heat. (Possible answer: *The friction between skate and ice may cause the ice to melt a little.*)

Elaborate

Fossil Fuels

Divide the class into small groups. Have students investigate where fossil fuels come from and what they are used for. Invite each group to make a poster about fossil fuels and present it to the class.

4 Read and underline three types of fossil fuels.

Changes of Other Energy to Heat

Where does heat come from? Sometimes, other forms of energy change and give off heat. This heat can be used in many different ways.

Fossil Fuel to Heat

Fossil fuels are a source of heat. Coal, natural gas, and oil are all fossil fuels. When these fuels are burned, they produce useful heat. People often use this heat in their homes, schools, and other buildings.

Light to Heat

Have you ever noticed that a black T-shirt gets warm on a hot, sunny day? Objects that are black absorb all the colors of visible light. When the light is absorbed, it becomes a form of heat. So, the black T-shirt warms up. Laser light produces a great deal of heat. A laser is a powerful beam of light. Lasers are often used to cut, drill, and bond materials. Doctors can use lasers to treat eyes, skin, and other organs.

The sun's energy also can change to heat. A solar panel absorbs energy from the sun. In a solar heat system, the sun's energy heats a liquid that flows through the solar panels. The heated liquid is pumped through pipes. It can then be used to heat homes and wash clothes, for example.

Think! What could be other important uses of solar panels?

5 Read. What evidence would show that the friction of the ice skates sliding on ice causes heat? Discuss with a partner.

Friction to Heat

Try rubbing your hands together quickly. The heat you feel is an effect of friction. Friction occurs when two surfaces rub together. As a result of friction, heat is produced. For example, when a person skates on ice, the ice and skates rub together. The skates of this ice skater cause friction. This friction gives off heat.

Lesson 3 Check Got it? 60-Second Video Unit 7 85

Think!

Direct students' attention to the picture of the solar panel. Have students read the *Light to Heat* text. Ask *What could be other important uses of solar panels?* Invite students to discuss the question as a class. (Possible answers: *to heat water in swimming pools, to cook, to produce electricity*)

Evaluate

Lesson 3 Check Assessment for Learning

Distribute the *Lesson 3 Check* and guide students as they complete it. Check answers as a class. Then ask students to grade their progress on the topic of thermal energy and heat 1 to 3: 3 = *I understand how thermal energy is transferred*; 2 = *I need to study more*; 1 = *I need help!* Encourage students giving themselves a 1 or 2 to describe what they found difficult and what they need to study more.

Review Key Words for Lesson 3 (see Student's Book page 83). Play the *Got it? 60-Second Video* to review the lesson material.

Let's Investigate!

In this unit, students learn how energy causes change. In this lab, students will perform an experiment to explore how heat is transferred through objects by conduction.

Let's Investigate! Lab **Which material is the better heat conductor?**

Objective: Students will observe whether metal or plastic is a better conductor of heat.

Materials: 2 small plastic beads, plastic spoon, metal spoon, 2 clear plastic cups (500 mL), very warm water, margarine (1 stick, whole class use), clock with second hand, pouring container (teacher use)

Digital Resources: *Let's Investigate!* Digital Lab, *Let's Investigate! Activity Card* (1 per group)

Advance Preparation: For each group, fill a plastic cup with about 60 mL (about 1/4 cup) of very warm water just before beginning the experiment.

- Divide students into small groups and distribute materials.
- Show students how to place 1/4 of a spoonful of margarine on the handle of each spoon and to stick a bead into the margarine on each spoon. Have students place both spoons in an empty cup.
- Ask students to pick up the cup filled halfway with very warm water and gently pour it into the cup with the spoons.
- Invite students to observe the beads closely and time how long it takes for each ball to fall.
- Have students record their data and write explanations for their observations.
- Ask groups to discuss which of the materials would be better for a cooking pot.
- At the end of the activity, have students share their inferences with the class.

Teacher Time-Saving Option: Show the *Let's Investigate!* Digital Lab as an alternative to the hands-on lab activity.

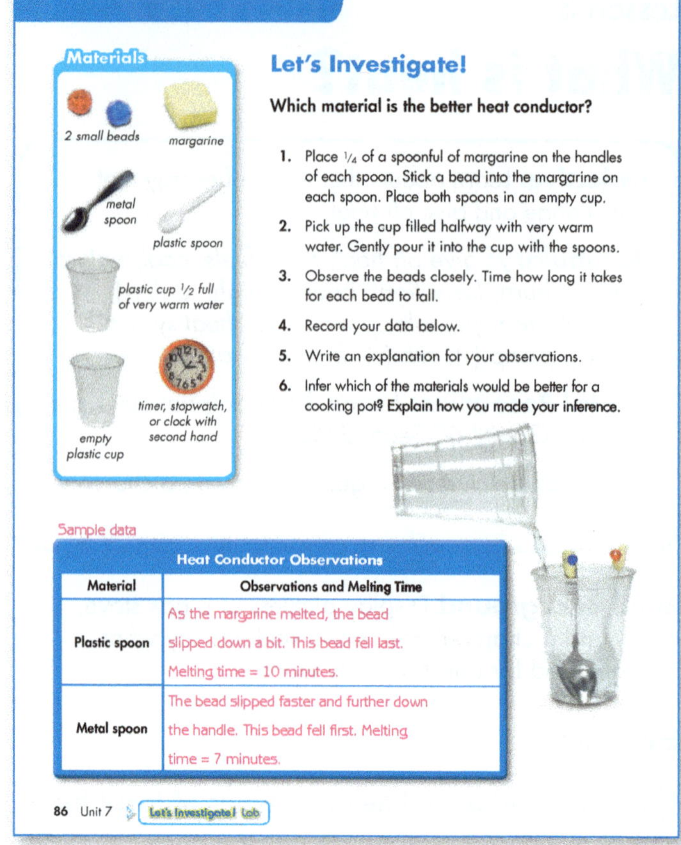

Materials

Let's Investigate!

Which material is the better heat conductor?

1. Place 1/4 of a spoonful of margarine on the handles of each spoon. Stick a bead into the margarine on each spoon. Place both spoons in an empty cup.
2. Pick up the cup filled halfway with very warm water. Gently pour it into the cup with the spoons.
3. Observe the beads closely. Time how long it takes for each bead to fall.
4. Record your data below.
5. Write an explanation for your observations.
6. Infer which of the materials would be better for a cooking pot? **Explain how you made your inference.**

Sample data

Heat Conductor Observations

Material	Observations and Melting Time
Plastic spoon	As the margarine melted, the bead slipped down a bit. This bead fell last. Melting time = 10 minutes.
Metal spoon	The bead slipped faster and further down the handle. This bead fell first. Melting time = 7 minutes.

86 Unit 7 Let's Investigate! Lab

Class Project: Energy Savers

Materials: large sheet of construction paper (1 per group), art supplies

Have students describe different forms of energy that are used in their homes. Remind students that forms of energy can be found indoors and outdoors.

- Divide the class into small groups. Have students research how energy can be saved. Have them list one example for each form of energy.
- Have students make a poster that illustrates five ways to save energy around the house and present it to the class.
- Invite them to tell which form of energy was the most difficult to identify. Ask students if there are any devices that cannot be turned off or used less to conserve energy. A fish tank's aerator would be one example because fish need a constant supply of air to breathe.
- Students should note that lights and heating or cooling systems use energy, for example. Suggestions for conserving energy might include turning off lights when no one is in the room, keeping the thermostat turned down in cooler weather or up in warmer weather, and so on.

Unlock the Big Question

Have students refer to the Big Question on the Unit Opener page. In pairs, have them recall how energy can change. Invite student pairs to share their answers to question 7 on the *Let's Investigate! Activity Card.*

Unit 7 Review

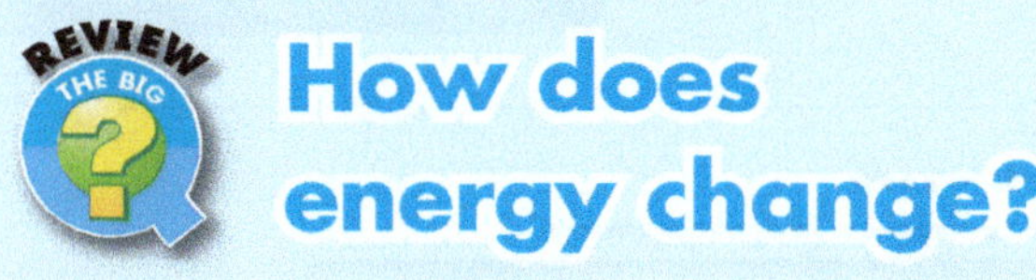

How does energy change?

Digital Resources: Print out 1 of each per student: *Got it? Self Assessment*, *Got it? Quiz*

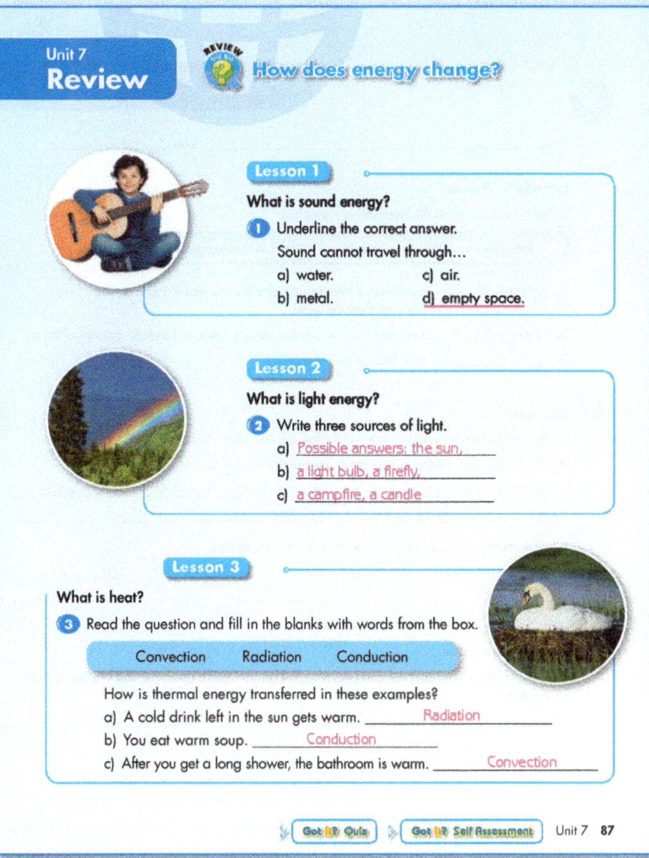

Unit 7 Review — How does energy change?

Lesson 1

What is sound energy?

1 Underline the correct answer.

Sound cannot travel through...

a) water. c) air.

b) metal. d) empty space.

Lesson 2

What is light energy?

2 Write three sources of light.

a) Possible answers: the sun,

b) a light bulb, a firefly,

c) a campfire, a candle

Lesson 3

What is heat?

3 Read the question and fill in the blanks with words from the box.

Convection Radiation Conduction

How is thermal energy transferred in these examples?

a) A cold drink left in the sun gets warm. Radiation

b) You eat warm soup. Conduction

c) After you get a long shower, the bathroom is warm. Convection

Got it? Quiz Got it? Self Assessment Unit 7 87

Evaluate

Strategies for Targeted Review

The following are strategies for providing targeted review for students if they encounter challenges with the content.

Lesson 1 What is sound energy?

Question 1

If... students are having difficulty underlining the correct answer, then... direct students to page 77. Have them find the information about what mediums sound travels in.

Lesson 2 What is light energy?

Question 2

If... students are having difficulty remembering three sources of light, then... direct students to page 81. Have students read the text to find the information they need.

Lesson 3 What is heat?

Question 3

If... students are having difficulty identifying examples of conduction, radiation, and convection, then... direct students' attention to pages 84 and 85. Have students find examples of how thermal energy is transferred through conduction, convection, and radiation.

ELL Language Support

Before students start working on the Review activities, have them read each question aloud along with you.

Got it? Self Assessment

Immediately after students have completed the Review activities, distribute a *Got it? Self Assessment* to each student. Have students complete the *Stop! Wait!* and *Go!* statements for each lesson, allowing them to look back through the lesson material if necessary.

Got it? Quiz

Distribute a Unit 7 *Got it? Quiz* to each student. Quizzes may be used for assessing students' understanding of unit concepts as well as for grading purposes.

Lesson 1 Check

Name ______________________ Date ____________

Words to Know

Write the word next to the description it matches.

frequency	pitch	wavelength

1. wavelength ______ the distance between a point on one wave and a similar point on the next wave
2. frequency ______ the number of waves that pass a point in a certain amount of time
3. pitch ______ how high or low a sound is

Explain

Write whether each statement is true or false. Explain your choice.

4. The higher the amplitude of a wave, the quieter it sounds.

   This statement is ___false___ because the higher the amplitude of a wave, the more energy it has and the louder it sounds.

5. Objects that vibrate more quickly have lower frequencies.

   This statement is ___false___ because objects that vibrate more quickly have higher frequencies.

Apply Concepts

6. How would a drum played on Earth sound different from a drum played in outer space? Explain your answer.

   A drum played on Earth would create sound because there is air through which vibrations can pass. However, outer space is empty, so a drum played there would create no sound.

Unit 7, *Lesson 1 Check* • What is sound energy?

Lesson 2 Check

Name ______________________ Date ____________

Words to Know

Write the word next to the description it matches.

absorption	prism	refraction

1. absorption ______ occurs when an object takes in a light wave
2. refraction ______ the bending of light when it passes into a new medium
3. prism ______ a piece of glass that separates white light into its different wavelengths

Explain

Write whether each statement is true or false. Explain your choice.

4. A piece of glass reflects more light than a piece of paper.

   This statement is ___true___ because the glass is smoother and shinier than the paper.

5. Only sound energy travels in waves.

   This statement is ___false___ because both light and sound travel in waves from their sources.

Apply Concepts

6. Does the color red have a lower frequency than the color green? Explain.

   Yes. Red is on the left side of the visible light spectrum, so it has the lowest frequency of all the colors.

Unit 7, *Lesson 2 Check* • What is light energy?

Lesson 3 Check

Name ______________________ Date ____________

Words to Know

Write the word next to the description it matches.

conduction	convection	radiation

1. convection ______ the transfer of heat as matter moves
2. radiation ______ energy that is sent out in waves
3. conduction ______ the transfer of heat that occurs when one thing touches another

Explain

Write whether the statement is true or false. Explain your choice.

4. When convection occurs, a solid moves from place to place.

   This statement is ___false___ because convenction involves a gas or liquid moving from place to place.

Write the answer to the question on the line.

5. What forms when gas or liquid transfers heat as it moves?

   A convection current forms.

Apply Concepts

6. Explain how a hot-water radiator heats a room's air. Include in your answer the way heat moves from the metal radiator to the room.

   A radiator uses conduction and convection to heat a room. Hot water moves through the inside of the radiator. Heat is transferred to the metal in the radiator, which is an example of conduction. The warm metal in the radiator transfers heat to nearby air particles in the room, and they move faster and rise upward. This allows cooler air to touch the hot metal of the radiator so heat can be transferred. Convection currents move the air through the room.

Unit 7, *Lesson 3 Check* • What is heat?

Lesson 1 Explore My Planet! Activity Card

Name ______________________ Date ____________

Fun Fact: Guitar

How does an acoustic guitar make sound? It starts with plucking the guitar's strings. Plucking causes the strings to vibrate. The vibrations first move through the saddle. Then they move through the bridge to the soundboard. The soundboard is the wooden piece that makes up the front of the guitar's body. The entire soundboard then vibrates. The body of the guitar is shaped in a way that makes the vibrations louder. These sounds come out through the sound hole, producing the sound you hear when someone plays a guitar.

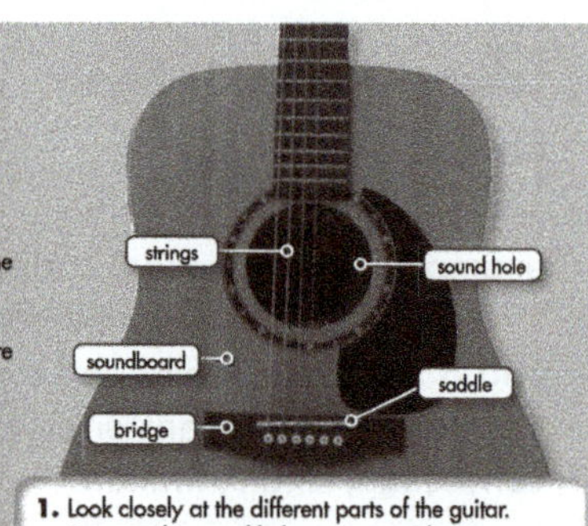

1. Look closely at the different parts of the guitar. Suppose the sound hole were covered up. How do you think the sound would change?

   Possible answer: The sound would be quieter.

How might tightening or loosening a guitar string affect the sound it makes?

Possible answer: Tightening or loosening a string would affect the frequency of the string's vibrations, so the pitch would become higher or lower.

Unit 7, Lesson 1 *Explore My Planet! Activity* • What is sound energy?

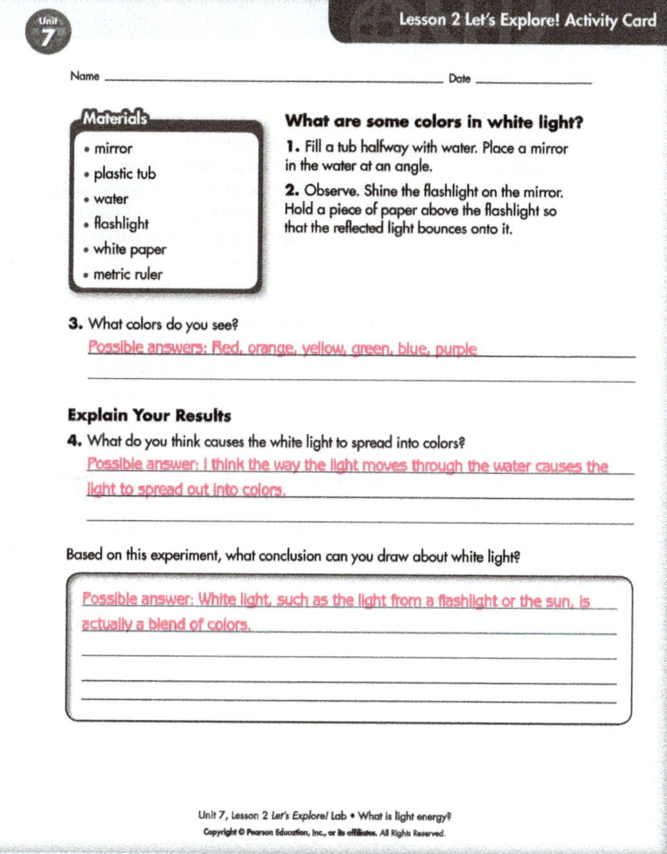

Unit 7

Lesson 2 Let's Explore! Activity Card

Name ______________________ Date ____________

Materials

- mirror
- plastic tub
- water
- flashlight
- white paper
- metric ruler

What are some colors in white light?

1. Fill a tub halfway with water. Place a mirror in the water at an angle.
2. Observe. Shine the flashlight on the mirror. Hold a piece of paper above the flashlight so that the reflected light bounces onto it.
3. What colors do you see?
   Possible answers: Red, orange, yellow, green, blue, purple

Explain Your Results

4. What do you think causes the white light to spread into colors?
   Possible answer: I think the way the light moves through the water causes the light to spread out into colors.

Based on this experiment, what conclusion can you draw about white light?

Possible answer: White light, such as the light from a flashlight or the sun, is actually a blend of colors.

Unit 7, Lesson 2 *Let's Explore!* Lab • What is light energy?

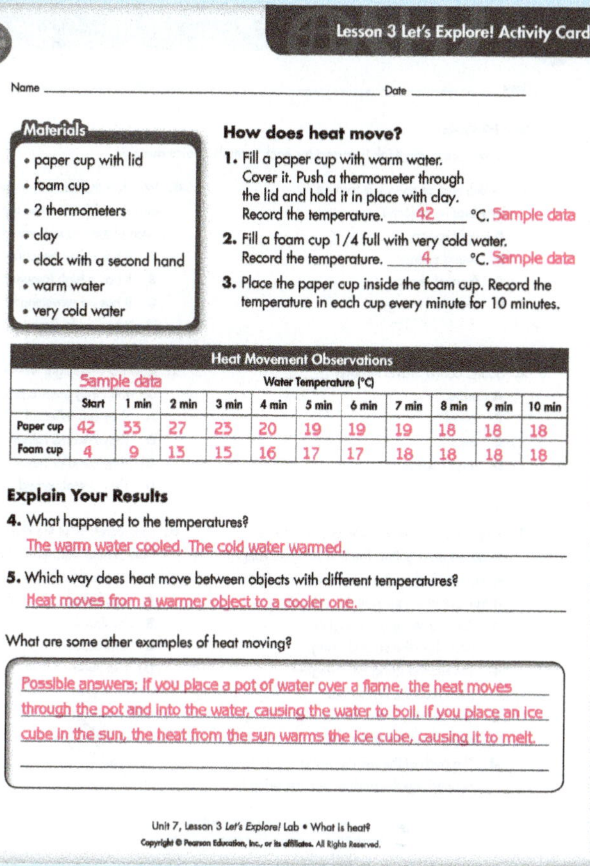

Unit 7

Lesson 3 Let's Explore! Activity Card

Name ______________________ Date ____________

Materials

- paper cup with lid
- foam cup
- 2 thermometers
- clay
- clock with a second hand
- warm water
- very cold water

How does heat move?

1. Fill a paper cup with warm water. Cover it. Push a thermometer through the lid and hold it in place with clay. Record the temperature. 42 °C. Sample data
2. Fill a foam cup 1/4 full with very cold water. Record the temperature. 4 °C. Sample data
3. Place the paper cup inside the foam cup. Record the temperature in each cup every minute for 10 minutes.

Heat Movement Observations

Sample data	Water Temperature (°C)										
	Start	1 min	2 min	3 min	4 min	5 min	6 min	7 min	8 min	9 min	10 min
Paper cup	42	33	27	23	20	19	19	19	18	18	18
Foam cup	4	9	13	15	16	17	17	18	18	18	18

Explain Your Results

4. What happened to the temperatures?
   The warm water cooled. The cold water warmed.
5. Which way does heat move between objects with different temperatures?
   Heat moves from a warmer object to a cooler one.

What are some other examples of heat moving?

Possible answers: If you place a pot of water over a flame, the heat moves through the pot and into the water, causing the water to boil. If you place an ice cube in the sun, the heat from the sun warms the ice cube, causing it to melt.

Unit 7, Lesson 3 *Let's Explore!* Lab • What is heat?

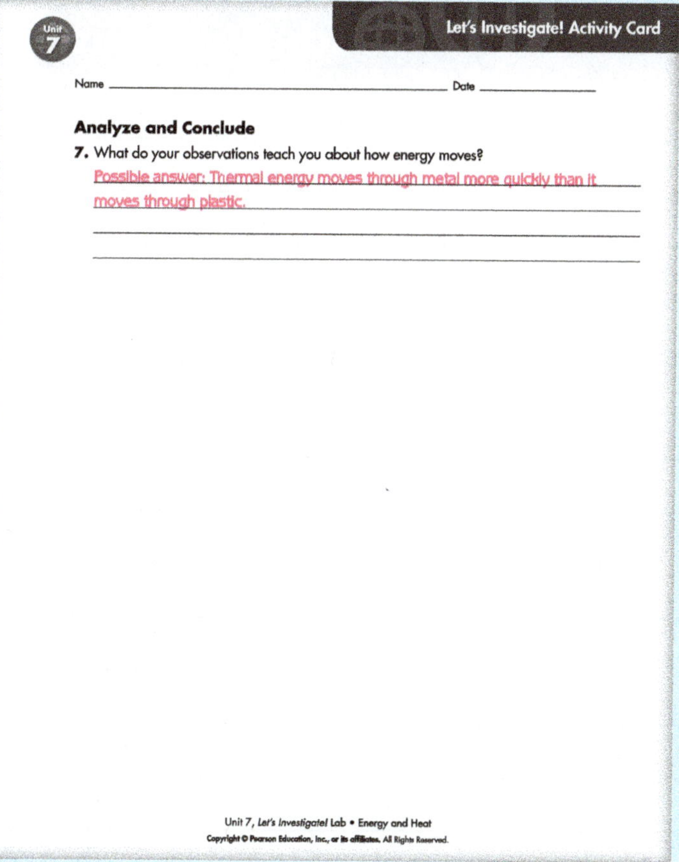

Unit 7

Let's Investigate! Activity Card

Name ______________________ Date ____________

Analyze and Conclude

7. What do your observations teach you about how energy moves?
   Possible answer: Thermal energy moves through metal more quickly than it moves through plastic.

Unit 7, *Let's Investigate!* Lab • Energy and Heat

Unit 7

Lessons 1–3 Got it? Self Assessment

Name ______________________ Date ____________

Got it? Self Assessment

Complete the statements for each lesson.

Lesson 1 What is sound energy?

Stop! I need help with ____________________

Wait! I have a question about ____________________

Go! Now I know ____________________

Lesson 2 What is light energy?

Stop! I need help with ____________________

Wait! I have a question about ____________________

Go! Now I know ____________________

Lesson 3 What is heat?

Stop! I need help with ____________________

Wait! I have a question about ____________________

Go! Now I know ____________________

Unit 7, *Got it? Self Assessment* • Energy and Heat

Unit 7

Got it? Quiz

Name ______________________ Date ____________

Got it? Quiz

Circle the choice you think is correct for each multiple choice question.

1. Which type of energy is formed by vibrating objects?
- **A** light energy
- **(B)** sound energy
- **C** thermal energy
- **D** electrical energy

2. Sound cannot travel through ________.
- **A** water
- **B** metal
- **C** air
- **(D)** empty space

3. A guitarist strums two strings of the same thickness on a guitar. The first string vibrates more quickly than the second string. What is true about the pitch of the strings?
- **(A)** The first string has a higher pitch than the second string.
- **B** The second string has a higher pitch than the first string.
- **C** The first string and the second string have the same pitch.
- **D** The pitch of the second string depends on the pitch of the first string.

4. Which is the best description of a sound wave when the pitch of the sound is very high?
- **A** It has a low frequency.
- **(B)** It has a high frequency.
- **C** It has no wavelength.
- **D** It has no frequency.

5. How do white light waves act when they enter a prism?
- **A** They are reflected.
- **B** They become raindrops.
- **(C)** They bend or refract.
- **D** They are absorbed.

6. Look carefully at the illustration. How is heat transferred from the campfire to the people?
- **A** conduction
- **B** insulation
- **C** convection
- **(D)** radiation

Unit 7, *Got it? Quiz* • Energy and Heat

Unit 7

Got it? Quiz

Name ______________________ Date ____________

7. Which of the following can act like a prism?
- **A** a radio
- **(B)** a raindrop
- **C** a piece of paper
- **D** a laser

8. Which of the following shows convection?
- **A** The sun warms your skin.
- **B** You rub your hands together quickly.
- **C** A metal spoon in soup becomes warm.
- **(D)** A mobile turns from a candle burning below it.

9. Suppose light waves refract as they pass through a raindrop. Describe what happens to the light.

Possible answer: The colors in the visible light spectrum separate as they leave the raindrop, and a rainbow is formed.

10. Think back to this morning as you were getting ready for school. Name two different forms of energy you used or saw. How did you use or how were they being use?

Answers will vary.

Unit 7, *Got it? Quiz* • Energy and Heat

Teacher's Notes

Unit 7 Study Guide

How does energy change?

Lesson 1
What is sound energy?

- Sound is produced by vibrating objects. The way an object is made and the way it vibrates affect the type of sound we hear.
- A sound wave's frequency and energy also affect the sound we hear.

Lesson 2
What is light energy?

- Visible light is made up of waves with different wavelengths and frequencies.
- Light can be refracted, reflected, or absorbed.

Lesson 3
What is heat?

- Heat is the transfer of thermal energy.
- Heat can move by conduction, convection, or radiation.
- Other forms of energy can change to heat.

Review the Big Question

How does energy change?

Have students use what they have learned from the unit to answer the question in their own words. *How has your answer to the Big Question changed since the beginning of the unit? What are some things you learned that caused your answer to change?*

Make a Concept Map

Have students make a concept map like the one shown on this page to help them organize key concepts.

Unit 7 Concept Map

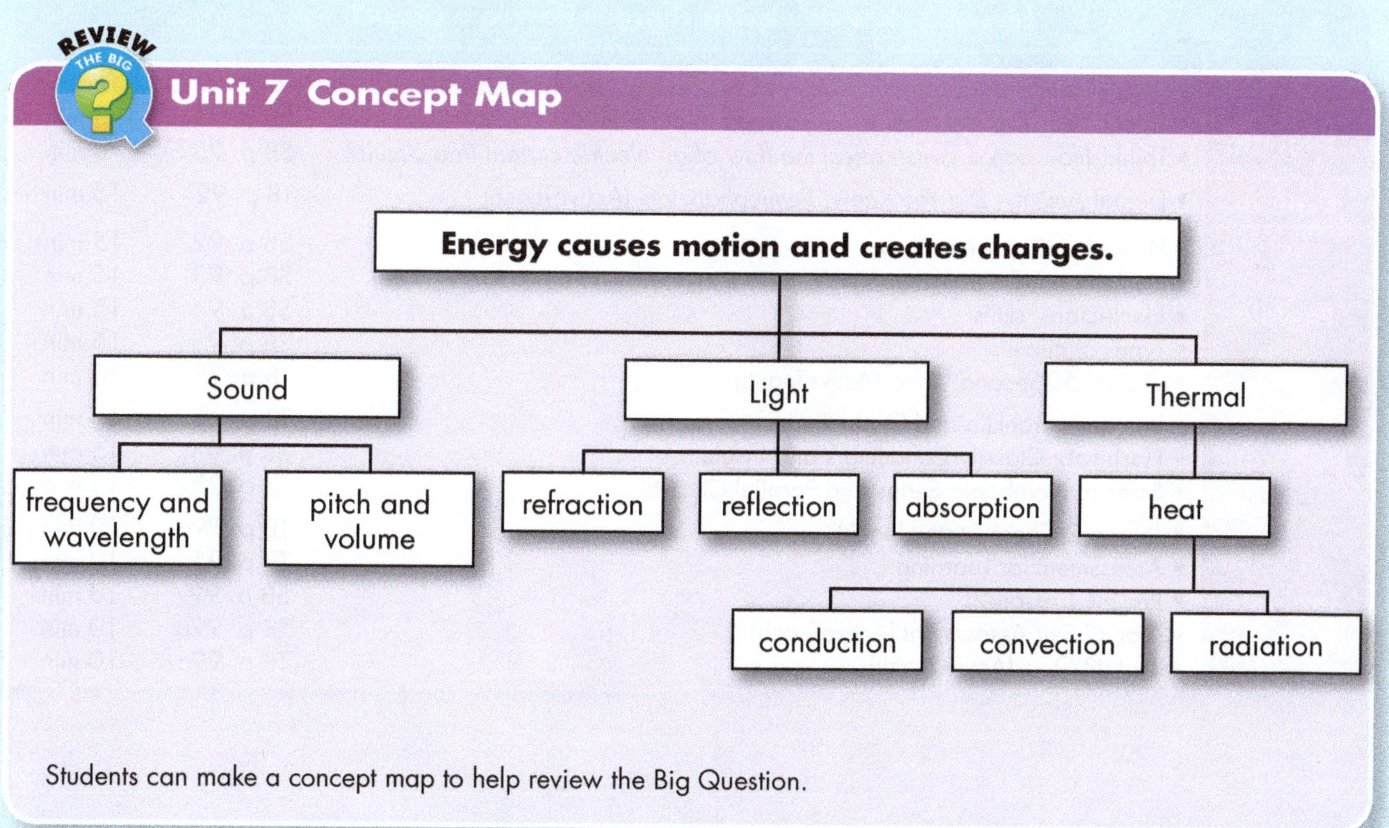

Students can make a concept map to help review the Big Question.

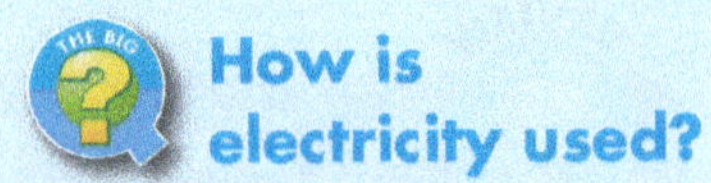

Lesson Plan

Unit Opener & Lesson 1 What is static electricity?			
	Activity	**Pages**	**Time**
Engage	• Unit Opener: Think! *What lights the night?* • Unit Opener: Identify what electronic devices need to work. • Unit Opener: Discuss how electric devices have improved human lives. • Think! *What causes the glow of a lightning bolt?*	SB p. 88 SB p. 88 SB p. 88 SB p. 91	5 min 10 min 10 min 5 min
Explore	• Digital Lab: *What is one effect of static electricity?* (ActiveTeach)	TB p. 89	15 min
Explain	• Atoms and static electricity • How charged objects behave • Effects of static electricity • *Got it? 60-Second Video* (ActiveTeach)	SB p. 89 SB p. 90 SB p. 91 TB p. 91	15 min 15 min 25 min 10 min
Elaborate	• Science Notebook: Electric Charge • At-Home Lab: Strength of Force • Science Notebook: Positive and Negative Charges • Science Notebook: How Lightning Forms	TB p. 89 SB p. 90 TB p. 90 TB p. 91	10 min 15 min 10 min 20 min
Evaluate	• *Lesson 1 Check* (ActiveTeach) • Assessment for Learning • Review (Lesson 1) • *Got it? Self Assessment* (ActiveTeach) • *Got it? Quiz* (ActiveTeach)	TB p. 99a TB p. 91 SB p. 99 TB p. 99b TB p. 99c	10 min 10 min 10 min 10 min 10 min

Lesson 2 How do electric charges flow in a circuit?			
	Activity	**Pages**	**Time**
Engage	• Think! *Why are electrical wires covered in plastic?* • Think! *How can a switch affect the flow of an electric current in a circuit?*	SB p. 93 SB p. 95	5 min 5 min
Explore	• Digital Activity: *Did You Know: Semiconductors* (ActiveTeach)	TB p. 92	15 min
Explain	• How electric currents flow • Insulators and conductors • Electricians' skills • Types of circuits • *Got it? 60-Second Video* (ActiveTeach)	SB p. 92 SB p. 93 SB p. 94 SB p. 95 TB p. 95	15 min 15 min 15 min 15 min 10 min
Elaborate	• Benjamin Franklin and Electricity • Flash Lab: Classify Conductors and Insulators • Science Notebook: Series and Parallel Circuits	TB p. 93 SB p. 94 TB p. 95	15 min 15 min 15 min
Evaluate	• *Lesson 2 Check* (ActiveTeach) • Assessment for Learning • Review (Lesson 2) • *Got it? Self Assessment* (ActiveTeach) • *Got it? Quiz* (ActiveTeach)	TB p. 99a TB p. 95 SB p. 99 TB p. 99b TB p. 99c	10 min 10 min 10 min 10 min 10 min

Lesson 3 How does electricity transfer energy?			
	Activity	**Pages**	**Time**
Engage	• Think! *Why do many hair dryers use heating coils as resistors?*	TB p. 97	5 min
Explore	• Digital Activity: *Voices from History: Thomas Edison* (ActiveTeach)	TB p. 96	15 min
Explain	• How energy changes form • Filaments, coils, and phantom energy • *Got it? 60-Second Video* (ActiveTeach)	SB p. 96 SB p. 97 TB p. 97	15 min 15 min 10 min
Elaborate	• Flash Lab: Motion and Heat • Electricity Every Day	TB p. 96 TB p. 97	20 min 15 min
Evaluate	• *Lesson 3 Check* (ActiveTeach) • Assessment for Learning • Review (Lesson 3) • *Got it? Self Assessment* (ActiveTeach) • *Got it? Quiz* (ActiveTeach)	TB p. 99a TB p. 97 SB p. 99 TB p. 99b TB p. 99c	10 min 10 min 10 min 10 min 10 min
Lab	• *Let's Investigate! How can a switch make a complete circuit?* (ActiveTeach)	SB p. 98	30 min

Flash Cards

atom

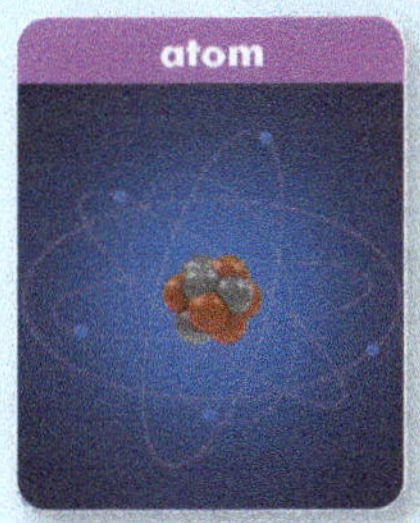

electric charge

static electricity

lightning

conductor

insulator

electrician

circuit

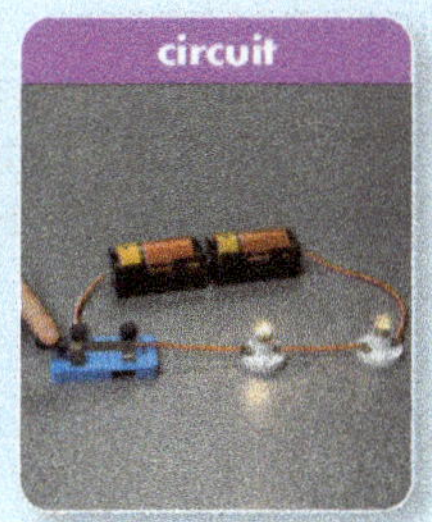

filament

Lesson 1

Key Words	ELL Support
atom, electric charge, static electricity, electric force, lightning	**Vocabulary:** *protons, neutrons, electrons, attract, push away, positive, negative, positively charged, negatively charged, pull toward, repel, stand on end, rub against, strands of hair, dash* (v), *carpet, doorknob, zap* (n), *thunderstorm, bolt of lightning, flash* (v/n), *drag your feet, spark* (n), *droplets, ground* (n), *glow* (v/n)

Lesson 2

Key Words	ELL Support
electric current, circuit, battery, conductor, insulator, series circuit, parallel circuit	**Vocabulary:** *plug* (v), *string of light bulbs, electrical outlet, light up, flow* (v/n), *loop* (n), *wires, switch, flashlight, copper, gold, silver, lead* (n), *graphite, electrical cord, electrician, gloves, rubber, glass, dry wood, eraser, chalk, coin, tire, scissors, jug, rubber boots, ring, gold conductors*

Lesson 3

Key Words	ELL Support
resistor, filament, phantom energy	**Vocabulary:** *tank, pump* (n/v), *light, sound, motion, pluck* (v), *guitar string, sound energy, wind turbine, wind energy, heat* (v), *light* (v/n), *incandescent light bulb, melt, pass through, give off heat, coil, appliances, plug into, power source, household, conserve, power strip, switch off, environment* **Word Formation:** *electricity, electric, electrician, electrically*

Unit 8 Electricity

Unit Objectives

Lesson 1: Students will explain what static electricity is and how charged objects behave.

Lesson 2: Students will describe how electric current flows in a circuit.

Lesson 3: Students will explain how energy changes form and how electricity is transformed into light and gives off heat.

Vocabulary: *light* (v), *television, computer, video games, plug* (n), *wires, socket, appliances, electric devices*

Materials: pictures of common appliances

Introduce the Big Question

How is electricity used?

Build Background Display pictures of common appliances. Elicit their names and what they are used for. Have students brainstorm other electric appliances they and their families use, and make a list on the board.

Engage

What lights the night?

Point to the photo on the bottom right. *This is what the sky above Mexico and a small part of the United States of America look like at night. What lights the night?* Ask volunteers to share their ideas with the class, along with their reasoning.

1 Complete the words to label the pictures.

Point to the pictures of the devices in the exercise and see if students can identify each item. Then invite students to complete the names by filling in the blanks. When they have finished, check answers and practice pronunciation by having volunteers name each item. Ask students how often they use each of the devices.

2 What do the three appliances above need to make them work? How do they get it? Discuss with a partner.

Explain the meaning of the word *appliance*. Then elicit the names of the three appliances above. Have pairs discuss what they need in order to make them work and how they get it. (Possible answer: *Electricity is transferred to our houses to power the electronics and appliances.*)

Unit 8 Electricity

How is electricity used?

I will learn

- what static energy is.
- how electric charges flow in a circuit.
- how electricity transfers energy.

1 Complete the words to label the pictures.

television computer video game

plug wires socket

2 What do the three appliances above need to make them work? How do they get it? Discuss with a partner.

3 How have electric devices improved human lives? Discuss as a class.

Think! What lights the night?

88 Unit 8

3 How have electric devices improved human lives? Discuss as a class.

On the board, make a two-column chart with the headings *Electric Devices* and *Usefulness*. *What electric devices do you find the most useful? Why do you think they are so useful?* Write students' ideas in the corresponding columns. Discuss with the class how such devices have improved human lives.

Think! Again!

What lights the night?

Ask students to describe what happens when they plug a lamp into an electrical outlet and turn the switch on. (*The lamp gives off light.*) Then ask students to describe what process takes place. (Possible answer: *Electrical energy is transformed into light energy.*) *How does electrical energy help to light the cities in your country at night? Electrical energy changes to light energy. What else can light the night?* Accept all logical answers.

ELL Content Support

Explain to students that the electricity for all the appliances we use may come from different sources, such as power plants, wind turbines, and solar panels. Ask students to investigate which source is the most powerful, which is the most expensive, and which is the most polluting.

Lesson 1

What is static electricity?

Objective: Learn what atoms are and how their particles determine electric charges.

Vocabulary: *atoms, building blocks, make up, matter, protons, neutrons, electrons, electric charge, attract, push away, positive, negative, balance* (v), *overall, neutral, gain* (v), *end up, lose, static electricity, positively charged, negatively charged*

Digital Resources: Flash Cards (*static electricity, atom*), *Let's Explore!* Digital Lab

Materials: picture of cinderblocks or bricks

Unlock the Big Question

Write the following on the board: *I will learn what static electricity is and how charged objects behave.*

Build Background Write the antonym pair *attract–repel* on the board. Have students hold their hands in front of them with palms facing each other. Say *attract* and show students how to move their hands together until their palms touch. Then say *repel* and show students how to move their hands away from each other.

Explore

Let's Explore! Lab What is one effect of static electricity?

Objective: Students will observe how a static electrical charge on paper strips affects them.

Digital Resources: *Let's Explore!* Digital Lab, *Let's Explore! Activity Card* (1 per student) (*Optional*: Do the lab in class; refer to the *Activity Card* for materials and steps.)

- Show the Digital Lab and have students complete the *Activity Card*. Have students check their answers in small groups or pairs. Provide support as needed.

Explain

1 What is causing the light between these wires? Discuss as a class.

Direct students' attention to the picture at the top of the page. Ask *What is causing the light between these wires?* Students may say that the light is caused by an electrical spark.

2 Read and circle the balloon that has a negative charge.

Display the *atom* Flash Card and the picture of some cinderblocks or bricks. Read the first paragraph aloud

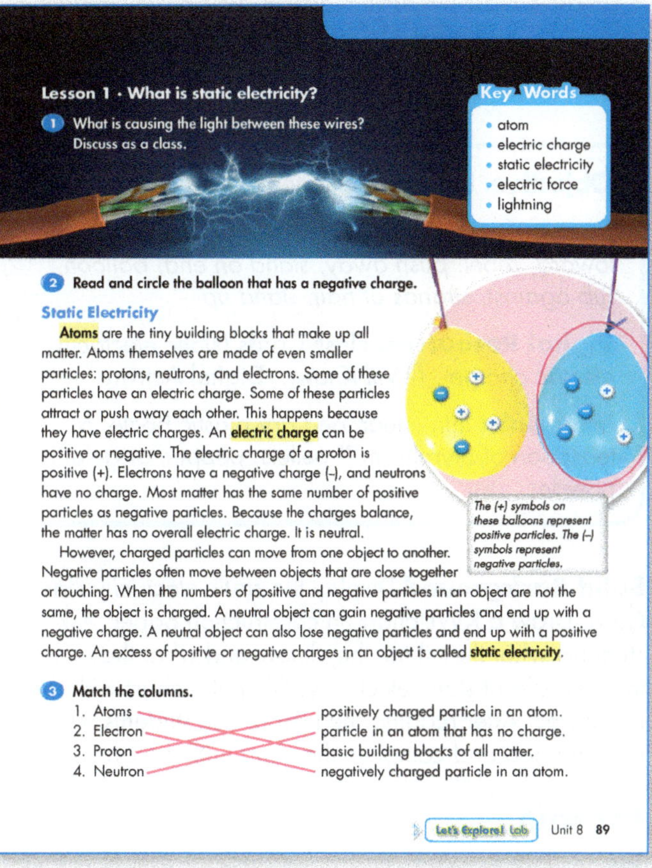

Lesson 1 • What is static electricity?

1 What is causing the light between these wires? Discuss as a class.

Key Words
- atom
- electric charge
- static electricity
- electric force
- lightning

2 Read and circle the balloon that has a negative charge.

Static Electricity

Atoms are the tiny building blocks that make up all matter. Atoms themselves are made of even smaller particles: protons, neutrons, and electrons. Some of these particles have an electric charge. Some of these particles attract or push away each other. This happens because they have electric charges. An **electric charge** can be positive or negative. The electric charge of a proton is positive (+). Electrons have a negative charge (–), and neutrons have no charge. Most matter has the same number of positive particles as negative particles. Because the charges balance, the matter has no overall electric charge. It is neutral.

However, charged particles can move from one object to another. Negative particles often move between objects that are close together or touching. When the numbers of positive and negative particles in an object are not the same, the object is charged. A neutral object can gain negative particles and end up with a negative charge. A neutral object can also lose negative particles and end up with a positive charge. An excess of positive or negative charges in an object is called **static electricity**.

The (+) symbols on these balloons represent positive particles. The (–) symbols represent negative particles.

3 Match the columns.

1. Atoms	positively charged particle in an atom.
2. Electron	particle in an atom that has no charge.
3. Proton	basic building blocks of all matter.
4. Neutron	negatively charged particle in an atom.

Let's Explore! Lab Unit 8 89

for students. *How can atoms and cinderblocks be similar? Because atoms make up all matter, the way cinderblocks make up buildings.* Read the caption beneath the two balloons with the class. Discuss with students which balloon has a negative charge.

ELL Content Support

Explain that the type of energy that is used in our homes, schools, and businesses is called *current electricity*. Current electricity is a stream of electrons that flows through a conductor. *An electrical conductor is a material that is good at allowing a current of electricity to flow through it. Power plants produce the electricity that is sent through wires, which are conductors, to most homes, schools, and businesses.*

3 Match the columns.

Have pairs read the text and match the columns. Check answers as a class.

Elaborate

Science Notebook: Electric Charge

Have each student make a word web with the vocabulary term *electric charge* in the middle. In three circles around it, have students find and fill in related words: *positive charge, negative charge,* and *no charge*. Ask students to read their word webs aloud to the rest of the class.

Lesson 1

What is static electricity?

Objective: Learn how charged objects behave.

Vocabulary: *opposite charges, attract, pull toward, repel, push away, stand on end, balloon, rub against, strands of hair, stand up*

Digital Resources: Flash Cards (*static electricity, electric charge*), *I Will Know...* Digital Activity

Materials: thin, clear, and dry plastic bottle, a teaspoon of Styrofoam™ pellets, pepper, or gelatin powder

Build Background Display the *static electricity* Flash Card. Draw a KWL chart on construction paper. Ask students what they think they know and what they want to know about static electricity. Fill in the chart with their responses. After reading the lesson, ask students what they learned and complete the chart with their answers.

Explain

4 Read and write a caption that describes what is causing the girl's hair to stand on end.

Display the *electric charge* Flash Card. Have students explain what it is. *Remember that all matter in the universe is made of atoms. This means that everything can have an electric charge. An electric charge can be positive or negative.*

Read the text aloud for students. Point at the *static electricity* Flash Card for pairs to write a descriptive caption. Check answers as a class. Make sure it is clear to students that opposite charges attract each other.

5 Read and draw positive (+) and negative (–) symbols on the red balloon and the wall.

Direct students' attention to the picture of the balloons and describe it. Ask students to read and draw symbols on the red balloon and the wall. Have students explain in their own words why this happens.

6 What might happen if you place a negatively charged balloon close to small paper scraps? Discuss as a class.

How can you charge a balloon negatively? Have students find the answer in the text. (*You can rub it against your hair.*) *What might happen if you place a negatively charged balloon close to small paper scraps?* (*The negatively charged balloon will attract and lift the paper scraps.*) Invite students to discuss why they think this happens.

4 **Read and write a caption that describes what is causing the girl's hair to stand on end.**

Positive charges on this girl's strands of hair repel one another and are attracted to the negative charges on the balloon.

How Charged Objects Behave

Charged objects behave in predictable ways. If two objects have opposite charges, they attract, or pull toward, each other. Objects with the same charge repel, or push away from, each other. Have you ever noticed that your hair stands on end after you rub a balloon against it? When you do this, negative charges move from your hair to the balloon. The strands of your hair are left with a positive charge. The hairs stand up as they try to push as far away from each other as possible.

5 **Read and draw positive (+) and negative (–) symbols on the red balloon and the wall.**

Electric Force

The pull or push between two charged objects is called an electric force. Electric force gets stronger when the charged objects are close together. **Electric force** can also exist between charged objects and neutral objects. For this reason, electric force can move lightweight neutral objects or lift them into the air.

If you rub a balloon against your hair and then place the balloon against a wall, the balloon will stick to the wall. Why does this happen? First, the balloon picks up negative particles from your hair. Then, the extra negative particles on the balloon push away some of the negative particles on the wall. The part of the wall close to the balloon is left with an excess of positive particles. These positive particles attract the negative particles on the balloon. An electric force holds the balloon against the wall until the balloon loses its charge.

6 **What might happen if you place a negatively charged balloon close to small paper scraps? Discuss as a class.**

At-Home Lab

Strength of Force
Gather lightweight objects of different sizes, such as scraps of paper and cloth. Set them on a table. Rub an inflated balloon against your hair. Hold the balloon close to the objects. Does the balloon attract any objects? Which ones? What can you conclude about the strength of the balloon's electric force?

90 Unit 8 I Will Know...

Elaborate

At-Home Lab

Strength of Force

Materials: a balloon, lightweight objects, such as scraps of paper and cloth

Students should notice that the balloon attracts small, very light objects. They should conclude that the strength of the balloon's electric force is greater than the force of gravity acting on the objects it attracts.

Science Notebook: Positive and Negative Charges

Get a thin, clear plastic bottle. Make sure the inside is dry. Add a teaspoon of Styrofoam™ pellets, pepper, or gelatin powder. Rub a balloon with a piece of cloth. Hold the charged balloon near the edge of the bottle, but not touching it. Have students observe and discuss what happens. (*The Styrofoam™ pellets, pepper, or gelatin powder moves to the section of the plastic bottle near the charged balloon.*)

I Will Know...

Have students do the *I Will Know...* Digital Activity.

Lesson 1

What is static electricity?

Objective: Learn about the similarities between small shocks and lightning.

Vocabulary: *dash* (v), *carpet, doorknob, zap* (n), *startle, thunderstorm, bolt of lightning, flash* (v/n), *static, release* (v/n), *static energy, build up, drag your feet, negatively charged, spark* (n), *droplets, rub against, ground* (n), *attract, cloud, heat up, glow* (v/n)

Digital Resources: Flash Cards (*static electricity, electric charge, lightning*), *Lesson 1 Check* (print out 1 per student), *Got it? 60-Second Video*

Build Background Display the *static electricity* and *electric charge* Flash Cards and the KWL chart to review what students have learned about static electricity so far. Tell students that, in today's class, they will learn new things about static electricity and, therefore, will be able to complete the chart with new information.

Explain

7 Read and underline the sentence that tells how a bolt of lightning and a small zap of electricity are similar.

Discuss with students what can happen when they drag their feet on the carpet and then touch a metal object. Ask students to read the first paragraph to confirm their answers. Display the *lightning* Flash Card. *How are a bolt of lightning and a small zap of electricity similar?* Ask students to read and underline the sentence that explains the similarity.

8 Read. With a partner, explain in your own words what can cause a small shock.

Imagine a girl touches a metal doorknob and feels a small shock. What probably caused the shock? Ask students to read and discuss the answer in pairs.

9 What do small shocks and lightning have in common? Discuss with a partner.

Have pairs discuss what small shocks and lightning have in common.

Think!

What causes the glow of a lightning bolt?

Display the *lightning* Flash Card. Discuss with the class what causes the glow of a lightning bolt. (Possible answer: *When static electricity is released in lightning, the electrical energy heats up the air and makes it glow.*)

7 Read and underline the sentence that tells how a bolt of lightning and a small zap of electricity are similar.

Effects of Static Electricity

You dash across a carpet and touch a metal doorknob. Ouch! A small zap of electricity startles you. During a thunderstorm, you see a big bolt of lightning flash across the sky. How are the small zap and the flash of lightning similar? Both result from the movement of charged particles.

The word *static* means *not moving*, and electric charge can stay on an object for some time. But eventually the charge does move. The negative particles on an object with negative charge may move to another object. This movement of charged particles releases static electricity.

8 Read. With a partner, explain in your own words what can cause a small shock.

Small Shocks

Static electricity can build up on all types of objects. It also can build up on you! You can pick up negative particles when you drag your feet on a carpet. You become negatively charged. When you reach for a metal doorknob, the negative particles travel from you to the doorknob. You might even see the spark caused by the release of static electricity!

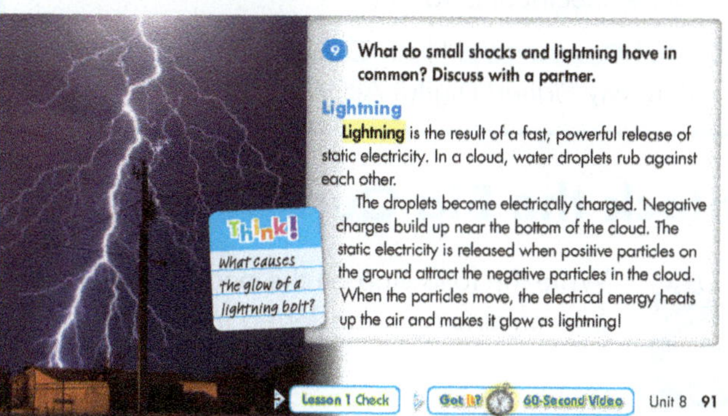

9 What do small shocks and lightning have in common? Discuss with a partner.
Lightning
Lightning is the result of a fast, powerful release of static electricity. In a cloud, water droplets rub against each other.
The droplets become electrically charged. Negative charges build up near the bottom of the cloud. The static electricity is released when positive particles on the ground attract the negative particles in the cloud. When the particles move, the electrical energy heats up the air and makes it glow as lightning!
Think!
What causes the glow of a lightning bolt?

Lesson 1 Check | Got it? 60-Second Video | Unit 8 91

Elaborate

Science Notebook: How Lightning Forms

Discuss with the class the series of events that produces lightning. Have students illustrate and describe the steps in their Science Notebooks. Students should start by drawing a cloud and the ground below it. Then they should draw minus signs at the bottom of the cloud to represent the negative charges that have built up in the cloud. Students should continue to illustrate until they have drawn a flash of lightning.

Evaluate

Lesson 1 Check **Assessment for Learning**

Distribute the *Lesson 1 Check* and guide students as they complete it. Check answers as a class. Then ask students to grade their progress on the topic of static electricity from 1 to 3: 3 = *I understand that small shocks and lightning are effects of static electricity*; 2 = *I need to study more*; 1 = *I need help!* Encourage students giving themselves a 1 or 2 to describe what they found difficult and what they need to study more.

Review Key Words for Lesson 1 (see Student's Book page 89). Play the *Got it? 60-Second Video* to review the lesson material.

Lesson 2

How do electric charges flow in a circuit?

Objective: Learn how electricity flows in a circuit. Learn that some materials conduct electricity.

Vocabulary: *plug* (v), *string of light bulbs, electric outlet, light up, flow* (v/n), *invisibly, electric current, electric charges, loop* (n), *circuit, battery, wires, path, gaps, breaks off, switch, flashlight, turn off, atoms, conductors, copper, gold, silver, lead* (n), *graphite, electrical cord*

Digital Resources: Flash Card (*conductor*), *Explore My Planet!* Digital Activity

Unlock the Big Question

Write the following on the board: *I will know how electricity is transferred in a circuit.*

Build Background Turn the classroom light off and on. *How does this kind of energy get to your house or your school whenever you need it? The poles and wires outside, above or below the ground, are the electrical transmission and distribution system.* Display the *conductor* Flash Card. *Electricity travels through the wires in your home to the outlets and switches so that there is always electricity for you when you need it.*

Explore

Explore My Planet! Did You Know: Semiconductors

Objective: Students will learn what semiconductors are made of and how they are used in computers.

Digital Resources: *Explore My Planet!* Digital Activity, *Explore My Planet! Activity Card* (1 per student)

- Display the *conductor* Flash Card. *Remember how we said that a conductor was a material that was good at transferring electricity? Materials that do not transfer electricity are called insulators. In this activity, we are going to learn about materials that transfer electricity, but only a little bit of it. They are called semiconductors.*
- Show the *Explore My Planet!* to the class.
- Have pairs complete the *Activity Card*. Provide support as needed.
- Ask volunteers to read their answers to the class.

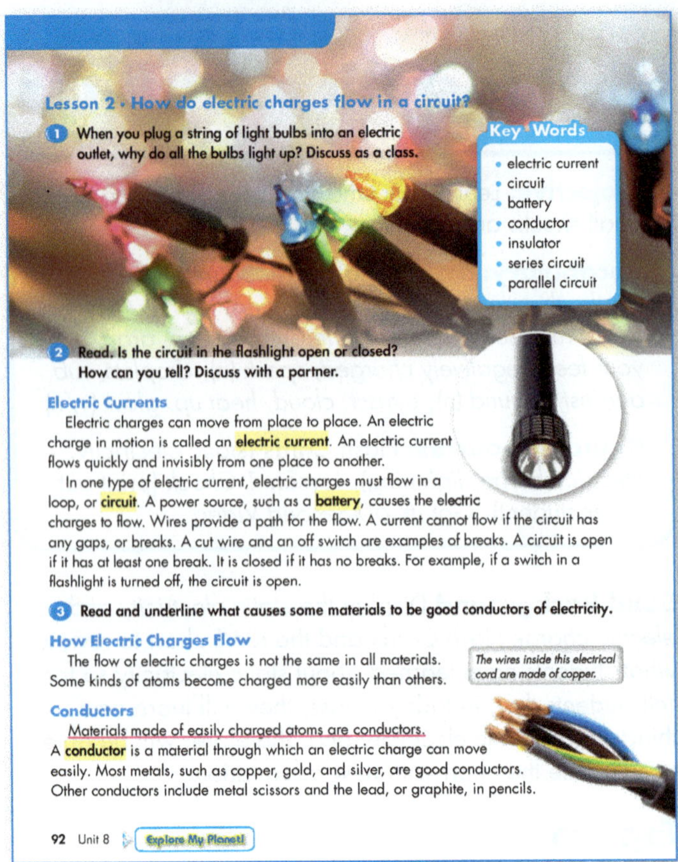

Lesson 2 • How do electric charges flow in a circuit?

1 When you plug a string of light bulbs into an electric outlet, why do all the bulbs light up? Discuss as a class.

Key Words
- electric current
- circuit
- battery
- conductor
- insulator
- series circuit
- parallel circuit

2 Read. Is the circuit in the flashlight open or closed? How can you tell? Discuss with a partner.

Electric Currents

Electric charges can move from place to place. An electric charge in motion is called an **electric current**. An electric current flows quickly and invisibly from one place to another.

In one type of electric current, electric charges must flow in a loop, or **circuit**. A power source, such as a **battery**, causes the electric charges to flow. Wires provide a path for the flow. A current cannot flow if the circuit has any gaps, or breaks. A cut wire and an off switch are examples of breaks. A circuit is open if it has at least one break. It is closed if it has no breaks. For example, if a switch in a flashlight is turned off, the circuit is open.

3 Read and underline what causes some materials to be good conductors of electricity.

How Electric Charges Flow

The flow of electric charges is not the same in all materials. Some kinds of atoms become charged more easily than others.

The wires inside this electrical cord are made of copper.

Conductors

Materials made of easily charged atoms are conductors. A **conductor** is a material through which an electric charge can move easily. Most metals, such as copper, gold, and silver, are good conductors. Other conductors include metal scissors and the lead, or graphite, in pencils.

92 Unit 8 Explore My Planet!

Explain

1 When you plug a string of light bulbs into an electric outlet, why do all the bulbs light up? Discuss as a class.

Look at the string of light bulbs. Have you ever seen lights like this? When you plug a string of light bulbs into an electric outlet, why do all the bulbs light up? Students may say electricity lights the bulbs.

2 Read. Is the circuit in the flashlight open or closed? How can you tell? Discuss with a partner.

Ask students to look at the flashlight and say if it is on or off and how they know. Have pairs read and discuss whether the circuit is open or closed. Check answers as a class. Ask questions to check comprehension. *What do the wires in the string of light bulbs in the picture do?* (Answer: *They provide a path for the energy to flow.*) *Do you think there are any gaps in the circuit?* (Answer: *No, because all the lights are on.*)

3 Read and underline what causes some materials to be good conductors of electricity.

Display the *conductor* Flash Card. Have students say what materials the wires are made of. Read the text for students to underline what causes some materials to be good conductors of electricity.

Lesson 2

How do electric charges flow in a circuit?

Objective: Identify what materials are conductors and insulators.

Vocabulary: *electrician, gloves, atoms, insulator, electric charge, plastic, rubber, glass, dry wood, eraser, chalk, coin, tire, scissors, jug, rubber boots, ring, gold conductors, flow* (v), *electrical wires*

Digital Resources: Flash Cards (*conductor, insulator*), *I Will Know...* Digital Activity

Build Background Display the *conductor* Flash Card. *What is a conductor?* Write on the board: *A conductor is a material through which electric charge moves easily.*

Explain

4 Read and look at the picture. Why do electricians often wear special gloves? Discuss with a partner.

Display the *insulator* Flash Card. *What material are the gloves made of?* (Possible answers: *plastic, rubber*) Write the following sentence frame on the board: *An insulator is a material through which an electric charge moves* ________. Elicit ways to complete the sentence. Have students read to confirm their predictions. Ask *Why do electricians often wear special gloves?* Students answer the question in pairs. (Possible answer: *Electric charge cannot move through the plastic or rubber and shock the electrician.*)

5 What is each object made of? Is it a conductor or an insulator? Discuss with a partner and write your answers below.

First have pairs discuss what material each object is made of and then decide whether each object is a conductor or an insulator.

6 The computer part below is made of gold conductors. Will gold allow electric charges to flow easily in this computer part? Why or why not? Discuss as a class.

Challenge students to explain whether gold will allow electric charges to flow easily in the computer in the picture. (Possible answer: *Yes. Gold is a conductor, so electric charges move easily through it.*) Explain to students that gold is a very good conductor and that oxygen does not chemically change it, like oxygen changes iron and other metals.

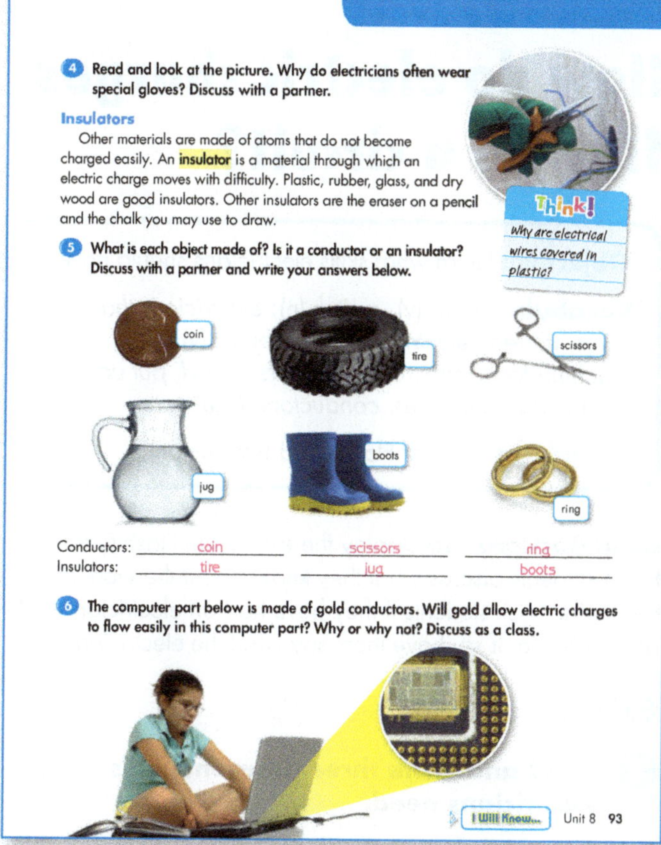
4 Read and look at the picture. Why do electricians often wear special gloves? Discuss with a partner.

Insulators

Other materials are made of atoms that do not become charged easily. An insulator is a material through which an electric charge moves with difficulty. Plastic, rubber, glass, and dry wood are good insulators. Other insulators are the eraser on a pencil and the chalk you may use to draw.

Think! Why are electrical wires covered in plastic?

5 What is each object made of? Is it a conductor or an insulator? Discuss with a partner and write your answers below.

coin, tire, scissors, jug, boots, ring

Conductors: coin, scissors, ring
Insulators: tire, jug, boots

6 The computer part below is made of gold conductors. Will gold allow electric charges to flow easily in this computer part? Why or why not? Discuss as a class.

I Will Know... Unit 8 93

ELL Content Support

Draw on the board a wide water pipe and a narrow water pipe. *A wide pipe has more room for water to flow through than a narrow pipe. Which has more resistance, a wide pipe or a narrow pipe? A narrow pipe!* Then have students remember the water pipe analogy and ask them which wire would have more resistance to the flow of electric current, a thin wire or a thick one.

Think!

Why are electrical wires covered in plastic?

Review what students have learned about conductors and insulators. Invite volunteers to name as many conductors and insulators as possible. Then ask the question and have students discuss. (Possible answer: *Plastic acts as an insulator, so current does not flow out of the wire.*)

Elaborate

Benjamin Franklin and Electricity

Divide the class into small groups. Have students find information about Benjamin Franklin and his contributions to the understanding of electricity. Students should focus on his experiments and what he used as a conductor. Invite students to make posters that illustrate their findings.

I Will Know...

Have students do the *I Will Know...* Digital Activity.

Lesson 2

How do electric charges flow in a circuit?

Objective: Learn what skills electricians should have.

Vocabulary: *flip* (v), *switch* (n), *electrician, thank, wires, current, electrical wiring, repair* (v), *fix, machine tools, safe, climb ladders, crawl, put up, electrical shocks, fires, conductors, insulators*

Digital Resources: Flash Card (*electrician*)

Build Background Display the *electrician* Flash Card. Have students discuss what they know about the work of an electrician. Ask students if they have had an electrician work at their home. If so, have them say what the electrician did.

Explain

7 Read and write three different skills electricians need.

Write *skill* on the board. *You are going to read about the skills electricians need. What does skill mean?* Write students' ideas on the board. *Skill is the ability to do something well.* Have students write three different skills electricians need. Accept all logical answers.

8 Why is it important for electricians to be good at problem solving? Discuss as a class and write your answer.

Electricians should like to work with their hands and be good at problem solving. Why? Encourage students to discuss freely. (Possible answer: *They need to be able to safely find out why a circuit is not working.*)

ELL Vocabulary Support

Write the words *electricity, electrician, electric,* and *electrically* on the board. Pronounce each word and have students repeat after you. Help them to recognize that the second *c* can be pronounced differently in *electricity, electrician* and *electric* and that *electric* and *electrically* have the same pronunciation of the second *c*. Guide students to identify if each word is a noun, an adjective, or an adverb. On the board, write the following sentences for students to copy and complete in their notebooks:

1. ________ appliances make our lives simpler.
2. An excess of positive or negative charges in an object is called static ________.
3. An __________ makes sure that electrical systems are safe.
4. Lightning happens when water droplets in a cloud become ________ charged.

7 Read and write three different skills electricians need.

An Electrician's Job

You flip a switch, and a light goes on. Every time this happens, you have an electrician to thank.

Electricians run the wires that carry current throughout your home, your school, and any building that has electricity. Some electricians work mainly in houses and other small buildings. Others work in office buildings, where they might install telephones and cables for computers as well as electrical wiring. Still others work in large factories, where they might repair robots or fix machine tools.

Electricians check to make sure that electrical systems are safe. They usually spend much of the workday on their feet. Sometimes they need to climb ladders or crawl into small spaces to put up or repair wires. They must work carefully because poor wiring can cause electrical shocks and fires.

If you like to work with your hands and are good at problem solving, you might like to become an electrician.

Skills:
1. the ability to spend much of the workday on their feet
2. the ability to work carefully
3. the ability to solve problems

8 Why is it important for electricians to be good at problem solving? Discuss as a class and write your answer.

Possible answers:
Because poor wiring can cause electrical shocks and fires. Because sometimes they have to fix things like robots.

Flash Lab

Classify Conductors and Insulators
Look at objects on your desk and around your classroom. Think about what material each object is made of. Make a list of objects that are conductors and a list of objects that are insulators.

94 Unit 8

Elaborate

Flash Lab

Classify Conductors and Insulators

Materials: classroom objects

- Students gather about ten objects they may have on their desks or from around the classroom.
- In pairs, students discuss what material each object is made of and make a list of objects that are conductors and a list of objects that are insulators.
- Point out to students that the erasers in their pencils are insulators, but the graphite, which makes the pencil mark, is a conductor.
- Ask students to explain their insulator and conductor choices. Students' lists of conductors should include metal objects. Their lists of insulators should include items made of plastic, rubber, glass, and wood.

Lesson 2

How do electric charges flow in a circuit?

Objective: Learn how electricity moves through series and parallel circuits.

Vocabulary: *circuit, energy source, electric charges, wires, electrical outlets, resistors, light bulbs, flow* (v), *switch* (n), *symbols, series circuit, loop, parallel circuit, paths, power source*

Digital Resources: Flash Card (*circuit*), *Lesson 2 Check* (print out 1 per student), *Got it? 60-Second Video*

Build Background Explain that a circuit is like a circular road for electricity. Draw a loop on the board to represent the road. Erase part of the loop and ask students what would happen if a car came to the gap in the road. Explain that electricity, like a car, cannot cross a gap. Tell students that a switch is like a bridge that allows electricity to keep flowing.

Explain

9 Read and look at the circuit symbols to the right. Then label the parts of the series circuit below.

Display the *circuit* Flash Card. Have students name the three objects in the circuit. (Answers: *batteries, switch, light bulbs*) Then direct students' attention to the circuit symbols diagram at the top of the page. Explain to students that those symbols are used by electricians to represent the parts of a circuit. Finally ask pairs to read and label the parts of the series circuit.

10 Read and look at the picture. Circle the loop that is broken.

Tell students that the circuit in the picture above is called a series circuit. Have them read and find out what happens when the circuit has a burned-out or missing light bulb. (*The electric current stops flowing.*) Direct students' attention to the picture at the bottom of the page. Have students discuss the difference between a series and a parallel circuit. Ask pairs to read and circle the loop that is broken.

ELL Content Support

Tell students that the movement of electrical charges through a circuit can be described in the same way as the movement of students through a cafeteria line. *If no one disrupts the line, then everyone moves along smoothly. If one person in the line stops, everyone behind must also stop. The same thing happens when you stop the flow of electric current with an open switch.*

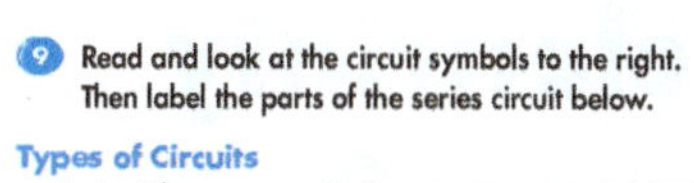

9 Read and look at the circuit symbols to the right. Then label the parts of the series circuit below.

CIRCUIT SYMBOLS	
Switch	
Energy Source	
Resistor	

Types of Circuits

A circuit has many parts. Its energy source provides the energy to move electric charges through its wires. Batteries and electrical outlets are energy sources. A circuit also has resistors such as light bulbs or machines. Resistors transform energy into other forms of energy. They use the energy that flows through the circuit. A circuit may also have a switch. The symbols are used to represent the parts of a circuit.

Think! How can a switch affect the flow of an electric current in a circuit?

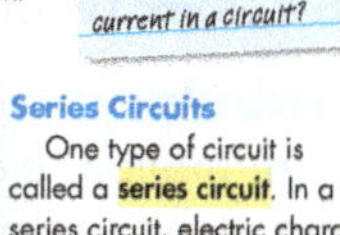

Series Circuits

One type of circuit is called a **series circuit**. In a series circuit, electric charge can only flow in a single loop. Any break in the loop, such as a burned-out or missing bulb, stops the current from flowing.

10 Read and look at the picture. Circle the loop that is broken.

Parallel Circuits

Another type of circuit is a parallel circuit. A **parallel circuit** has two or more paths through which electric charges may flow. Each path leaves from the power source and returns to it. The current that flows through one path does not have to flow through the other paths. Therefore, if one loop in the circuit is broken, the current will still flow through the other loops.

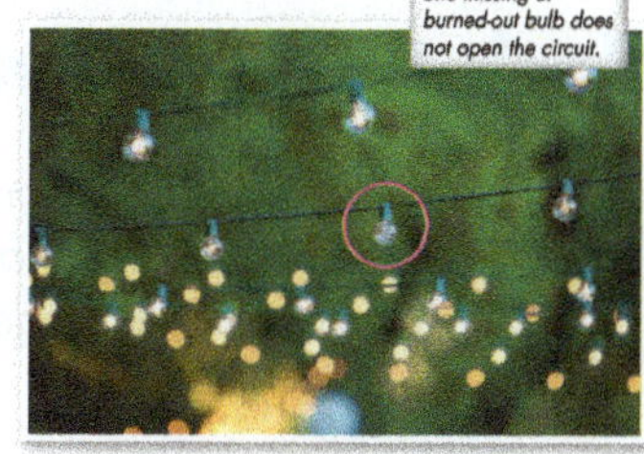

In a parallel circuit, one missing or burned-out bulb does not open the circuit.

Lesson 2 Check | Got it? 60-Second Video

Elaborate

Science Notebook: Series and Parallel Circuits

Have students draw a picture of a series circuit and a parallel circuit. Then have them label each drawing. Prompt students to identify and explain the differences between them.

Direct students' attention to the *circuit* Flash Card. Ask *How can a switch affect the flow of an electric current in a circuit?* (Possible answer: *If the switch is turned off or is open, the circuit is broken, and the current stops flowing.*)

Evaluate

Lesson 2 Check Assessment for Learning

Distribute the *Lesson 2 Check* and guide students as they complete it. Check answers as a class. Then ask students to grade their progress on the topic of how electricity flows in a circuit from 1 to 3: 3 = *I understand how electricity flows in a circuit*; 2 = *I need to study more*; 1 = *I need help!* Encourage students giving themselves a 1 or 2 to describe what they found difficult and what they need to study more.

Got it? 60-Second Video

Review Key Words for Lesson 2 (see Student's Book page 92). Play the *Got it? 60-Second Video* to review the lesson material.

Lesson 3

How does electricity transfer energy?

Objective: Learn how energy changes form.

Vocabulary: *organisms, tank, pump (n/v), light, sound, motion, electrical energy, pluck (v), guitar string, sound energy, wind turbine, wind energy*

Digital Resources: *Explore My Planet!* Digital Activity, *I Will Know...* Digital Activity

Materials: picture of a light bulb

Unlock the Big Question

Write the following on the board: *I will learn how energy changes form. I will know how electricity changes into light and gives off heat.*

Build Background Make connections to students' lives by having them name ways they use electricity every day. Brainstorm electric devices that students would not like to live without.

Explore

Explore My Planet! **Voices from History: Thomas Edison**

Objective: Students will learn about Thomas Edison's improvement of the light bulb.

Digital Resources: *Explore My Planet!* Digital Activity, *Explore My Planet! Activity Card* (1 per student)

- Display a picture of a light bulb. Have students discuss what people's lives were like without light bulbs.
- Show the *Explore My Planet!* to the class. Have students form pairs and complete the *Activity Card.* Provide support as needed.
- Review answers as a class.
- Then have students discuss how Edison's improvement of the light bulb has been important to their lives.

Explain

1 Why is electricity important to the organisms in this tank? Discuss with a partner and list three reasons.

Direct students' attention to the picture at the top of the page. *What organisms live in this tank? Fish and plants. What are three parts of a fish tank that are powered by electricity?* (Possible answers: *the air pump, the filter, the heater, the lamp*) Ask *Why is electricity important to the organisms in this tank?* Have pairs discuss and list three reasons.

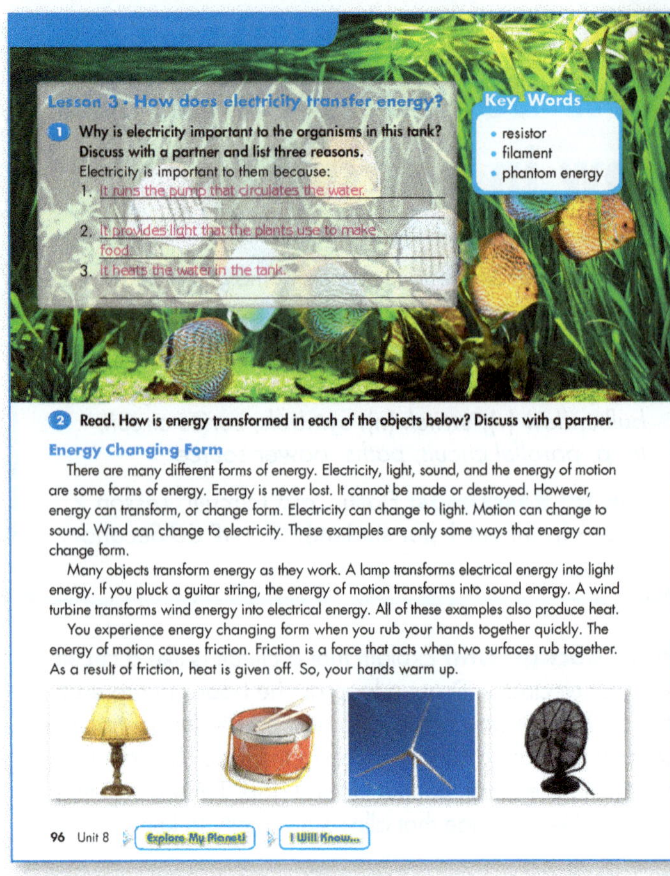
Lesson 3 • How does electricity transfer energy?

1 Why is electricity important to the organisms in this tank? Discuss with a partner and list three reasons.

Electricity is important to them because:

1. It runs the pump that circulates the water.
2. It provides light that the plants use to make food.
3. It heats the water in the tank.

Key Words
- resistor
- filament
- phantom energy

2 Read. How is energy transformed in each of the objects below? Discuss with a partner.

Energy Changing Form

There are many different forms of energy. Electricity, light, sound, and the energy of motion are some forms of energy. Energy is never lost. It cannot be made or destroyed. However, energy can transform, or change form. Electricity can change to light. Motion can change to sound. Wind can change to electricity. These examples are only some ways that energy can change form.

Many objects transform energy as they work. A lamp transforms electrical energy into light energy. If you pluck a guitar string, the energy of motion transforms into sound energy. A wind turbine transforms wind energy into electrical energy. All of these examples also produce heat.

You experience energy changing form when you rub your hands together quickly. The energy of motion causes friction. Friction is a force that acts when two surfaces rub together. As a result of friction, heat is given off. So, your hands warm up.

96 Unit 8 Explore My Planet! I Will Know...

2 Read. How is energy transformed in each of the objects below? Discuss with a partner.

Have students read the first paragraph and underline the four different forms of energy (*electricity, light, sound, and motion*). *How can energy change form?* Ask pairs to read the second paragraph and discuss how energy is transformed in each of the objects pictured at the bottom of the page. (Answers: *A lamp transforms electricity into light. When you play the drum, the energy of motion transforms into sound energy. A wind turbine transforms wind energy into electrical energy. An electric fan changes electrical energy into motion.*)

Elaborate

Motion and Heat

Have students rub an eraser quickly across their desk several times. Ask them to touch the eraser and describe how it feels. Students should note that the eraser is warmer after rubbing than it was before rubbing. Discuss with the class how energy transformed.

I Will Know...

Have students do the *I Will Know...* Digital Activity.

Lesson 3

How does electricity transfer energy?

Objective: Learn how electrical energy transforms into heat and understand what phantom energy is.

Vocabulary: *light bulb, resistor, heat* (v), *light* (v/n), *filament, incandescent light bulb, melt, pass through, give off heat, coil, appliances, devices, plug into, power source, phantom energy, make up, household, conserve, plug* (v), *power strip, switch off, step* (n), *environment*

Digital Resources: Flash Cards (*circuit, filament*), *Lesson 3 Check* (print out 1 per student), *Got it? 60-Second Video*

Build Background Brainstorm sources of energy that can be used to heat food. (Possible answers: *fossil fuels that give off heat as they are burned, electricity that can produce heat, burning firewood*)

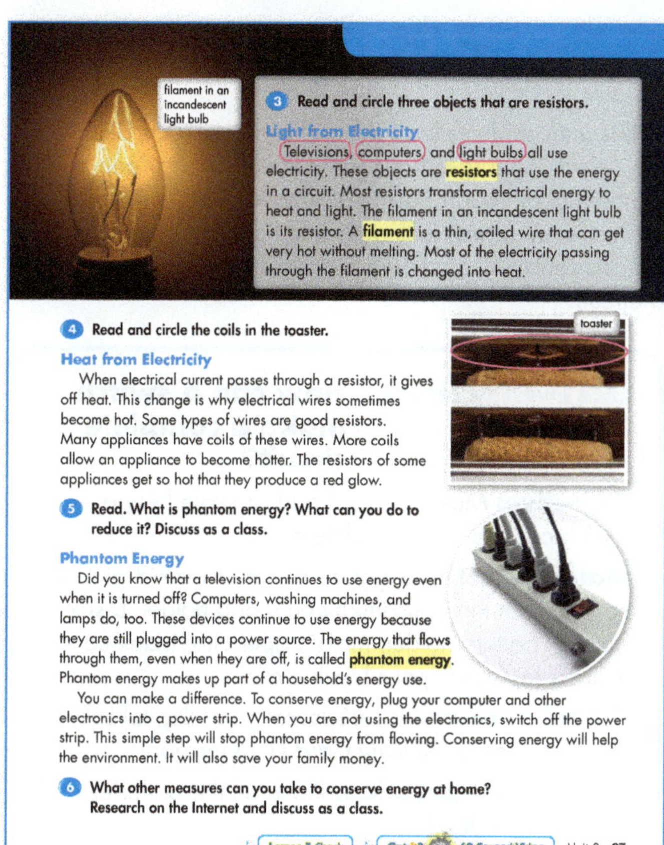
filament in an incandescent light bulb

3 Read and circle three objects that are resistors.

Light from Electricity

Televisions, computers, and light bulbs all use electricity. These objects are **resistors** that use the energy in a circuit. Most resistors transform electrical energy to heat and light. The filament in an incandescent light bulb is its resistor. A **filament** is a thin, coiled wire that can get very hot without melting. Most of the electricity passing through the filament is changed into heat.

4 Read and circle the coils in the toaster.

Heat from Electricity

When electrical current passes through a resistor, it gives off heat. This change is why electrical wires sometimes become hot. Some types of wires are good resistors. Many appliances have coils of these wires. More coils allow an appliance to become hotter. The resistors of some appliances get so hot that they produce a red glow.

toaster

5 Read. What is phantom energy? What can you do to reduce it? Discuss as a class.

Phantom Energy

Did you know that a television continues to use energy even when it is turned off? Computers, washing machines, and lamps do, too. These devices continue to use energy because they are still plugged into a power source. The energy that flows through them, even when they are off, is called **phantom energy**. Phantom energy makes up part of a household's energy use.

You can make a difference. To conserve energy, plug your computer and other electronics into a power strip. When you are not using the electronics, switch off the power strip. This simple step will stop phantom energy from flowing. Conserving energy will help the environment. It will also save your family money.

6 What other measures can you take to conserve energy at home? Research on the Internet and discuss as a class.

Lesson 3 Check | Got it? 60-Second Video | Unit 8 97

Explain

3 Read and circle three objects that are resistors.

Display the *circuit* Flash Card. Have students name the parts of the circuit. *Resistors transform energy into other forms of energy. Resistors use the electrical energy that flows through the circuit.* Elicit examples of resistors. Have students read and circle three objects that are resistors. Discuss with the class how they can tell resistors transform electricity into heat and light.

4 Read and circle the coils in the toaster.

Read the text aloud for the class. Have students explain in their own words what coils are. Then ask students to circle the coils in the toaster.

5 Read. What is phantom energy? What can you do to reduce it? Discuss as a class.

Have students read and answer the first question individually. Make sure it is clear to students that, even if electronic devices are turned off, if they are still plugged in, they continue to use energy. Discuss with the class what they can do to reduce phantom energy.

6 What other measures can you take to conserve energy at home? Research on the Internet and discuss as a class.

Give students time to research what other measures they can take to conserve energy at home. Then invite students to discuss the question as a class.

Ask *Why do many hair dryers use heating coils as resistors?* Invite students to discuss the question as a class. (Possible answer: *The coils become hot to help dry the hair.*)

Elaborate

Electricity Every Day

Have students brainstorm different ways that they use electricity in the course of a typical day. Have students write a short story describing the life of a student who uses electricity all day long.

Evaluate

Lesson 3 Check Assessment for Learning

Distribute the *Lesson 3 Check* and guide students as they complete it. Check answers as a class. Then ask students to grade their progress on the topic of how electricity transfers energy from 1 to 3: 3 = *I understand how energy is transferred and transformed.* 2 = *I need to study more;* 1 = *I need help!* Encourage students giving themselves a 1 or 2 to describe what they found difficult and what they need to study more.

Review Key Words for Lesson 3 (see Student's Book page 96). Play the *Got it? 60-Second Video* to review the lesson material.

Let's Investigate!

In this unit, students learn how charged objects behave, how electric charges flow, what materials they can and cannot flow through easily, and how energy can change into different forms, including into light and heat. In this lab, students will observe how a switch opens and closes a circuit.

Let's Investigate! Lab How can a switch make a complete circuit?

Objective: Make a circuit and observe how a switch can make a complete circuit.

Materials: safety goggles, insulated wire (3 pieces, each 20 cm long), flashlight bulb and holder, D-size battery, battery holder, metal paper clip, index card, 2 brass fasteners, pencil with eraser

Digital Resources: *Let's Investigate!* Digital Lab, *Let's Investigate! Activity Card* (1 per group)

Advance Preparation: Place all the batteries and bulbs in their holders.

- Divide the class into small groups and distribute materials.
- Show students how to make a circuit and a switch using the pictures as a guide.
- Have groups discuss what they think will happen when the paper clip touches the other fastener.
- Demonstrate how students should use the pencil eraser to move the paper clip so that it touches the other fastener.
- Students will observe that the bulb lights up when the paper clip touches both fasteners.
- Ask students to complete the *Activity Card* and share their results with the class.

Teacher Time-Saving Option: Show the *Let's Investigate!* Digital Lab as an alternative to the hands-on lab activity.

Unlock the Big Question

Have students refer to the Big Question on the Unit Opener page. In pairs, have them recall what they have learned about how electricity can be used. Invite student pairs to share their answers to question 6 on the *Let's Investigate! Activity Card*.

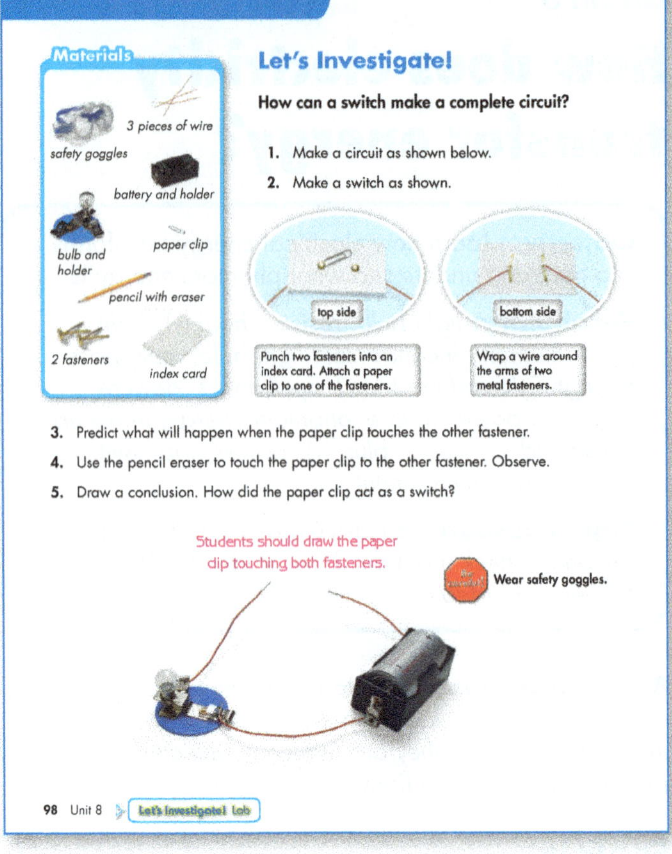

Materials

3 pieces of wire

safety goggles

battery and holder

paper clip

bulb and holder

pencil with eraser

2 fasteners

index card

Let's Investigate!

How can a switch make a complete circuit?

1. Make a circuit as shown below.
2. Make a switch as shown.

top side

bottom side

Punch two fasteners into an index card. Attach a paper clip to one of the fasteners.

Wrap a wire around the arms of two metal fasteners.

3. Predict what will happen when the paper clip touches the other fastener.
4. Use the pencil eraser to touch the paper clip to the other fastener. Observe.
5. Draw a conclusion. How did the paper clip act as a switch?

Students should draw the paper clip touching both fasteners.

Wear safety goggles.

98 Unit 8 Let's Investigate! Lab

Class Project: History of the Light Bulb

Materials: large sheet of construction paper (1 per group), art supplies

Divide the class into small groups. Ask each group to research improvements that have been made to light bulbs since they were invented, including improvements to filaments and glass casings. You may wish to have groups research different types of light bulbs, such as incandescent, fluorescent, tungsten-halogen, and LED bulbs. Have each group make a poster with pictures that show the light bulbs' parts and the improvements. Encourage students to include pictures and to write captions that describe how the improvements made the light bulbs better. Have groups present their posters to the class.

Unit 8 Review

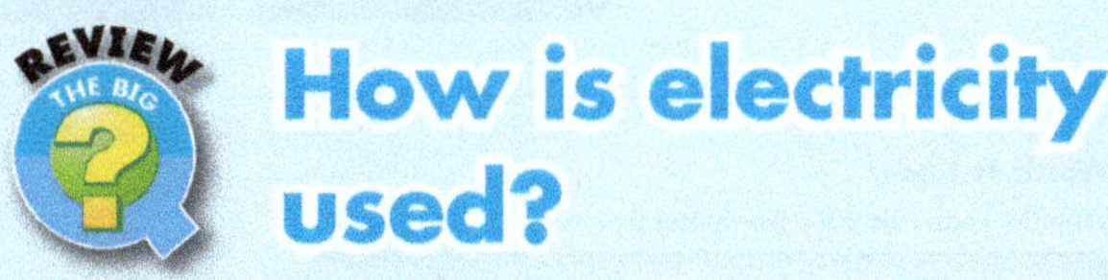

How is electricity used?

Digital Resources: Print out 1 of each per student: *Got it? Self Assessment, Got it? Quiz*

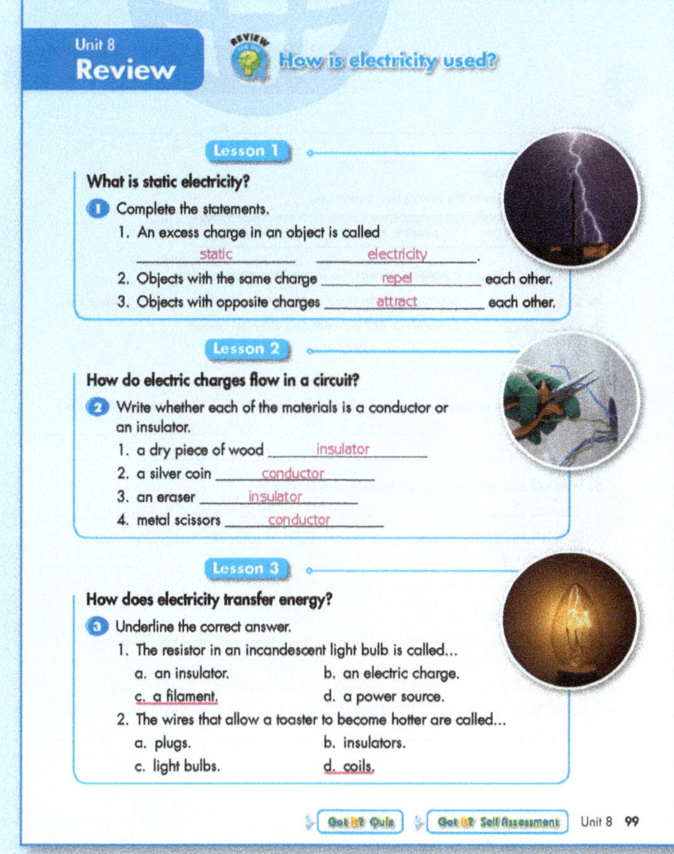

Unit 8
Review

How is electricity used?

Lesson 1

What is static electricity?

1 Complete the statements.

1. An excess charge in an object is called static electricity.
2. Objects with the same charge repel each other.
3. Objects with opposite charges attract each other.

Lesson 2

How do electric charges flow in a circuit?

2 Write whether each of the materials is a conductor or an insulator.

1. a dry piece of wood insulator
2. a silver coin conductor
3. an eraser insulator
4. metal scissors conductor

Lesson 3

How does electricity transfer energy?

3 Underline the correct answer.

1. The resistor in an incandescent light bulb is called...
   a. an insulator. b. an electric charge. c. a filament. d. a power source.
2. The wires that allow a toaster to become hotter are called...
   a. plugs. b. insulators. c. light bulbs. d. coils.

Got it? Quiz Got it? Self Assessment Unit 8 99

Evaluate

Strategies for Targeted Review

The following are strategies for providing targeted review for students if they encounter challenges with the content.

Lesson 1 What is static electricity?

Question 1

If... students are having difficulty completing the statements, then... direct students to page 89 and have them reread to find the key words.

Lesson 2 How do electric charges flow in a circuit?

Question 2

If... students are having difficulty identifying whether the objects are conductors or insulators, then... direct students to pages 92 and 93 to find out what materials are conductors and what materials are insulators.

Lesson 3 How does electricity transfer energy?

Question 3

If... students are having difficulty choosing the correct words to complete the sentences, then... direct students' attention to the pictures of the light bulb and the toaster on page 97 and have students identify the resistors.

ELL Language Support

Before students start working on the Review activities, have them read each question aloud along with you.

Got it? Self Assessment

Immediately after students have completed the Review activities, distribute a *Got it? Self Assessment* to each student. Have students complete the *Stop! Wait!* and *Go!* statements for each lesson, allowing them to look back through the lesson material if necessary.

Got it? Quiz

Distribute a Unit 8 *Got it? Quiz* to each student. Quizzes may be used for assessing students' understanding of unit concepts as well as for grading purposes.

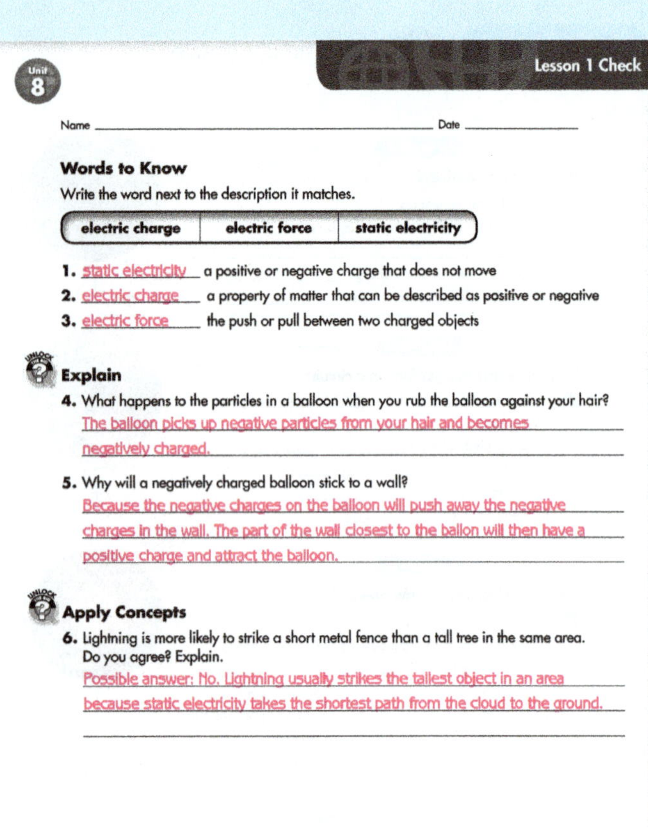

Unit 8

Lesson 1 Check

Name ______________________ Date ____________

Words to Know

Write the word next to the description it matches.

electric charge	electric force	static electricity

1. static electricity a positive or negative charge that does not move
2. electric charge a property of matter that can be described as positive or negative
3. electric force the push or pull between two charged objects

Explain

4. What happens to the particles in a balloon when you rub the balloon against your hair?
The balloon picks up negative particles from your hair and becomes negatively charged.

5. Why will a negatively charged balloon stick to a wall?
Because the negative charges on the balloon will push away the negative charges in the wall. The part of the wall closest to the ballon will then have a positive charge and attract the balloon.

Apply Concepts

6. Lightning is more likely to strike a short metal fence than a tall tree in the same area. Do you agree? Explain.
Possible answer: No. Lightning usually strikes the tallest object in an area because static electricity takes the shortest path from the cloud to the ground.

Unit 8, *Lesson 1 Check* • What is static electricity?

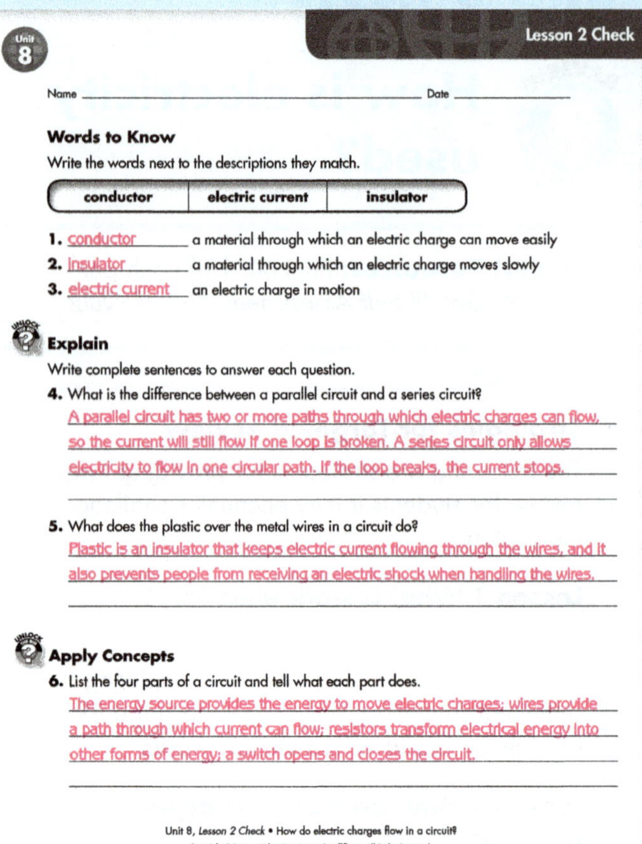

Unit 8

Lesson 2 Check

Name ______________________ Date ____________

Words to Know

Write the words next to the descriptions they match.

conductor	electric current	insulator

1. conductor a material through which an electric charge can move easily
2. insulator a material through which an electric charge moves slowly
3. electric current an electric charge in motion

Explain

Write complete sentences to answer each question.

4. What is the difference between a parallel circuit and a series circuit?
A parallel circuit has two or more paths through which electric charges can flow, so the current will still flow if one loop is broken. A series circuit only allows electricity to flow in one circular path. If the loop breaks, the current stops.

5. What does the plastic over the metal wires in a circuit do?
Plastic is an insulator that keeps electric current flowing through the wires, and it also prevents people from receiving an electric shock when handling the wires.

Apply Concepts

6. List the four parts of a circuit and tell what each part does.
The energy source provides the energy to move electric charges; wires provide a path through which current can flow; resistors transform electrical energy into other forms of energy; a switch opens and closes the circuit.

Unit 8, *Lesson 2 Check* • How do electric charges flow in a circuit?

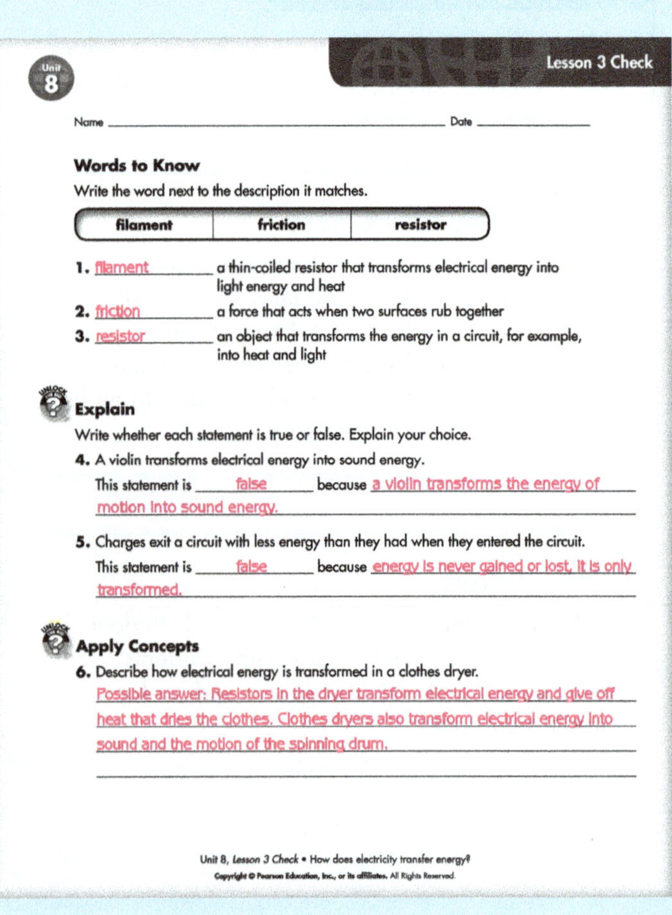

Unit 8

Lesson 3 Check

Name ______________________ Date ____________

Words to Know

Write the word next to the description it matches.

filament	friction	resistor

1. filament a thin-coiled resistor that transforms electrical energy into light energy and heat
2. friction a force that acts when two surfaces rub together
3. resistor an object that transforms the energy in a circuit, for example, into heat and light

Explain

Write whether each statement is true or false. Explain your choice.

4. A violin transforms electrical energy into sound energy.
This statement is false because a violin transforms the energy of motion into sound energy.

5. Charges exit a circuit with less energy than they had when they entered the circuit.
This statement is false because energy is never gained or lost, it is only transformed.

Apply Concepts

6. Describe how electrical energy is transformed in a clothes dryer.
Possible answer: Resistors in the dryer transform electrical energy and give off heat that dries the clothes. Clothes dryers also transform electrical energy into sound and the motion of the spinning drum.

Unit 8, *Lesson 3 Check* • How does electricity transfer energy?

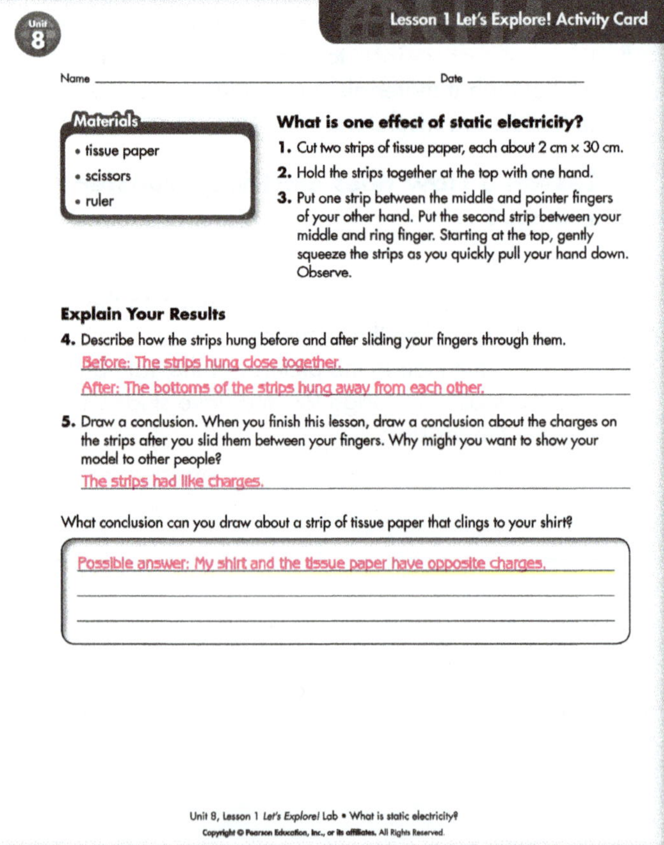

Unit 8

Lesson 1 Let's Explore! Activity Card

Name ______________________ Date ____________

Materials
- tissue paper
- scissors
- ruler

What is one effect of static electricity?

1. Cut two strips of tissue paper, each about 2 cm × 30 cm.
2. Hold the strips together at the top with one hand.
3. Put one strip between the middle and pointer fingers of your other hand. Put the second strip between your middle and ring finger. Starting at the top, gently squeeze the strips as you quickly pull your hand down. Observe.

Explain Your Results

4. Describe how the strips hung before and after sliding your fingers through them.
Before: The strips hung close together.
After: The bottoms of the strips hung away from each other.

5. Draw a conclusion. When you finish this lesson, draw a conclusion about the charges on the strips after you slid them between your fingers. Why might you want to show your model to other people?
The strips had like charges.

What conclusion can you draw about a strip of tissue paper that clings to your shirt?

Possible answer: My shirt and the tissue paper have opposite charges.

Unit 8, Lesson 1 *Let's Explore!* Lab • What is static electricity?

Unit 8

Lesson 2 Explore My Planet! Activity Card

Name ______________________ Date ______________

Did You Know: Semiconductors

The most important part of a computer uses semiconductors. As you will be able to tell from the name, semiconductors are not true conductors. They only conduct a small amount of electricity. Semiconductors share properties with conductors and with insulators.

The main element of semiconductors is silicon, which is also the main element of sand. Pure silicon acts in the same way as an insulator. To make a semiconductor, silicon is combined with a small amount of other substances like arsenic and boron. These substances convert the silicon into a material that is able to conduct some electrical current.

Different kinds of semiconductor materials can be joined together to make a diode. A diode is a device that usually allows current to flow in one direction only. Information can be generated by controlling whether current flows or does not flow through a diode. Computers use diodes and other devices to process information.

1. Explain why a semiconductor cannot be made out of pure silicon.
   Possible answer: Pure silicon acts as an insulator. Engineers need to add small amounts of other substances to the silicon so that it can pass a small amount of current.

Glass is another substance that is made out of silicon. Do you think it is a conductor or an insulator?

Possible answers: Glass is probably an insulator because silicon acts like an insulator unless it is mixed with some other chemical.

Unit 8, Lesson 2 *Explore My Planet!* Activity • How do electric charges flow in a circuit?

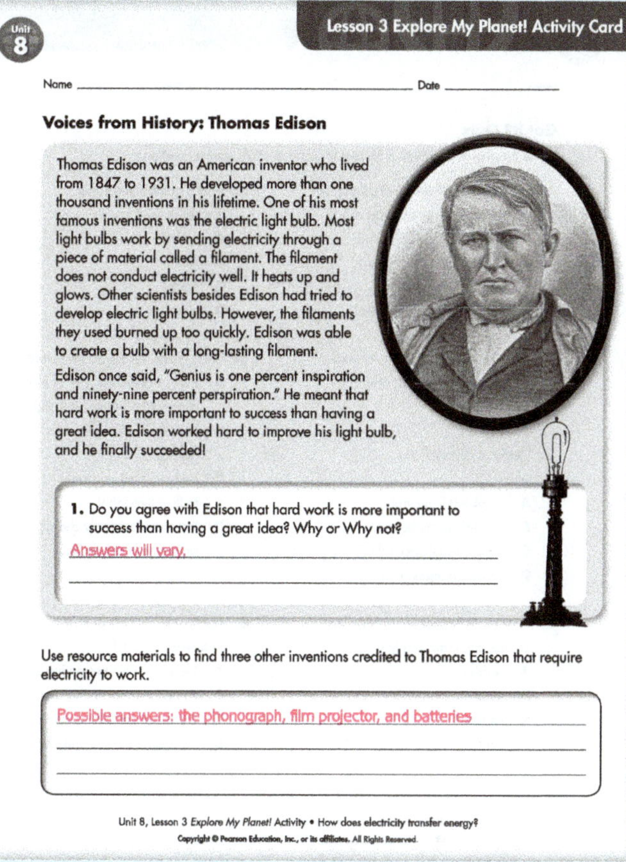

Unit 8

Lesson 3 Explore My Planet! Activity Card

Name ______________________ Date ______________

Voices from History: Thomas Edison

Thomas Edison was an American inventor who lived from 1847 to 1931. He developed more than one thousand inventions in his lifetime. One of his most famous inventions was the electric light bulb. Most light bulbs work by sending electricity through a piece of material called a filament. The filament does not conduct electricity well. It heats up and glows. Other scientists besides Edison had tried to develop electric light bulbs. However, the filaments they used burned up too quickly. Edison was able to create a bulb with a long-lasting filament.

Edison once said, "Genius is one percent inspiration and ninety-nine percent perspiration." He meant that hard work is more important to success than having a great idea. Edison worked hard to improve his light bulb, and he finally succeeded!

1. Do you agree with Edison that hard work is more important to success than having a great idea? Why or Why not?
   Answers will vary.

Use resource materials to find three other inventions credited to Thomas Edison that require electricity to work.

Possible answers: the phonograph, film projector, and batteries

Unit 8, Lesson 3 *Explore My Planet!* Activity • How does electricity transfer energy?

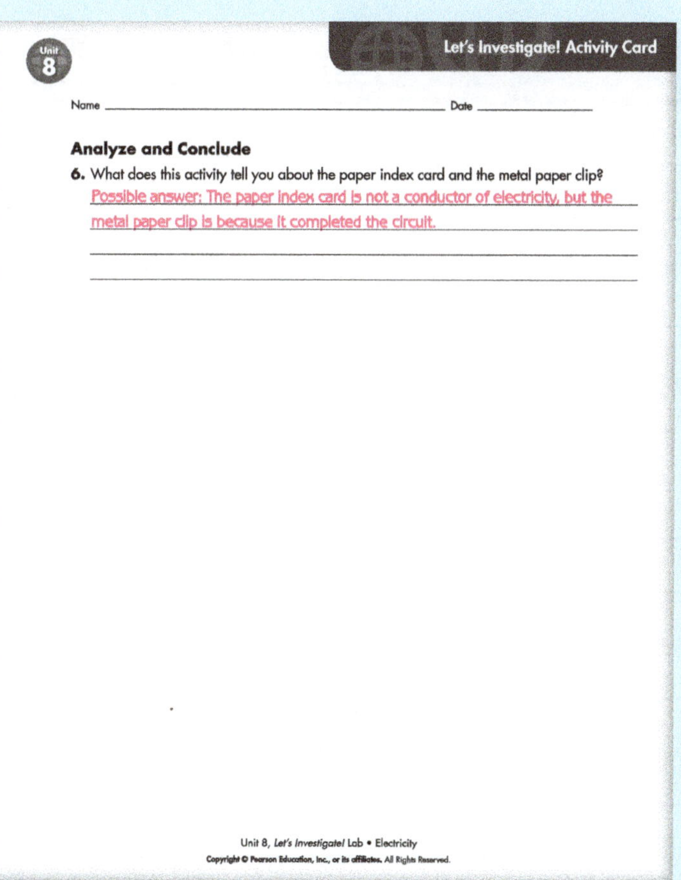

Unit 8

Let's Investigate! Activity Card

Name ______________________ Date ______________

Analyze and Conclude

6. What does this activity tell you about the paper index card and the metal paper clip?
   Possible answer: The paper index card is not a conductor of electricity, but the metal paper clip is because it completed the circuit.

Unit 8, *Let's Investigate!* Lab • Electricity

Unit 8

Lessons 1–3 Got it? Self Assessment

Name ______________________ Date ______________

Got it? Self Assessment

Complete the statements for each lesson.

Lesson 1 What is static electricity?

Stop! I need help with ______________________

Wait! I have a question about ______________________

Go! Now I know ______________________

Lesson 2 How do electric charges flow in a circuit?

Stop! I need help with ______________________

Wait! I have a question about ______________________

Go! Now I know ______________________

Lesson 3 How does electricity transfer energy?

Stop! I need help with ______________________

Wait! I have a question about ______________________

Go! Now I know ______________________

Unit 8, *Got it? Self Assessment* • Electricity

Unit 8 **Got it? Quiz**

Name ______________________ Date ____________

Got it? Quiz

Circle the choice you think is correct for each multiple choice question.

1. You see two balloons that attract one another. What can you conclude about the balloons?

- **(A)** The balloons do not have the same charge.
- **B** Both balloons are neutral.
- **C** The electric force between the balloons is weak.
- **D** Electric force is not affecting the balloons.

2. Electrical current flows when __________ .

- **(A)** an object is neutral
- **B** charged particles move
- **C** objects repel each other
- **D** matter changes state

3. A material through which an electric charge can move easily is a(n)

- **A** resistor.
- **B** insulator.
- **C** filament.
- **(D)** conductor.

4. Which of these materials is a conductor?

- **A** rubber
- **B** chalk
- **(C)** gold
- **D** glass

5. What is one use for an insulator?

- **A** It slows down the flow of electric current to produce heat and light.
- **B** It stores electric energy for later use.
- **C** It allows electric current to flow through it.
- **(D)** It makes electrically charged objects safe to handle.

6. Electric current cannot flow in a series circuit if

- **A** the circuit contains an insulator.
- **(B)** there is a break in the circuit.
- **C** the circuit contains a resistor.
- **D** he circuit has only one loop in it.

Unit 8, *Got it? Quiz* • Electricity

Unit 8 **Got it? Quiz**

Name ______________________ Date ____________

7. A washing machine is an example of transforming

- **A** electricity into light.
- **B** motion energy into electric energy.
- **(C)** electricity into motion energy.
- **D** heat into motion energy.

8. One example of a resistor is __________.

- **A** a power cord
- **B** a parallel circuit
- **C** a battery
- **(D)** a filament

9. Why might you feel a shock when you walk across a carpet and then touch a metal knob? Explain.

Possible answer: When I walk on the carpet, I pick up negative particles and become negatively charged. The negative particles move from me to the knob. Their movement releases electrical energy I feel as a shock.

10. List three appliances that transform electrical energy into another form of energy. Identify the form or forms of energy each produces.

Possible answer: Toaster, porch light, computer, microwave oven, cell phone. Each changes electrical energy into light and gives of heat.

Unit 8, *Got it? Quiz* • Electricity

Teacher's Notes

Unit 8 Study Guide

How is electricity used?

Lesson 1

What is static electricity?

- An excess charge in an object is called static electricity.
- Objects with the same charge repel. Objects with opposite charges attract.

Lesson 2

How do electric charges flow in a circuit?

- An electric charge flows through conductors easily.
- An electric charge flows through insulators with difficulty.
- For a current to flow, electric charges must complete a circuit.

Lesson 3

How does electricity transfer energy?

- Electrical energy, light energy, and the energy of motion are some forms of energy.
- Electrical energy can change to light energy and give off heat.

Review the Big Question

How is electricity used?

Have students use what they have learned from the unit to answer the question in their own words. *How has your answer to the Big Question changed since the beginning of the unit? What are some things you learned that caused your answer to change?*

Make a Concept Map

Have students make a concept map like the one shown on this page to help them organize key concepts.

Unit 8 Concept Map

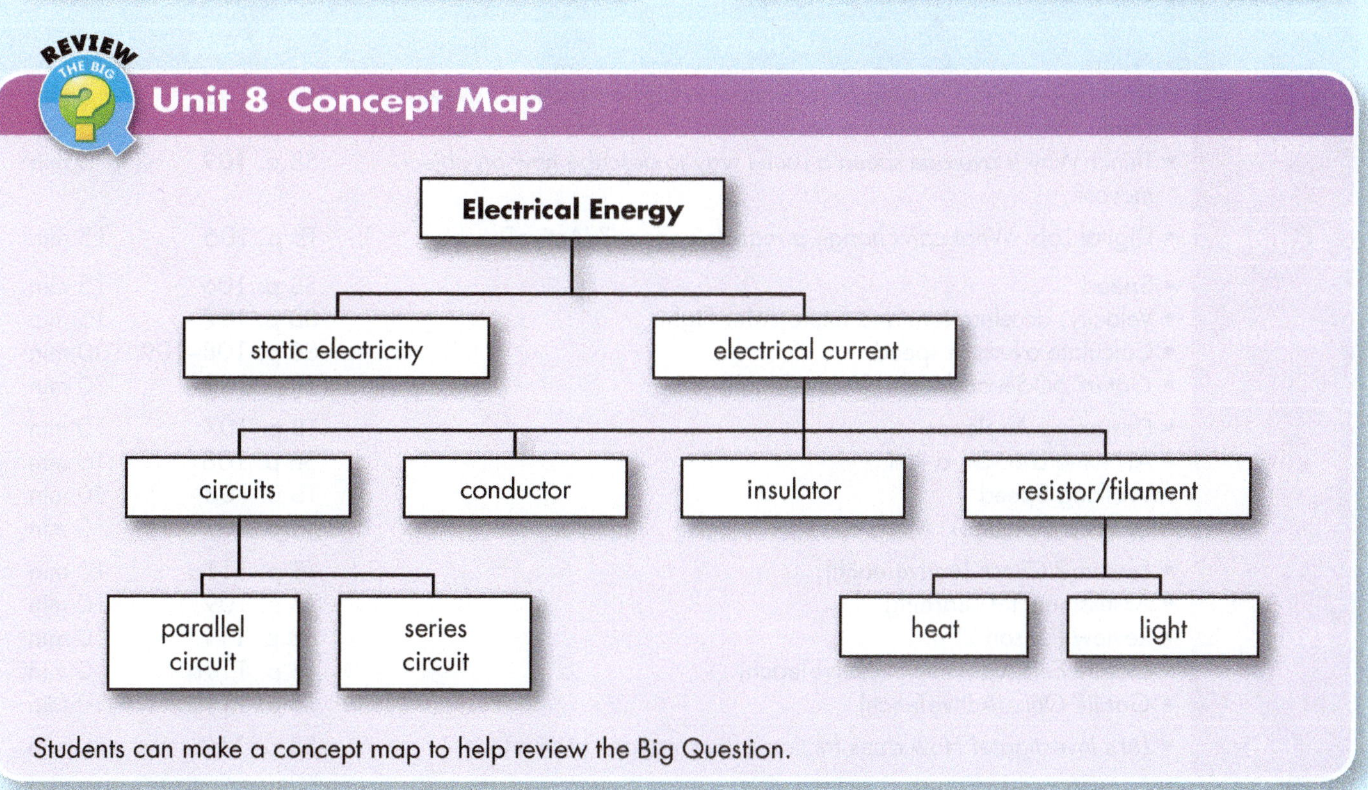

Students can make a concept map to help review the Big Question.

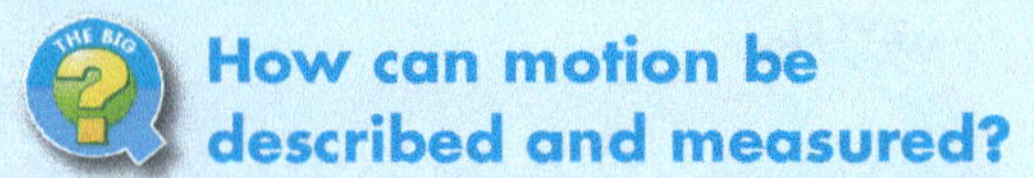

How can motion be described and measured?

Lesson Plan

Unit Opener & Lesson 1 What is motion?

	Activity	Pages	Time
Engage	• Unit Opener: Think! *What affects motion?* • Unit Opener: Identify what it takes to make objects move. • Unit Opener: Discuss conditions to ride a bike fast. • Think! *What will happen if a moving bumper car is hit from the side by another bumper car?*	SB p. 100 SB p. 100 SB p. 100 SB p. 103	5 min 10 min 5 min 5 min
Explore	• Digital Activity: *Misconception: Motion Sickness* (ActiveTeach)	TB p. 101	15 min
Explain	• Describe motion • Relative motion and reference points • Contact forces and motion • How forces affect motion • Force, mass, gravity, and weight • *Got it? 60-Second Video* (ActiveTeach)	SB p. 101 SB p. 102 SB p. 103 SB p. 104 TB p. 105 TB p. 105	15 min 15 min 15 min 15 min 10 min 10 min
Elaborate	• Sequence of Motion • Who is the reference point? • Science Notebook: Forces I Use • Flash Lab: The Wrecking Ball • Science Notebook: Less Force and Greater Force	TB p. 101 TB p. 102 TB p. 103 SB p. 104 TB p. 105	10 min 15 min 30 min 30 min 30 min
Evaluate	• *Lesson 1 Check* (ActiveTeach) • Assessment for Learning • Review (Lesson 1) • *Got it? Self Assessment* (ActiveTeach) • *Got it? Quiz* (ActiveTeach)	TB p. 111a TB p. 105 SB p. 111 TB p. 111b TB p. 111b	10 min 10 min 10 min 10 min 10 min

Lesson 2 What is speed?

	Activity	Pages	Time
Engage	• Think! *How can a moving object increase and decrease its speed?* • Think! *What happens to the acceleration of a jet plane as it takes off?* • Think! *Why is average speed a useful way to describe how an object moves?*	TB p. 106 SB p. 107 SB p. 109	5 min 5 min 5 min
Explore	• Digital Lab: *What can change a marble's speed?* (ActiveTeach)	TB p. 106	15 min
Explain	• Speed • Velocity, acceleration, and future safer flights • Calculate average speed • *Got it? 60-Second Video* (ActiveTeach)	SB p. 106 SB p. 107 SB p. 108–109 TB p. 109	15 min 15 min 30 min 30 min
Elaborate	• Designing Airplanes • At-Home Lab: On a Roll • Walking Speed • Science Notebook: Marathon	TB p. 107 SB p. 108 TB p. 108 TB p. 109	10 min 10 min 20 min 10 min
Evaluate	• *Lesson 2 Check* (ActiveTeach) • Assessment for Learning • Review (Lesson 2) • *Got it? Self Assessment* (ActiveTeach) • *Got it? Quiz* (ActiveTeach)	TB p. 111a TB p. 109 SB p. 111 TB p. 111b TB p. 111b	10 min 10 min 10 min 10 min 10 min
Lab	• *Let's Investigate! How does friction affect motion?* (ActiveTeach)	SB p. 110	30 min

Flash Cards

Lesson 1

Key Words	ELL Support
motion, relative motion, reference point, force, balanced forces, gravity, mass, weight	**Vocabulary:** *track, straight line, curved path, back, forth, vibrate, rotate, position, ride your bike, walk down a street, pass (v), fixed, stand still, bikers, water slide, frame of reference, ride in a car, push (n/v), pull (n/v), slow down, change direction, contact force, pedals, handlebars, pedal (v), speed up, unbalanced, opposite directions, mass, shopping cart, groceries, gain, strength, ground, drop (v)* **Sequencers:** *first, next, finally* **Homograph:** *relative*

Lesson 2

Key Words	ELL Support
speed, velocity, direction, acceleration, average speed	**Vocabulary:** *gazelle, peregrine falcon, cheetah, hare, speed (v/n), land mammal, swoop (n), prey, speed up, slow down, accelerate, roller coaster, curved path, engineers, damaged, mid-flight, software, engine, thrust, trip (n), roll (v), calculate, leg, entire trip* **Words to Describe Direction and Speed:** direction: *north, south, east, west, left, right, up, down;* speed: *fast, slow, speed up, slow down, accelerate*

Unit Objectives

Lesson 1: Students will know how an object's mass affects the amount of force needed to move it and how Earth's gravity affects objects.

Lesson 2: Students will calculate and describe the speed of an object.

Vocabulary: *power* (v), *soccer ball, wagon, bicycle, boat, shopping cart, toy car, ride* (v), *motion*

How can motion be described and measured?

Build Background Have students close their eyes and imagine what you tell them. *It is a bright and sunny morning, and you have just found the perfect spot to watch the bike race. The racers are crouched over their handlebars and moving at top speed. Their knees pump up and down as they pedal furiously. The wheels of their bikes spin so rapidly that the spokes are a blur.* Ask students comprehension questions to be sure they understood the story. *Your favorite racer gets behind! You start to think about what he can do to move faster.* Elicit ideas from the class.

Engage

Think!

What affects motion?

Point to the photo on the bottom right. Ask students to think about the skills needed to ride a bike. Guide students to conclude that the legs power the bike forward and the upper body is used to steer. Have students brainstorm what forces are involved.

1 Label the pictures.

Point to the pictures and allow students to say any words they already know. Ask students to work in pairs and write the words. Review the answers by pointing to the pictures for students to say the words.

2 What can you do to make each of the objects above move? Discuss with a partner.

What do all these objects have in common? They all need a certain kind of force to make them move. Have pairs discuss what they can do to make each object move. Invite volunteers to describe the step-by-step processes they can follow to make the objects move.

Unit 9 Motion

How can motion be described and measured?

I will learn
- what motion is.
- what speed is.

1 Label the pictures.

soccer ball | wagon | bicycle

boat | shopping cart | toy car

2 What can you do to make each of the objects above move? Discuss with a partner.

3 When you ride a bike, what conditions are necessary to ride as fast as you can? Discuss as a class.

Think! What affects motion?

100 Unit 9

ELL Content Support

Write and say the word *push* and then demonstrate examples, like pushing a chair. Repeat with the word *pull*. Remind students that pushes and pulls are forces that affect motion. Use board drawings to demonstrate the following. *Most of the forces you use are contact forces. When you hit a ball with a bat, the bat's force changes the speed and direction of the ball. If the bat does not make contact with the ball, these changes cannot occur.*

3 When you ride a bike, what conditions are necessary to ride as fast as you can? Discuss as a class.

Point to the picture of the bike at the top of the page. Have students name its parts. (Possible answers: *pedals, handlebar, brakes, wheels, seat*) Ask students to describe what they do to ride a bike. Discuss with the class what conditions are necessary to ride as fast as possible. Have students justify their answers.

Think! Again!

This cyclist is riding on a steep road. What affects the cyclist's motion? (Possible answer: *The pedaling speed and gravity affect how fast the cyclist moves.*) Ask volunteers to share their ideas with the class along with their reasoning.

Lesson 1

What is motion?

Objective: Learn what motion is.

Vocabulary: *toy car, track, straight line, curved path, back, forth, vibrate, rotate, measure* (v), *motion, position*

Digital Resources: Flash Card (*motion*), *Explore My Planet!* Digital Activity

Materials: pictures of different rides at an amusement park

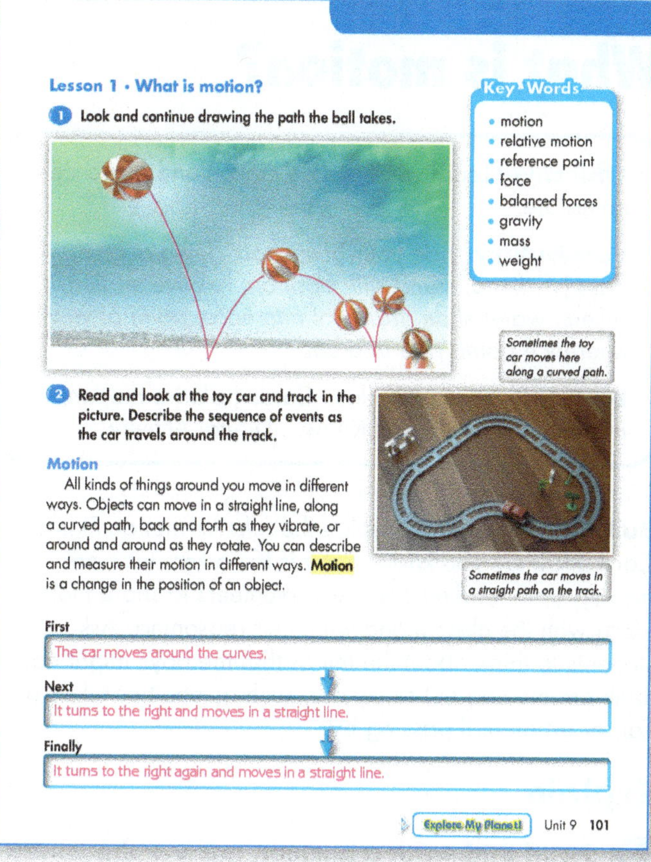

Lesson 1 • What is motion?

1 Look and continue drawing the path the ball takes.

Key Words
- motion
- relative motion
- reference point
- force
- balanced forces
- gravity
- mass
- weight

Sometimes the toy car moves here along a curved path.

2 Read and look at the toy car and track in the picture. Describe the sequence of events as the car travels around the track.

Motion

All kinds of things around you move in different ways. Objects can move in a straight line, along a curved path, back and forth as they vibrate, or around and around as they rotate. You can describe and measure their motion in different ways. **Motion** is a change in the position of an object.

Sometimes the car moves in a straight path on the track.

First
The car moves around the curves.

Next
It turns to the right and moves in a straight line.

Finally
It turns to the right again and moves in a straight line.

Explore My Planet! Unit 9 101

Unlock the Big Question

Write the following on the board: *I will understand what motion is. I will know that motion is relative and is affected by forces.*

Build Background Display pictures of different rides at an amusement park. Have students say which they like and dislike. Discuss with the class how they feel when they go on these or similar rides.

Explore

Explore My Planet! Misconception: Motion Sickness

Objective: Students will learn what motion sickness is.

Digital Resources: *Explore My Planet!* Digital Activity, *Explore My Planet! Activity Card* (1 per student)

- Show the *Explore My Planet!* Ask students to describe the picture.
- Read the text aloud for students. Have them complete the *Activity Card*.
- Ask students to check their answers in small groups or pairs. Provide support as needed.
- Then invite volunteers to share their responses with the class.

Explain

1 Look and continue drawing the path the ball takes.

Call students' attention to the picture at the top of the page. Be sure students recognize that the time-lapse photo shows a bouncing ball slowing down as it moves toward the right side of the illustration. Remind students to use what they have just noticed about the bouncing ball as they draw the path of the ball on the right side of the photo. *What do you notice about how the ball bounces each time as it moves to the right? The height of the ball decreases with each bounce.*

ELL Vocabulary Support

Draw a straight railway track and a curved highway on the board. Point to the straight railway track and say *straight*. Point to the curved highway and say *curved*. Demonstrate moving along a straight path and along a curved path. Say *I am walking in a straight line. I am walking along a curve.* Ask volunteers to show you each of these kinds of movement and to say how they are moving.

2 Read and look at the toy car and track in the picture. Describe the sequence of events as the car travels around the track.

Sequence refers to the order in which events happen. It can also refer to steps that are followed. Words such as first, next, after, *and* finally *tell you when each event or step begins. Understanding these words can help you recognize the sequence of events.* Have pairs read, describe, and write the sequence of events as the car travels around the track.

Elaborate

Sequence of Motion

Have students work in pairs. Have them make a ramp by leaning a grooved ruler on a book. Ask one partner to let go of a marble at the top of the ramp. Encourage the other partner to use the sequence words *first, next,* and *finally* to describe what happens. Have partners switch roles and repeat the activity.

Lesson 1

What is motion?

Objective: Learn what relative motion and reference points are.

Vocabulary: *ride your bike, walk down a street, pass (v), fixed, stand still, position, relative motion, bikers, water slide, frame of reference, point of view, reference point, ride in a car*

Digital Resources: Flash Cards (*relative motion, reference point*), *I Will Know...* Digital Activity

Build Background Display the *relative motion* Flash Card. *How do you know that it is the bikers who are moving and not the trees?* Ask volunteers to share their ideas with the class along with their reasoning. Ask students to think about the things that the boys might not think are moving but really are, such as another biker, a car, or an animal passing by.

Explain

3 Read and, with a partner, answer the question below.

Remind students that the word *position* means *where something is located.* Challenge them to name other synonyms for the word *position.* (Possible answers: *point, spot, place*) Read the text aloud for the class. Have pairs discuss the question and write the answer.

ELL Vocabulary Support

Point to the *relative motion* Flash Card. Ask students what they think the word *relative* means. (Possible answer: *family member*) Remind students that relatives are connected to one another because they belong to the same family. Explain that, in the term *relative motion*, the word *relative* is used to describe the connection between the motions of two or more objects.

4 Read and underline the three steps you take to tell if a person on a water slide moves.

Have students read the first paragraph and underline the three steps they can take to tell if a person on a water slide is moving. Check answers with the class.

5 Circle the moving object. Then draw an ✗ on two reference points for the girl.

Display the *reference point* Flash Card. Ask students to read the second paragraph and explain in their own words what a reference point is. Have students read the text on their own. *When you ride in a car, how can you tell your car is moving?* *By using a tree or a sign as a reference point.* Have students circle the moving object and draw an ✗ on two reference points for the girl.

---

3 Read and, with a partner, answer the question below.

Relative Motion

As you ride your bicycle or walk down a street, you pass trees, buildings, and other things that do not move. They are fixed in place. When you pass a fixed object, you know you are moving. When you stand still, you can tell that a car you see moves if it changes position. The change in one object's position compared with another object's position is called **relative motion**.

In the picture to the left, the three bikers are passing many trees. They also have different motions compared to each other because they probably do not always have the same speed.

If the bikers move at the same speed and the biker in the back uses only the biker in front as a reference, what might the biker in the back conclude about his or her own motion?
The biker might conclude that he or she is not moving at all.

4 Read and underline the three steps you take to tell if a person on a water slide moves.

Frame of Reference

How do you know if a person on a water slide moves? You look at the changing positions of the person. You compare the person's changing positions with the fixed position of the slide. You use the relative motion of the objects to decide what is moving and what is not moving.

Objects that do not seem to move define your frame of reference. Your frame of reference is like your point of view. How an object seems to move depends on your frame of reference.

One way to help you describe your motion is to find a reference point. A **reference point** is a place or object used to determine if an object is in motion. For example, when you ride in a car, you can tell your car is moving by observing a sign, a tree, or a building. Many objects can be reference points.

5 Circle the moving object. Then draw an ✗ on two reference points for the girl.

102 Unit 9 I Will Know...

ELL Content Support

Is it moving or not moving?
Students may think of motion as a simple matter of determining whether something is moving or not moving. To understand the motion of most objects we see on Earth, we need to observe any changes in the object's speed and/or direction. Motion is what happens to the object from one resting point to another resting point, including everything in between.

Elaborate

Who is the reference point?
Work with students in groups of three to model and identify a moving object and a reference point. Assign one student to serve as the reference point and another as the moving object. The third student identifies which student is the reference point. Ask *How do you know which student is the reference point?* (Possible answer: *Reference points do not seem to move, so the student who does not move can be the reference point.*)

I Will Know...
Have students do the *I Will Know...* Digital Activity.

Lesson 1

What is motion?

Objective: Learn how forces affect objects.

Vocabulary: *force, push* (n/v), *pull* (n/v), *stand still, direction, slow down, change direction, contact force, marble, level surface, size, strength, swing* (n), *balanced, bumper car*

Digital Resources: Flash Card (*force*)

Materials: pictures of bumper cars

6 **Read and list the five ways a force can affect motion.**

Forces Affect Objects

Forces make objects move or stop. A **force** is any push or pull. A force can make an object that is standing still start to move in the direction of the force. It can also make a moving object move faster, slow down, stop, or change direction.

Some forces act only with contact. A contact force must touch an object to affect it. A **marble** on a level surface will not move until you hit it with your finger or another object.

Pushing or pulling can change both the position and motion of an object. The size of the change depends on the strength of the push or pull. For example, the harder you push a swing, the higher and faster it will move.

All forces have size and direction. Notice the dogs pulling on the toy. They are pulling in opposite directions, but with the same amount of force. As long as they pull with forces that are the same size, the forces are **balanced**, and the toy will not move. If one dog pulls with more force, the forces will be unbalanced. The toy will move toward the dog pulling with greater force.

Five ways a force can affect motion:

1. It can make an object start to move in the direction of the force,
2. move faster,
3. slow down,
4. stop,
5. or change direction

7 **Look at the picture below. With a partner, describe what you think will happen.**

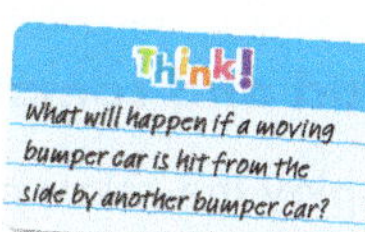

Unit 9 103

Build Background Place a chair in front of you. *What force is needed to move the chair?* (*A push or a pull.*) Pull and push the chair. *Why are many of the forces we use called contact forces?* (Possible answer: *Because something has to be in contact with, or touching, the object to move it.*)

Explain

6 Read and list the five ways a force can affect motion.

Ask *How can a force affect motion?* Accept all logical answers. Have students read and list the five ways force can affect motion. Check answers as a class.

7 Look at the picture below. With a partner, describe what you think will happen.

Display the *force* Flash Card. Have students describe what is happening in their own words. Then draw students' attention to the picture at the bottom of the page. Have pairs predict what might happen. (Possible answer: *The toy will move toward the larger dog because it seems to be pulling with greater force.*)

ELL Vocabulary Support

Write the words *balanced* and *unbalanced* on the board. Explain that *unbalanced* means *not balanced*. Show students that the prefix *un-* means *not* with examples such as *unclear, unfriendly,* and *unlucky*. Tell students that *un-* is just one example of a negative prefix. (You can encourage students to think of other examples of negative prefixes, such as *in-* and *dis-*.)

Elaborate

Science Notebook: Forces I Use

Invite students to choose an activity or sport that they enjoy doing. Choose one activity and have students identify the forces used in that activity. Write students' ideas on the board. Then have students brainstorm in groups and identify the forces that they use for the activity each student in their group chooses. (Students may mention swinging a bat to hit a baseball or kicking a soccer ball. Or, they may mention pushing their fingers against the keys of a piano.) Have students write their descriptions in their Science Notebooks. Invite them to illustrate their actions and use arrows to show how the motion of objects, such as a soccer ball that is kicked, changes.

Balanced and Unbalanced Forces

Have advanced learners work in pairs to become experts on balanced and unbalanced forces. Tell them to observe forces that affect everyday objects and events. Have students determine if the forces are balanced or unbalanced. Encourage students to make labeled drawings that show the objects and events and include arrows to represent the forces. Have students organize what they learn into a presentation for the class.

What will happen if a moving bumper car is hit from the side by another bumper car?

Display a picture of a bumper car. Ask *Have you ever driven a bumper car? What will happen if a moving bumper car is hit from the side by another bumper car?* (Possible answer: *The first car will probably change direction.*) Discuss the answers as a class.

Lesson 1

What is motion?

Objective: Learn how a force can change an object's position or the direction of its motion.

Vocabulary: *force, ride a bike, push the pedals, handlebars, pedal (v), slow down, speed up, unbalanced, opposite directions, collide*

Digital Resources: Flash Cards (*relative motion, force*)

Materials: picture of a person in a wheelchair

Build Background Draw a big bicycle on the board. Have volunteers come to the front and label its parts.

Explain

8 Read and complete the statements.

Display the *relative motion* Flash Card. Ask *How can a biker affect a bike's motion?* Accept all logical answers. Read the text and the pictures' captions aloud for students. Invite them to complete the statements. Check answers as a class.

9 With a partner, describe an example of balanced forces.

Display the *force* Flash Card. Discuss with the class why the forces are balanced. Use board drawings to explain the following situation: *You and a friend are each holding on to one end of a rope. If both of you pull on the rope in opposite directions with an equal amount of force, will the rope move?* (*No. The rope will not move. Since the forces on the rope are equal and acting in opposite directions, the forces are balanced. When balanced forces act on an object, it does not move.*)

Divide the class into pairs. Have students describe an example of balanced forces. (Possible answer: *My friend and I pushed a table in opposite directions. It did not move.*) Invite volunteers to share their answers with the class.

Elaborate

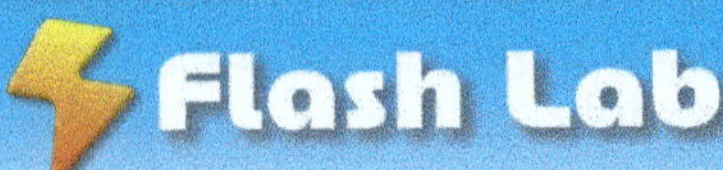

Flash Lab

The Wrecking Ball

Materials: 2 medium-sized sports balls, such as soccer balls

Students may take several tries to get the balls to collide. Students should note that both balls' positions and directions change when they hit each other.

8 Read and complete the statements.

Force and Motion

Force causes a change in motion in an object. The amount of force acting on an object affects how that object changes speed, direction, or both. When you ride a bike, you push the pedals. If you push harder, the bike goes faster. You turn the handlebars. The bike changes direction. Pedaling and turning change the bike's motion.

A moving object changes its motion only when a force acts on it. If balanced forces are applied to a moving object, it will keep moving at the same speed and in the same direction. The moving object will not slow down, speed up, or turn until the forces acting on it become unbalanced. An example is when you continue to pedal your bike with the same force. The bike will continue to move at the same speed because the same force is acting on it.

Balanced forces that act in opposite directions cancel each other out. For example, if you apply the same force to the brake and pedal at the same time, the bike's motion will not change.

Flash Lab

The Wrecking Ball

Work with a partner. Roll a ball across the floor. Note what is changing—the ball's position, its direction, or both. Roll the ball again. Have your partner roll another ball at it so that they collide. Note what is changing—the first ball's position, its direction, or both.

A person applies a force to bike pedals. The pedals transfer this force to the chain and then to the tires. This causes the bike to move.

The brakes produce a force that slows down or stops the bike when necessary.

1. A bike goes faster when you push the pedals harder.
2. If you stop pedaling, the bike will slow down.
3. It changes direction when you turn the handlebars.

9 With a partner, describe an example of balanced forces.

104 Unit 9

Motion on a School Day

Have students write a paragraph that describes three examples of motion that they observed during the school day. Challenge students to use one of the following phrases in their paragraphs: *contact force, balanced forces, unbalanced forces.*

ELL Content Support

Friction as a Contact Force

Ask students if they have ever dragged their foot on the ground while riding a bike. Ask a volunteer to describe what can happen as a result of that action. (*The bike will most likely slow down.*) Explain that friction between the ground and the foot will cause the bike to slow down.

Display a picture of someone in a wheelchair. Explain that a person operating a wheelchair can use friction to control the motion of the wheelchair. *When the wheelchair is moving, the person can put his or her hands on the hand rims that rotate along with the wheels. Friction between their hands and the moving rims of the wheels will cause the wheelchair to slow down.* Elicit from the students other examples of how friction can cause a moving object to slow down.

Lesson 1

What is motion?

Objective: Learn how mass, gravity, and weight can affect motion.

Vocabulary: *mass, force* (n), *shopping cart, groceries, gain, pull* (v), *gravity, strength, ground, drop* (v), *weight, Earth, moon*

Digital Resources: Flash Cards (*mass, gravity, weight*), *Lesson 1 Check* (print out 1 per student), *Got it? 60-Second Video*

Build Background Display the *mass, gravity,* and *weight* Flash Cards. Invite students to discuss how they think mass, gravity, and weight can affect motion.

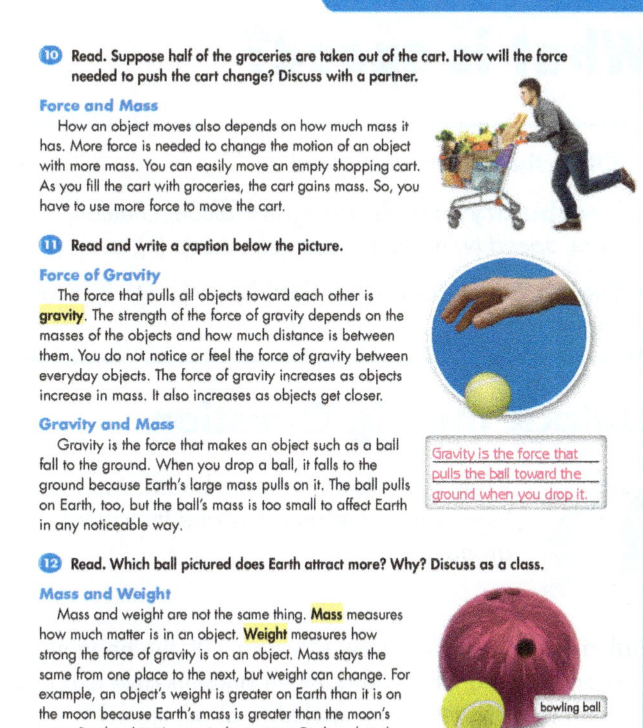

10 Read. Suppose half of the groceries are taken out of the cart. How will the force needed to push the cart change? Discuss with a partner.

Force and Mass

How an object moves also depends on how much mass it has. More force is needed to change the motion of an object with more mass. You can easily move an empty shopping cart. As you fill the cart with groceries, the cart gains mass. So, you have to use more force to move the cart.

11 Read and write a caption below the picture.

Force of Gravity

The force that pulls all objects toward each other is **gravity**. The strength of the force of gravity depends on the masses of the objects and how much distance is between them. You do not notice or feel the force of gravity between everyday objects. The force of gravity increases as objects increase in mass. It also increases as objects get closer.

Gravity and Mass

Gravity is the force that makes an object such as a ball fall to the ground. When you drop a ball, it falls to the ground because Earth's large mass pulls on it. The ball pulls on Earth, too, but the ball's mass is too small to affect Earth in any noticeable way.

Gravity is the force that pulls the ball toward the ground when you drop it.

12 Read. Which ball pictured does Earth attract more? Why? Discuss as a class.

Mass and Weight

Mass and weight are not the same thing. **Mass** measures how much matter is in an object. **Weight** measures how strong the force of gravity is on an object. Mass stays the same from one place to the next, but weight can change. For example, an object's weight is greater on Earth than it is on the moon because Earth's mass is greater than the moon's mass. But the object's mass is the same on Earth and on the moon because the amount of matter does not change.

Lesson 1 Check | Got it? 60-Second Video | Unit 9 105

Explain

10 Read. Suppose half of the groceries are taken out of the cart. How will the force needed to push the cart change? Discuss with a partner.

Have pairs read the text and discuss how the force needed to push the cart will change if half of the groceries are taken out. (Possible answer: *Less force will be needed to push the cart.*)

11 Read and write a caption below the picture.

Read the first paragraph aloud for students. Stand up, hold an eraser out in front of them, and let go of the eraser. After students watch the eraser drop, elicit or explain that Earth's large mass caused the eraser to fall to the ground. *The force that pulled the eraser toward Earth is called gravity.* Ask students to read the second paragraph individually and write a caption for the picture. Have students compare their answers with a partner.

12 Read. Which ball pictured does Earth attract more? Why? Discuss as a class.

Write *mass* and *weight* on the board. Read the text aloud to students. Have them explain in their own words the difference between mass and weight. (*Weight is the measure of the pull of gravity on an object. Mass is the amount of matter in an object.*) Discuss as a class which ball Earth attracts more and why. (Answer: *The bowling ball because it is heavier.*)

Elaborate

Science Notebook: Less Force and Greater Force

Have students make a two-column chart in their Science Notebooks to compare the amount of force needed to move certain objects in their daily lives. Students should label the columns *Less Force* and *Greater Force*. Examples may include pushing a stroller and pulling their wheeled backpacks; sliding a pencil and a dictionary across a table; pulling open an empty drawer and a drawer filled with items. In each case, students should recognize that the greater the mass of the object, the greater the force needed to move it.

What is motion?

Draw a graphic organizer for main idea and details with the title *What is motion?* Include a main idea box for each blue heading in the lesson. Have students recall and add details under each heading.

Evaluate

Lesson 1 Check **Assessment for Learning**

Distribute the *Lesson 1 Check* and guide students as they complete it. Check answers as a class. Then ask students to grade their progress on the topic of motion from 1 to 3: 3 = *I understand what motion is and how different forces affect motion*; 2 = *I need to study more*; 1 = *I need help!* Encourage students giving themselves a 1 or 2 to describe what they found difficult and what they need to study more.

Got it?  **60-Second Video**

Review Key Words for Lesson 1 (see Student's Book page 101). Play the *Got it? 60-Second Video* to review the lesson material.

Lesson 2

What is speed?

Objective: Learn what speed is.

Vocabulary: *gazelle, peregrine falcon, cheetah, hare, speed* (v/n), *land mammal, swoop* (n), *prey*

Digital Resources: Flash Card (*speed*), *Let's Explore!* Digital Lab

Unlock the Big Question

Write the following on the board: *I will know how to measure and describe the speed of an object.*

Build Background Draw a T-chart with the headings *Fast* and *Slow* on the board. Have students give examples of each to activate prior knowledge about speed.

Explore

Let's Explore! Lab What can change a marble's speed?

Objective: Students will measure distance and time to determine the speed of a marble that rolls down a ramp. Students will also observe how the steepness of the ramp affects the speed of the marble.

Digital Resources: *Let's Explore!* Digital Lab, *Let's Explore! Activity Card* (1 per student) (*Optional*: Do the lab in class; refer to the *Activity Card* for materials and steps.)

- Show the Digital Lab and have students complete the *Activity Card*.
- Ask students to check their answers in pairs or small groups. Provide support as needed.
- Review answers as a class. Guide students to conclude that the less time it takes an object to move a certain distance, the greater speed the object has.

Explain

1 Label the animals. Then, with a partner, number them (1–4) from the fastest to the slowest.

Elicit names of fast animals and write them on the board. Invite pairs to label the animals in the pictures and number them (1–4) from the fastest to the slowest.

2 With a partner, use the Internet to check your answers.

Allow some class time for students to go online to research the answers of the previous activity.

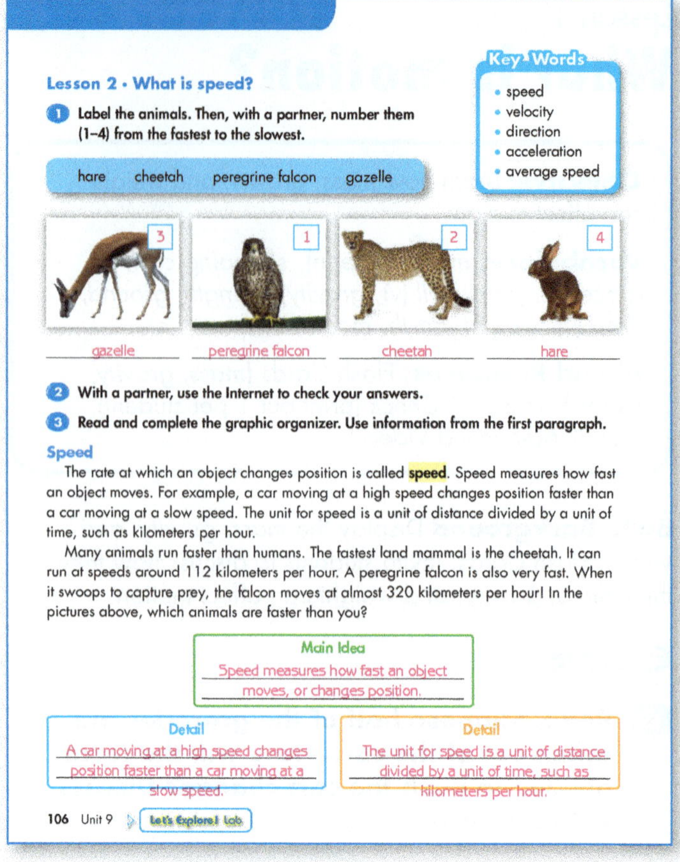

Lesson 2 • What is speed?

Key Words
- speed
- velocity
- direction
- acceleration
- average speed

1 Label the animals. Then, with a partner, number them (1–4) from the fastest to the slowest.

hare cheetah peregrine falcon gazelle

3 gazelle | 1 peregrine falcon | 2 cheetah | 4 hare

2 With a partner, use the Internet to check your answers.

3 Read and complete the graphic organizer. Use information from the first paragraph.

Speed

The rate at which an object changes position is called **speed**. Speed measures how fast an object moves. For example, a car moving at a high speed changes position faster than a car moving at a slow speed. The unit for speed is a unit of distance divided by a unit of time, such as kilometers per hour.

Many animals run faster than humans. The fastest land mammal is the cheetah. It can run at speeds around 112 kilometers per hour. A peregrine falcon is also very fast. When it swoops to capture prey, the falcon moves at almost 320 kilometers per hour! In the pictures above, which animals are faster than you?

Main Idea: Speed measures how fast an object moves, or changes position.

Detail: A car moving at a high speed changes position faster than a car moving at a slow speed.

Detail: The unit for speed is a unit of distance divided by a unit of time, such as kilometers per hour.

106 Unit 9 Let's Explore! Lab

ELL Language Support

Have students research the fastest average speed of each of the animals pictured on this page. Elicit the average speeds and write them on the board. Write on the board the following sentence frame for students to practice making comparative sentences: *A _____ can move fast, but a ______ can move faster.*

3 Read and complete the graphic organizer. Use information from the first paragraph.

Explain to students that using graphic organizers helps them to identify the main idea of a text and the details that support that idea. Have students read the text and complete the graphic organizer using information from the first paragraph. Ask students to compare their answers in pairs.

Think!

How can a moving object increase and decrease its speed?

Invite volunteers to write a definition of *speed* on the board. (*Speed is how fast an object changes position.*) Then discuss with the class how a moving object can increase and decrease its speed. (Possible answers: *An object can increase its speed when more force is applied to it in the direction it is moving. It can decrease its speed when force is applied in the opposite direction in which it is moving.*)

Lesson 2

What is speed?

Objective: Learn the difference between speed and velocity.

Vocabulary: *speed, direction, velocity, north, south, east, west, left, right, up, down, acceleration, speed up, slow down, accelerate, roller coaster, curved path, engineers, damaged, mid-flight, software, engine thrust*

Digital Resources: Flash Card (*velocity*), *I Will Know...* Digital Activity

Build Background Display the *velocity* Flash Card. *Have you ever ridden a roller coaster? Did you enjoy it? Why or why not? Why are they so exciting?* Allow students to respond freely.

Explain

4 Read. What are two things that must be measured in order to find an object's velocity? Discuss as a class.

Read the text aloud for students. Have them discuss what two things must be measured in order to know an object's velocity. Help students understand the difference between speed and velocity by having students describe your speed and then your velocity as you walk from one place to another in the room.

ELL Content Support

Direction and Speed

Draw a T-chart on the board with the headings *Direction* and *Speed*. Elicit words that can describe each concept and write them below the corresponding heading. (direction: *north, south, east, west, left, right, up, down;* speed: *fast, slow, speed up, slow down, accelerate*)

5 Read and complete the caption with a partner.

Divide the class into pairs. Ask students to describe how the roller coaster is moving. Tell students that they may use the words in the T-chart.

6 Read. How do you think engineers might use science and math to develop the software? Discuss in small groups.

Direct students' attention to the picture of the airplane. *A plane like this can fly steadily at 900 kilometers per hour, at altitudes between 11,000 and 12,000 meters. What if an airplane became damaged mid-flight?* Encourage students to discuss freely. *Did you know that there are programs used by computers that can help control the damaged airplane?* Read the text aloud to students. *How do you think engineers might use science and math to develop the software?* (Possible answer: *They might use science and math to calculate how airplanes respond in different scenarios.*)

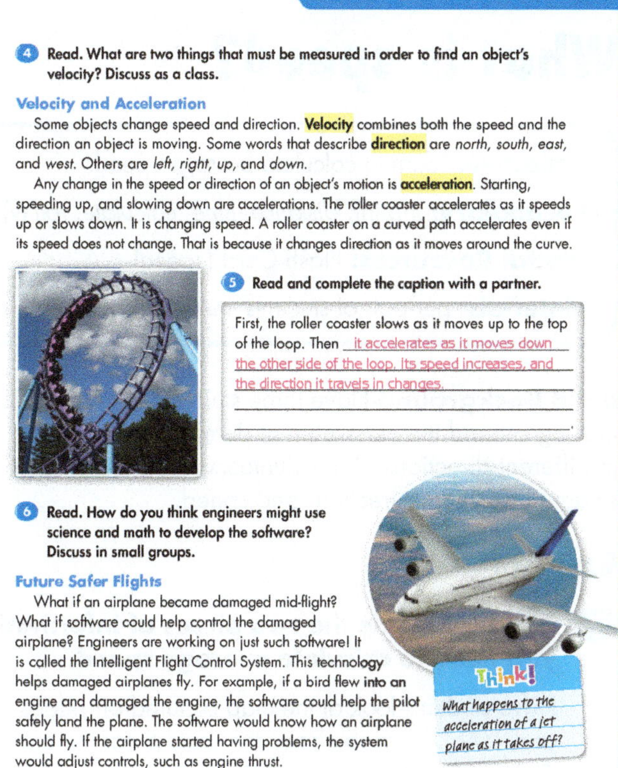

4 Read. What are two things that must be measured in order to find an object's velocity? Discuss as a class.

Velocity and Acceleration

Some objects change speed and direction. **Velocity** combines both the speed and the direction an object is moving. Some words that describe **direction** are *north, south, east,* and *west*. Others are *left, right, up,* and *down*.

Any change in the speed or direction of an object's motion is **acceleration**. Starting, speeding up, and slowing down are accelerations. The roller coaster accelerates as it speeds up or slows down. It is changing speed. A roller coaster on a curved path accelerates even if its speed does not change. That is because it changes direction as it moves around the curve.

5 Read and complete the caption with a partner.

First, the roller coaster slows as it moves up to the top of the loop. Then it accelerates as it moves down the other side of the loop. Its speed increases, and the direction it travels in changes.

6 Read. How do you think engineers might use science and math to develop the software? Discuss in small groups.

Future Safer Flights

What if an airplane became damaged mid-flight? What if software could help control the damaged airplane? Engineers are working on just such software! It is called the Intelligent Flight Control System. This technology helps damaged airplanes fly. For example, if a bird flew into an engine and damaged the engine, the software could help the pilot safely land the plane. The software would know how an airplane should fly. If the airplane started having problems, the system would adjust controls, such as engine thrust.

Think!

What happens to the acceleration of a jet plane as it takes off?

I Will Know... Unit 9 107

Elaborate

Designing Airplanes

Have students discuss what they know about the invention of airplanes that have engines. Explain that the Wright brothers tried many things that did not work before making the first controlled flight in an aircraft with an engine. Invite students to research in books or on the Internet how building bicycles might have helped the Wright brothers build airplanes.

Think!

What happens to the acceleration of a jet plane as it takes off?

Have students discuss when the acceleration of a jet plane usually changes. Draw a plane taking off on the board. Ask *What happens to the acceleration of a jet plane as it takes off?* (Possible answers: *Its velocity changes. Its speed increases without a change in direction on the runway. As the plane lifts off the ground, its velocity changes. Its direction changes, and its speed might change, too.*)

I Will Know...

Have students do the *I Will Know...* Digital Activity.

Lesson 2

What is speed?

Objective: Learn to calculate average speed.

Vocabulary: *trip* (n), *distance, time, calculate, roll* (v)

Digital Resources: Flash Card (*speed*)

Materials: per pair of students: meterstick, stopwatch

Build Background Have three students take turns moving around the classroom at different speeds and in different directions. Ask volunteers to describe each student's position, direction, and speed.

Explain

7 Read and circle the two pieces of information you need to calculate average speed.

Have students look at the map pictured on pages 108 and 109. *This map shows the route a car traveled. Today we will learn what information we need to calculate its average speed.* Display the *speed* Flash Card. *What does speed measure? Speed measures how fast an object moves.* Invite students to read and circle the two pieces of information needed to calculate average speed.

8 Read and calculate the average speed. Then complete the statement below.

What two pieces of information do you need to have before you can calculate average speed? The distance an object moves and the time it takes the object to move that distance. What do we have to do to find an object's average speed? Divide the distance the object moves by the total time spent moving. Have students read, calculate the car's average speed for the part of the trip in red, and complete the statement.

On a Roll

Materials: ball, masking tape

Before rolling the ball, ask students to record their predictions of what they might notice when they roll the ball each time. (Possible answer: *The ball will move faster and farther when rolled with more force.*) The following day in class, ask students if their observations matched their predictions.

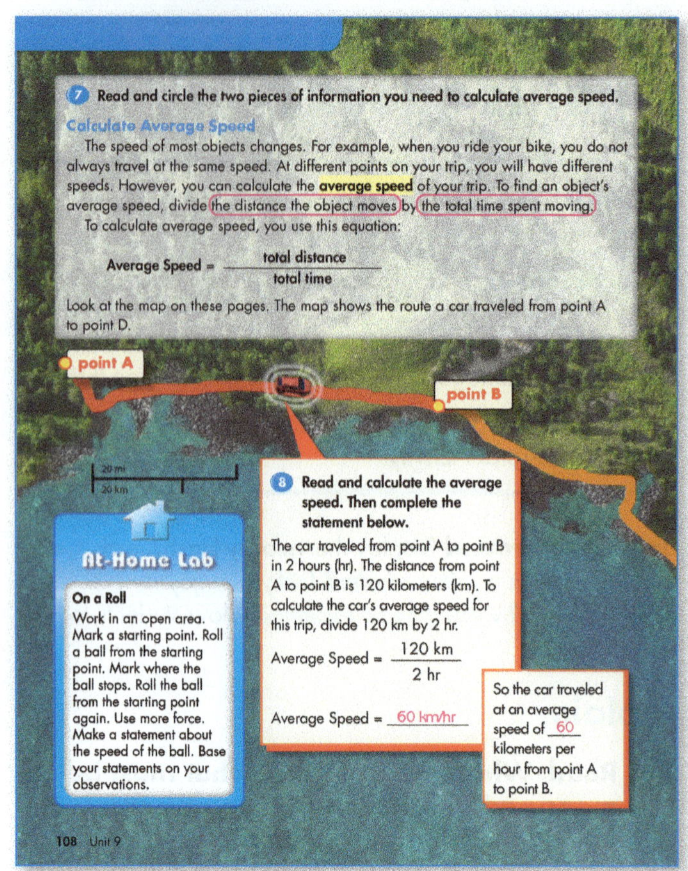

7 Read and circle the two pieces of information you need to calculate average speed.

Calculate Average Speed

The speed of most objects changes. For example, when you ride your bike, you do not always travel at the same speed. At different points on your trip, you will have different speeds. However, you can calculate the **average speed** of your trip. To find an object's average speed, divide the distance the object moves by the total time spent moving.

To calculate average speed, you use this equation:

$$\text{Average Speed} = \frac{\text{total distance}}{\text{total time}}$$

Look at the map on these pages. The map shows the route a car traveled from point A to point D.

8 Read and calculate the average speed. Then complete the statement below.

The car traveled from point A to point B in 2 hours (hr). The distance from point A to point B is 120 kilometers (km). To calculate the car's average speed for this trip, divide 120 km by 2 hr.

$$\text{Average Speed} = \frac{120\text{ km}}{2\text{ hr}}$$

Average Speed = 60 km/hr

So the car traveled at an average speed of 60 kilometers per hour from point A to point B.

At-Home Lab

On a Roll

Work in an open area. Mark a starting point. Roll a ball from the starting point. Mark where the ball stops. Roll the ball from the starting point again. Use more force. Make a statement about the speed of the ball. Base your statements on your observations.

108 Unit 9

Elaborate

Walking Speed

Have pairs of students use a meterstick to measure a straight line ten meters long on the ground. Have them mark the line with tape. Then ask one student to walk the ten meters at a normal pace as a second student uses a stopwatch to time the first student. Next, have the first student walk the ten meters taking very small steps while the second student times again. Then have students trade roles. Ask *What is the average normal walking speed of student 1? Of student 2? What is the average small-step walking speed of student 1? Of student 2?* Have students record their answers.

Science Notebook: A Walking Problem

Divide the class into pairs. Write the following problem for students to copy and solve in their Science Notebooks: *You walk four kilometers in 30 minutes. If you keep walking at the same average speed, how far will you walk in one hour? What is your average speed per hour?* (Answers: *4 kilometers × 2 = 8 kilometers; 8 kilometers/ 1 hour = 8 km/hr.*) Provide support as needed. Check answers by solving the problem on the board.

Lesson 2

What is speed?

Objective: Calculate average speed.

Vocabulary: *calculate, average speed, leg, entire trip, marathon*

Digital Resources: *Lesson 2 Check* (print out 1 per student), *Got it? 60-Second Video*

Materials: toy car, picture of a marathon

Build Background Have three volunteers come to the front. Give the first volunteer a toy car. Have the second student mark three locations: Point A, Point B, and Point C. Have the first student move the car at different speeds from one point to another, according to the third student's directions. Provide an example by saying: *Move the car slowly from Point A to Point B. Move the car at a faster speed from Point B to Point C.* Repeat with different volunteers.

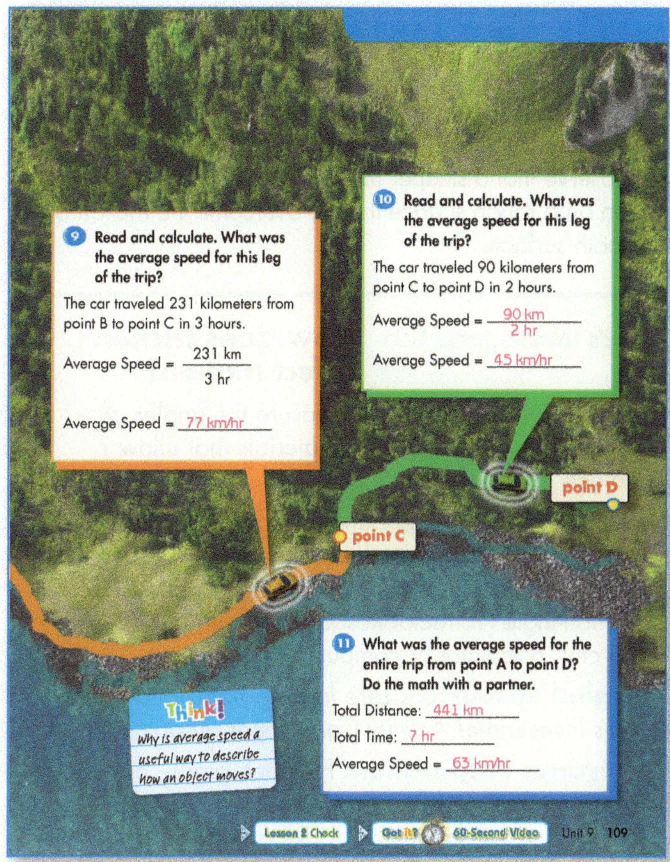

Explain

9 Read and calculate. What was the average speed for this leg of the trip?

How do you calculate average speed? Divide the distance an object moves by the time it takes to move that distance. Write the equation on the board for students to use as a guide. Have students calculate the average speed for the second leg of the trip.

10 Read and calculate. What was the average speed for this leg of the trip?

Have students calculate the average speed for the third leg of the trip.

11 What was the average speed for the entire trip from point A to point D? Do the math with a partner.

Have students calculate the average speed for the whole trip. Check answers as a class.

Elaborate

Science Notebook: Marathon

Display a picture of a marathon. Tell students that a marathon is an event in which runners follow a course that is about 41 kilometers long. *Cities all over the world hold annual marathons. No two courses are the same. Some are mostly flat, others have small hills, and still others are held in mountainous regions over very rough terrain. Weather is also a factor that affects these races. Because of these differences, the same runner can have different average speeds from one marathon to the next.*

Have students track a marathoner and note his or her finishing time for three or four marathons. Then have students calculate the marathoner's average running speed on the different courses. Have students show their work in their Science Notebooks.

Think!

Ask *Why is average speed a useful way to describe how an object moves?* Invite students to discuss the question as a class. (Possible answers: *Because the speed of most objects changes from one moment to the next. The average speed can help you say how long a different trip would take.*)

Evaluate

Lesson 2 Check Assessment for Learning

Distribute the *Lesson 2 Check* and guide students as they complete it. Check answers as a class. Then ask students to grade their progress on the topic of speed from 1 to 3: 3 = *I understand the difference between speed and velocity, and I can calculate average speed;* 2 = *I need to study more;* 1 = *I need help!* Encourage students giving themselves a 1 or 2 to describe what they found difficult and what they need to study more.

Got it? 60-Second Video

Review Key Words for Lesson 2 (see Student's Book page 106). Play the *Got it? 60-Second Video* to review the lesson material.

Let's Investigate!

In this unit, students learn that friction is a force that acts when the surfaces of objects rub together. In this lab, students will observe that a steeper ramp is needed to overcome the friction on a rough surface than to overcome the friction on a smooth surface.

Let's Investigate! Lab How does friction affect motion?

Objective: Students will measure the angles of ramps covered in different materials that allow objects to move down them.

Materials: sandpaper (medium grit, 23 × 28 cm), masking tape, corrugated cardboard (23 × 28 cm), small toy car, eraser, scissors, copy of Pattern for a Ramp Angle Protractor Resource, waxed paper (23 × 28 cm), calculator or computer (*optional*)

Digital Resources: *Let's Investigate!* Digital Lab, *Let's Investigate! Activity Card* (1 per group)

Advance Preparation: For each group, cut a piece of cardboard and a piece of waxed paper, each 23 × 28 cm in size. Make a copy of the Pattern for a Ramp Angle Protractor Resource for each group.

- Divide students into small groups and distribute materials.
- Show students how to tape the sandpaper to the piece of cardboard.
- Ask students to put a toy car and eraser at the top of the ramp. Have another student hold the Ramp Angle Protractor.
- Students should slowly raise the ramp by hand and record the angle when each object reaches the bottom of the ramp.
- Students should make three trials.
- Have students repeat the same procedure using waxed paper.
- Ask groups to discuss how friction affected the motion of the objects.
- At the end of the activity, have students share their inferences with the class.

Teacher Time-Saving Option: Show the *Let's Investigate!* Digital Lab as an alternative to the hands-on lab activity.

Unlock the Big Question

Have students refer to the Big Question on the Unit Opener page. In pairs, have them recall how motion can be described and measured. Invite student pairs to share their answers to questions 6, 7, and 8 on the *Let's Investigate! Activity Card*.

Materials

sandpaper, tape, scissors, cardboard, waxed paper, pattern for a Ramp Angle Protractor, calculator or computer (optional), eraser, toy car

Let's Investigate!

How does friction affect motion?

1. Tape sandpaper to a piece of cardboard.
2. Put a toy car and eraser at the top of the ramp. Have another student hold the Ramp Angle Protractor.
3. Slowly raise the ramp by hand. When each object reaches the bottom of the ramp, record the angle. Repeat 2 more times.
4. Discuss what would happen if you used waxed paper instead of sandpaper. Test your prediction 3 times. Record your results.
5. Describe how friction affected the motion of the objects on each surface.

Sample data

Effect of Friction on Motion

Trial	Angle When Object Reached Bottom of Ramp (degrees)			
	Sandpaper Surface		Waxed Paper Surface	
	Car	Eraser	Car	Eraser
1	5	52	3	32
2	2	54	1	31
3	1	50	2	30
Average	3	52	2	31

110 Unit 9 Let's Investigate! Lab

Class Project: Fastest Means of Transportation

Materials: large sheet of construction paper (1 per group), art supplies

Have students describe different means of transportation. Have them share their experiences with slow and fast forms of water, air, or land transportation.

- Divide the class into small groups. Have students research the fastest means of transportation and how they work. Have them list the fastest forms of water, air, and land transportation.
- Have each group make a poster that illustrates these means of transportation and present it to the class.

Unit 9 Review

How can motion be described and measured?

Digital Resources: Print out 1 of each per student: *Got it? Self Assessment*, *Got it? Quiz*

Evaluate

Strategies for Targeted Review

The following are strategies for providing targeted review for students if they encounter challenges with the content.

Lesson 1 What is motion?

Question 1

If... students are having difficulty underlining the correct answers, then... direct students to pages 102 and 103. Have them find the information about forces and relative motion.

Lesson 2 What is speed?

Question 2

If... students are having difficulty identifying what speed and velocity are, then... direct students to pages 106 and 107. Have students read the text to find the information they need.

Question 3

If... students are having difficulty identifying what data are needed to calculate average speed, then... direct students' attention to page 108. Have students read the text to find the information they need.

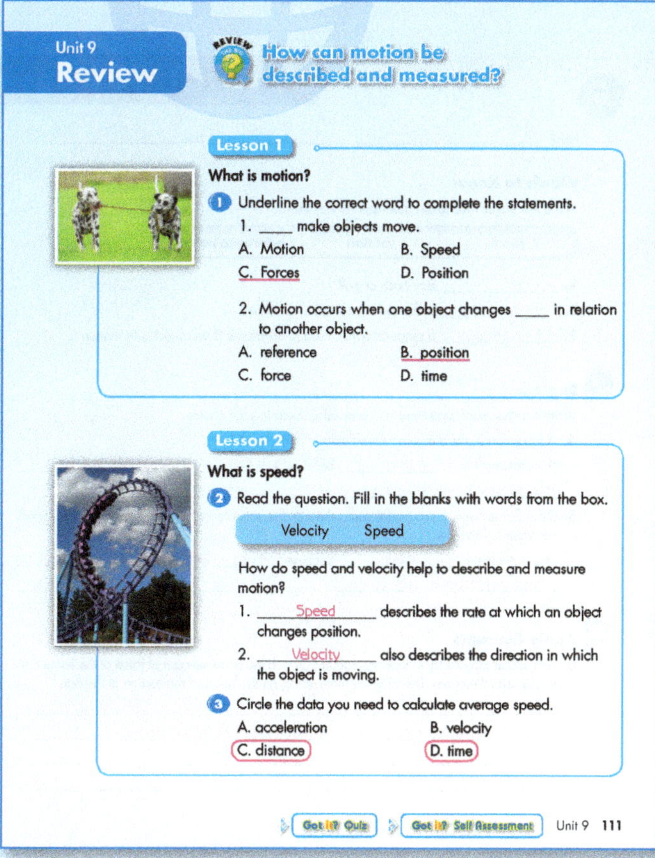

Unit 9 Review

How can motion be described and measured?

Lesson 1

What is motion?

1 Underline the correct word to complete the statements.

1. ____ make objects move.

A. Motion B. Speed
C. Forces D. Position

2. Motion occurs when one object changes ____ in relation to another object.

A. reference B. position
C. force D. time

Lesson 2

What is speed?

2 Read the question. Fill in the blanks with words from the box.

Velocity Speed

How do speed and velocity help to describe and measure motion?

1. ___Speed___ describes the rate at which an object changes position.

2. ___Velocity___ also describes the direction in which the object is moving.

3 Circle the data you need to calculate average speed.

A. acceleration B. velocity
C. distance D. time

Got it? Quiz Got it? Self Assessment Unit 9 111

ELL Language Support

Before students start working on the Review activities, have them read each question aloud along with you.

Got it? Self Assessment

Immediately after students have completed the Review activities, distribute a *Got it? Self Assessment* to each student. Have students complete the *Stop! Wait!* and *Go!* statements for each lesson, allowing them to look back through the lesson material if necessary.

Got it? Quiz

Distribute a Unit 9 *Got it? Quiz* to each student. Quizzes may be used for assessing students' understanding of unit concepts as well as for grading purposes.

Name ______________________ Date __________

Words to Know

Write the words next to the descriptions they match.

force	motion	reference point

1. force ________ any push or pull
2. motion ________ a change in position of an object
3. reference point ____ a place or object used to determine if an object is in motion

Explain

Write whether each statement is true or false. Explain your choice.

4. A force can make a moving object stop.

This statement is ____ true ____ because you could stop and object that was moving toward you by pushing on it.

5. On a flat surface, if you pedal your bike with the same force for 5 minutes, your speed will continue to increase.

This statement is ____ false ____ because the speed of an object will not change if the same force acts on it.

Apply Concepts

6. Two soccer players try to kick one soccer ball with the same amount of force at the same time in opposite directions. Describe the force acting on the ball and the motion of the ball.

Possible answer: The forces acting on the soccer ball will be balanced since equal and opposite forces cancel each other out. Because the forces acting on the object are balanced, it will have no motion.

Name ______________________ Date __________

Words to Know

Write the word next to the description it matches.

acceleration	speed	velocity

1. velocity ________ a measure of how fast and in what direction an object moves
2. speed ________ a measure of how fast an object moves
3. acceleration ____ a change in the speed or direction of a moving object

Explain

Write whether each statement is true or false. Explain your choice.

4. To find the average speed of a moving object, divide the time by the distance the object travels.

This statement is ____ false ____ because the average speed of a moving object is found by dividing the distance by the time.

5. An object is accelerating only if it is speeding up.

This statement is ____ false ____ because acceleration is any change in speed or direction, including slowing down.

Apply Concepts

6. A roller coaster track is 3,000 meters long. It takes 100 seconds to travel once around the roller coaster. What is the average speed of the roller coaster? Is the velocity constant throughout the trip? Explain.

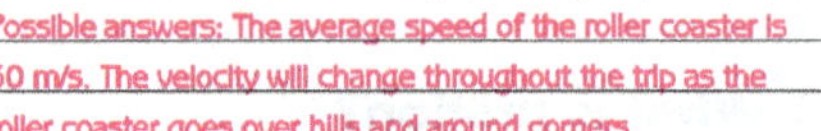

Possible answers: The average speed of the roller coaster is 30 m/s. The velocity will change throughout the trip as the roller coaster goes over hills and around corners.

Name ______________________ Date __________

Misconception: Motion Sickness

Have you ever felt sick in a car, boat, train, or airplane? You may have had motion sickness. Some people think that motion sickness is a problem related to the stomach. However, motion sickness happens when a person's sense of balance is thrown off. Balance is controlled by the inner ear. Sometimes the inner ear and the eyes process riding in something, such as a car or airplane, in different ways. This can cause a person to get pale, to get sweaty, or to vomit.

1. What do you think people with motion sickness could do to feel better?

Possible answers: They could take medicine to feel better. They could close their eyes. They could avoid going on rides that make them sick.

What are some other things that can affect your sense of balance?

Possible answers: spinning around, ear infections

Some people feel symptoms of motion sickness when watching movies or playing video games. Why might this be?

Possible answer: The picture makes it seem like you're moving, but the inner ear tells you you're not.

Name ______________________ Date __________

Materials

- 2 books
- ruler
- metal marble
- meterstick
- calculator
- timer

What can change a marble's speed?

1. Make a ramp with a book and a ruler. Use tape to label the start and the finish. Place a marble at the end of the ruler. Roll the marble down the ramp. Time how long it takes the marble to move 180 cm.

3.6 sec

Find the speed (speed = distance ÷ time).

50 cm/sec

2. Predict how raising the ramp would change the speed.

Possible answer: It will increase the speed.

Test your prediction by adding 1 book.

Time to move 180 cm = 2.0 sec

Speed = 90 cm/sec

Explain Your Results

3. How did raising the ramp change the marble's speed?

Possible answer: It increased the speed.

Why does raising the ramp increase the marble's speed?

Possible answers: The marble can gather more speed because it rolls from a greater height.

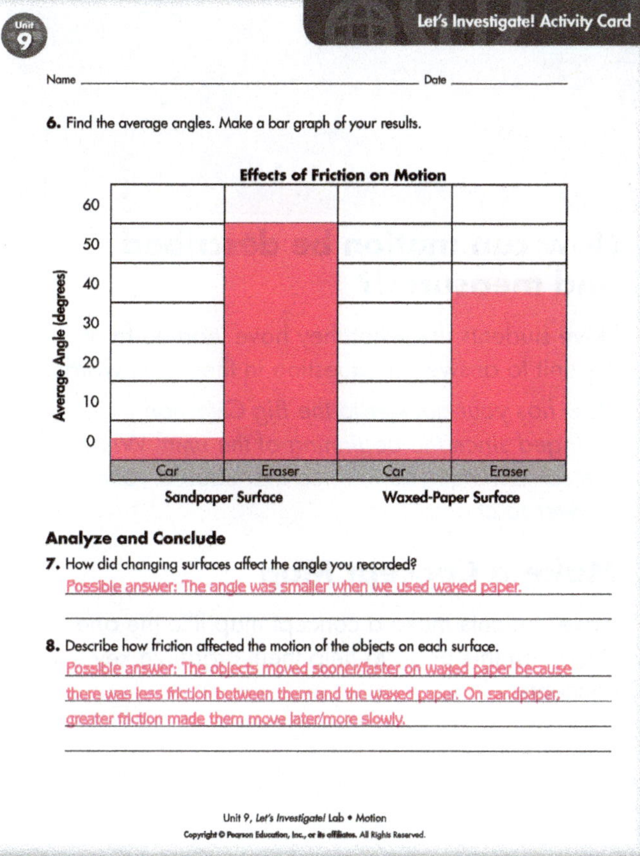
Unit 9

Let's Investigate! Activity Card

Name ______________________ Date ____________

6. Find the average angles. Make a bar graph of your results.

Analyze and Conclude

7. How did changing surfaces affect the angle you recorded?

Possible answer: The angle was smaller when we used waxed paper.

8. Describe how friction affected the motion of the objects on each surface.

Possible answer: The objects moved sooner/faster on waxed paper because there was less friction between them and the waxed paper. On sandpaper, greater friction made them move later/more slowly.

Unit 9, *Let's Investigate! Lab* • Motion

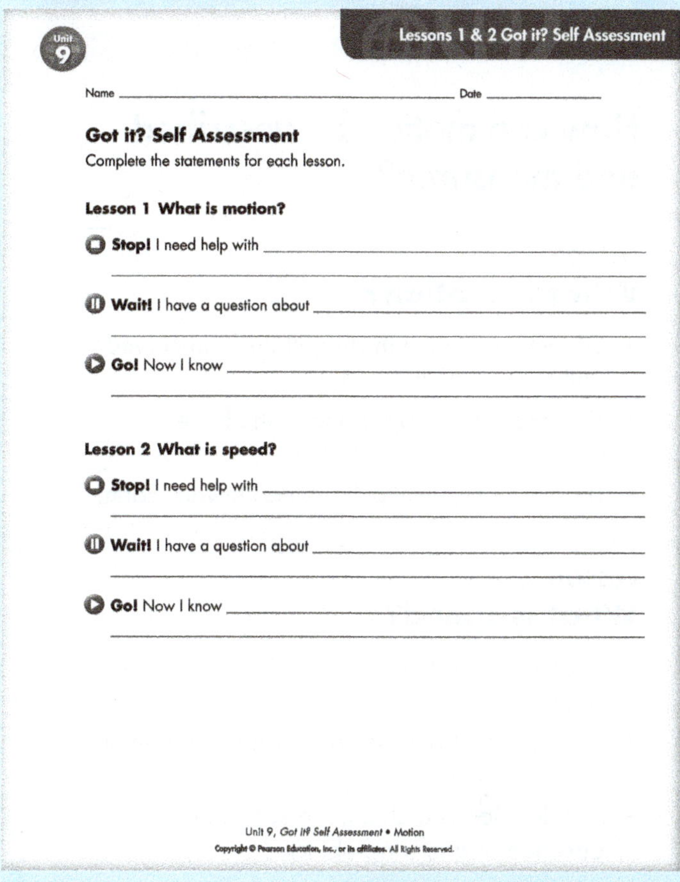
Unit 9

Lessons 1 & 2 Got it? Self Assessment

Name ______________________ Date ____________

Got it? Self Assessment

Complete the statements for each lesson.

Lesson 1 What is motion?

Stop! I need help with ______________________

Wait! I have a question about ______________________

Go! Now I know ______________________

Lesson 2 What is speed?

Stop! I need help with ______________________

Wait! I have a question about ______________________

Go! Now I know ______________________

Unit 9, *Got it? Self Assessment* • Motion

Unit 9

Got it? Quiz

Name ______________________ Date ____________

Got it? Quiz

Circle the choice you think is correct for each multiple choice question.

1. A teacher walks by a student's desk. The desk was in front of the teacher and is now behind the teacher. What term describes the change in the teacher's position compared to the desk?

- **A** velocity
- **(B)** relative motion
- **C** relative position
- **D** frame of reference

2. Which object requires the least amount of force to move it 5 m?

- **A** golf ball
- **B** soccer ball
- **C** bowling ball
- **(D)** table tennis ball

3. A car travels 160 km in 2 hours. What is the average speed of the car?

- **A** 0.01 km/hr
- **(B)** 80 km/hr
- **C** 160 km/hr
- **D** 320 km/hr

4. On which object would the force of gravity be strongest?

- **A** an apple
- **B** a book
- **C** a basketball
- **(D)** a truck

5. A student wants to describe the velocity of a moving car. What information does the student need?

- **A** speed and position
- **(B)** speed and direction
- **C** direction and position
- **D** direction and relative motion

6. A train goes around a corner at the same speed it was traveling on a straight track. What is this an example of?

- **A** speed
- **(B)** acceleration
- **C** relative motion
- **D** frame of reference

Unit 9, *Got it? Quiz* • Motion

Unit 9

Got it? Quiz

Name ______________________ Date ____________

7. Two students push boxes with the same mass for 10 seconds. Which student moved her box a greater distance?

- **A** the student who pushed with less force
- **B** the student who had less body mass
- **(C)** the student who pushed with more force
- **D** the student who had more body mass

8. Which of the following is NOT an example of an acceleration?

- **A** A train slowing down to a stop.
- **B** A train moving at the same speed going around a corner.
- **(C)** A train moving at the same speed in the same direction.
- **D** A train increasing its speed.

9. Describe what happens to the motion of a baseball after a pitcher throws it.

Possible answer: The pitcher throws the ball in the direction of the batter. The ball makes contact with the bat, and the force makes the ball change speed and direction.

10. You are moving a heavy box across a room. You ask a friend to help you lift it. Why is it easier when your friend helps?

Possible answer: The forces that you and your friend use are combined, so there is now a greater amount of force to move the box.

Unit 9, *Got it? Quiz* • Motion

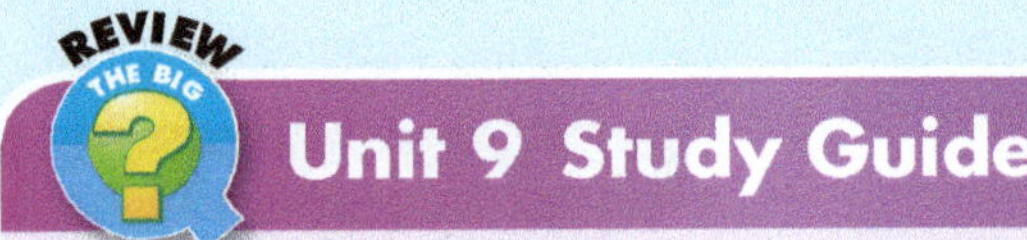

Unit 9 Study Guide

How can motion be described and measured?

Lesson 1
What is motion?

- Objects can move in straight lines, in curved paths, or back and forth.
- The mass of an object affects the force needed to change its motion.
- The force of gravity pulls objects toward Earth.

Lesson 2
What is speed?

- Speed is the rate at which an object changes position.
- Velocity describes the speed and direction of a moving object.
- Acceleration is a change in speed or direction of a moving object.

Review the Big Question

How can motion be described and measured?

Have students use what they have learned from the unit to answer the question in their own words.

How has your answer to the Big Question changed since the beginning of the unit? What are some things you learned that caused your answer to change?

Make a Concept Map

Have students make a concept map like the one shown on this page to help them organize key concepts.

Unit 9 Concept Map

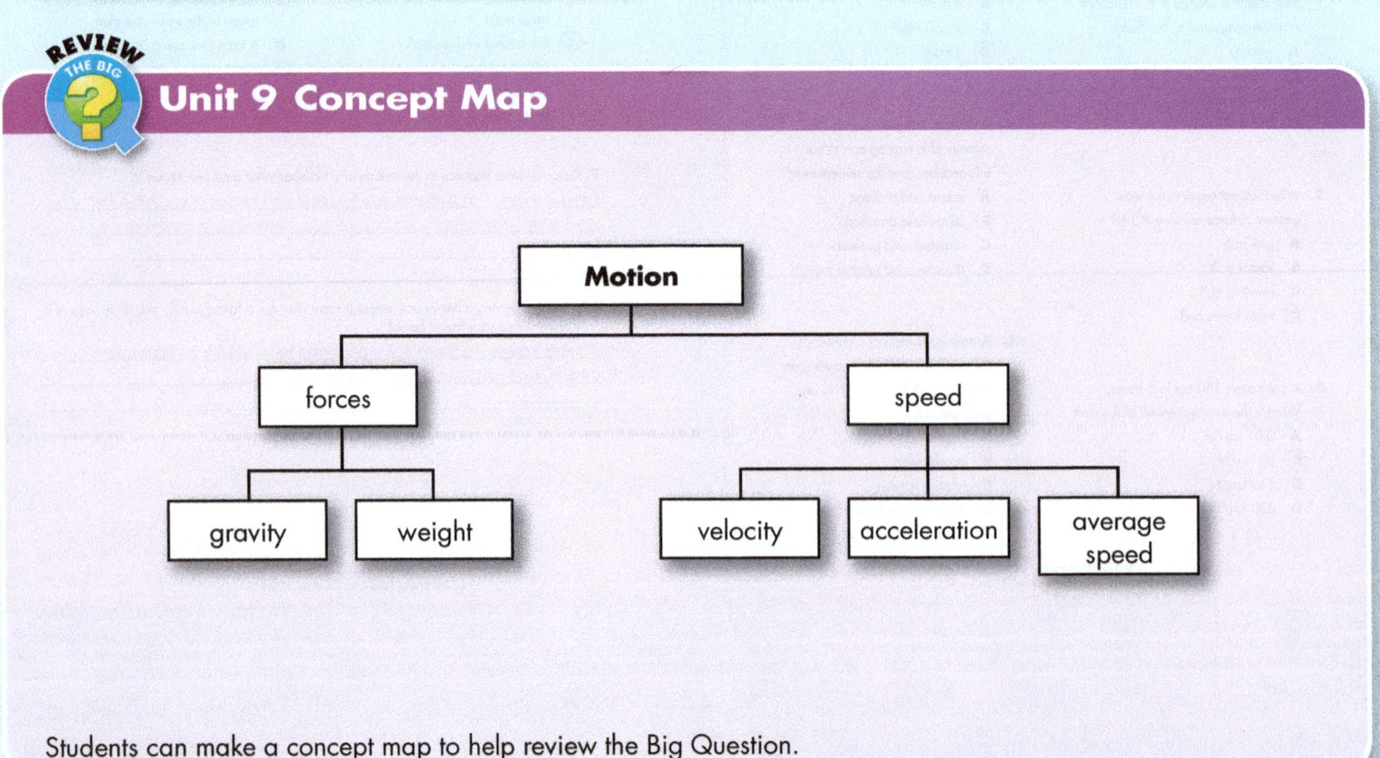

Students can make a concept map to help review the Big Question.